Praise

Animal, Vegeta

"Kingsolver's reveries on the joys of canning tomatoes and raising turkeys are every bit as transporting as—and more ecologically relevant than—any *Year in Provence*–style escapism. After reading Kingsolver's earthy lyricism, her husband's informative sidebars, and her daughter's enlightened teen perspective, you may never be able to eat a (fossil fuel–chomping) banana again." —*Washington Post*

"A hybrid book . . . part memoir . . . part call to action, part education, part recipe collection. . . . *Animal, Vegetable, Miracle* makes an important contribution to the chorus of voices calling for change."

—*Chicago Tribune*

"It's a lovely book. One wants with all one's heart to sit with [Kingsolver] on the porch at the end of the day and shell peas."

—*Los Angeles Times*

"Kingsolver, who writes evocatively about our connection to place, does so here with characteristic glowing prose. She provides the rapture, and Steven Hopp, her environmental biologist husband, provides pithy sidebars of facts and figures." —*Miami Herald*

"Homespun, unassuming, informed, positive, inspiring, zealously devoted to home and hearth . . . often wisecracking humorous. . . . The winning volume is unstinting in its concerns about this imperiled planet and the impact of extravagant American lifestyles."

—*Seattle Post-Intelligencer*

"Kingsolver has blessed us with a story as small as her Appalachian kitchen and as big as global climate change. . . . This novelist paints a compelling big picture of twenty-first-century America's national eating disorder—broad and ambitious, with nary an extraneous stroke." —*Denver Rocky Mountain News*

"Charming memoir and persuasive journalism. . . . Each season—and chapter—unfolds with a natural rhythm and mouth-watering appeal." —*Milwaukee Journal Sentinel*

"Equal parts folk wisdom and political activism. . . . This family effort instructs as much as it entertains." —*St. Louis Post-Dispatch*

"As satisfying and complete as a down-home supper. . . . Barbara Kingsolver is nothing less than a national treasure." —*Tucson Citizen*

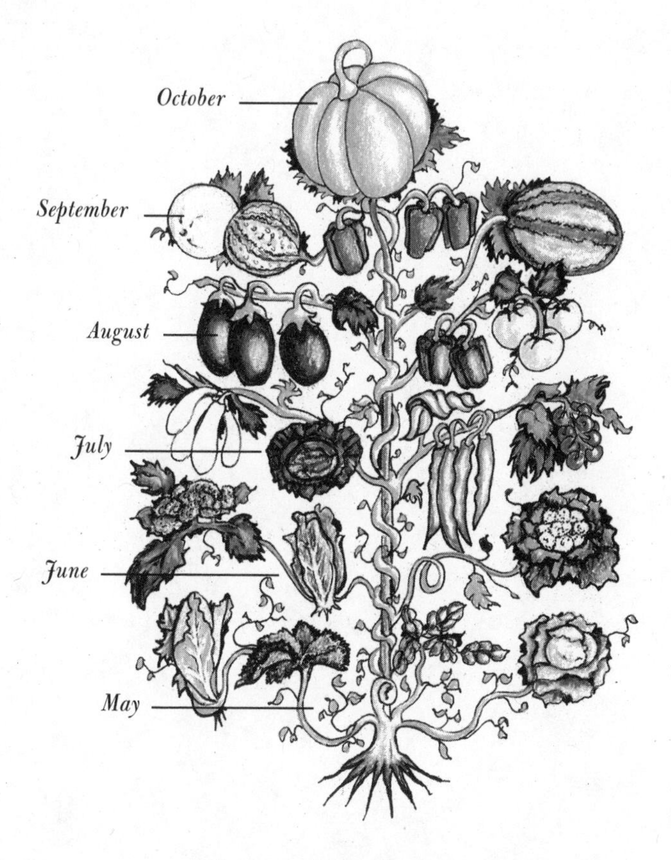

"Picture a single imaginary plant,
bearing throughout one season all the
different vegetables we harvest . . .
we'll call it a vegetannual."

Animal, Vegetable, Miracle

By the Same Author

FICTION

Prodigal Summer

The Poisonwood Bible

Pigs in Heaven

Animal Dreams

Homeland and Other Stories

The Bean Trees

ESSAYS

Small Wonder

High Tide in Tucson: Essays from Now or Never

POETRY

Another America

NONFICTION

Last Stand: America's Virgin Lands
(with photographs by Annie Griffiths Belt)

Holding the Line:
Women in the Great Arizona
Mine Strike of 1983

Animal, Vegetable, Miracle

TENTH ANNIVERSARY EDITION

A Year of Food Life

BARBARA
KINGSOLVER

with Steven L. Hopp, Camille Kingsolver, and Lily Hopp Kingsolver

ORIGINAL DRAWINGS BY RICHARD A. HOUSER

HARPER PERENNIAL

NEW YORK • LONDON • TORONTO • SYDNEY • NEW DELHI • AUCKLAND

A hardcover edition of this book was published in 2007 by HarperCollins Publishers.

HarperCollins books may be purchased for educational, business, or sales promotional use. For information, please e-mail the Special Markets Department at SPsales@harpercollins.com.

Original Drawings by Richard A. Houser

FIRST HARPER PERENNIAL EDITION PUBLISHED 2008, 2017.

Designed by Fritz Metsch

This book was printed on paper containing 100% postconsumer recycled fiber, processed chlorine-free.

The Library of Congress has catalogued the hardcover edition as follows:

Kingsolver, Barbara.
Animal, vegetable, miracle : a year of food life / Barbara Kingsolver, with Steven L. Hopp and Camille Kingsolver ; original drawings by Richard A. Houser.—1st ed.
370 p. : ill. ; 24 cm.
Includes bibliographical references (p. [355]–357).
ISBN: 978-0-06-085255-9
ISBN-10: 0-06-085255-0
1. Kingsolver, Barbara. 2. Hopp, Steven L., 1954–. 3. Farm life—Appalachian Region, Southern—Anecdotes. 4. Country life—Appalachian Region, Southern—Anecdotes. 5. Agriculture—Appalachian Region, Southern—Anecdotes. 6. Food habits—Appalachian Region, Southern—Anecdotes. I. Hopp, Steven L. II. Kingsolver, Camille. III. Title.

S521.5.A67 K56 2007
641.0973 22 2006053516

ISBN 978-0-06-265305-5

23 24 25 26 27 LBC 14 13 12 11 10

In memory of Jo Ellen

CONTENTS

1 · CALLED HOME

This story about good food begins in a quick-stop convenience market. It was our family's last day in Arizona, where I'd lived half my life and raised two kids for the whole of theirs. Now we were moving away forever, taking our nostalgic inventory of the things we would never see again: the bush where the roadrunner built a nest and fed lizards to her weird-looking babies; the tree Camille crashed into learning to ride a bike; the exact spot where Lily touched a dead snake. Our driveway was just the first tributary on a memory river sweeping us out.

One person's picture postcard is someone else's normal. This was the landscape whose every face we knew: giant saguaro cacti, coyotes, mountains, the wicked sun reflecting off bare gravel. We were leaving it now in one of its uglier moments, which made good-bye easier, but also seemed like a cheap shot—like ending a romance right when your partner has really bad bed hair. The desert that day looked like a nasty case of prickly heat caught in a long, naked wince.

This was the end of May. Our rainfall since Thanksgiving had measured less than *one inch.* The cacti, denizens of deprivation, looked ready to pull up roots and hitch a ride out if they could. The prickly pears waved good-bye with puckered, grayish pads. The tall, dehydrated saguaros stood around all teetery and sucked-in like very prickly supermodels. Even in the best of times desert creatures live on the edge of survival, getting by mostly on vapor and their own life savings. Now, as the southern tier of U.S. states came into a third consecutive year of drought, people elsewhere debated how seriously they should take global warming. We were staring it in the face.

Away went our little family, like rats leaping off the burning ship. It hurt to think about everything at once: our friends, our desert, old home, new home. We felt giddy and tragic as we pulled up at a little gas-and-go market on the outside edge of Tucson. Before we set off to seek our fortunes we had to gas up, of course, and buy snacks for the road. We did have a cooler in the back seat packed with respectable lunch fare. But we had more than two thousand miles to go. Before we crossed a few state lines we'd need to give our car a salt treatment and indulge in some things that go crunch.

This was the trip of our lives. We were ending our existence outside the city limits of Tucson, Arizona, to begin a rural one in southern Appalachia. We'd sold our house and stuffed the car with the most crucial things: birth certificates, books-on-tape, and a dog on drugs. (Just for the trip, I swear.) All other stuff would come in the moving van. For better or worse, we would soon be living on a farm.

For twenty years Steven had owned a piece of land in the southern Appalachians with a farmhouse, barn, orchards and fields, and a tax zoning known as "farm use." He was living there when I met him, teaching college and fixing up his old house one salvaged window at a time. I'd come as a visiting writer, recently divorced, with something of a fixer-upper life. We proceeded to wreck our agendas in the predictable fashion by falling in love. My young daughter and I were attached to our community in Tucson; Steven was just as attached to his own green pastures and the birdsong chorus of deciduous eastern woodlands. My father-in-law to be, upon hearing the exciting news about us, asked Steven, "Couldn't you find one closer?"

Apparently not. We held on to the farm by renting the farmhouse to another family, and maintained marital happiness by migrating like birds: for the school year we lived in Tucson, but every summer headed back to our rich foraging grounds, the farm. For three months a year we lived in a tiny, extremely crooked log cabin in the woods behind the farmhouse, listening to wood thrushes, growing our own food. The girls (for another child came along shortly) loved playing in the creek, catching turtles, experiencing real mud. I liked working the land, and increasingly came to think of this place as my home too. When all of us were ready, we decided, we'd go there for keeps.

We had many conventional reasons for relocation, including extended family. My Kingsolver ancestors came from that county in Virginia; I'd grown up only a few hours away, over the Kentucky line. Returning now would allow my kids more than just a hit-and-run, holiday acquaintance with grandparents and cousins. In my adult life I'd hardly shared a phone book with anyone else using my last name. Now I could spend Memorial Day decorating my ancestors' graves with peonies from my backyard. Tucson had opened my eyes to the world and given me a writing career, legions of friends, and a taste for the sensory extravagance of red hot chiles and five-alarm sunsets. But after twenty-five years in the desert, I'd been called home.

There is another reason the move felt right to us, and it's the purview of this book. We wanted to live in a place that could feed us: where rain falls, crops grow, and drinking water bubbles right up out of the ground. This might seem an abstract reason for leaving beloved friends and one of the most idyllic destination cities in the United States. But it was real to us. As it closes in on the million-souls mark, Tucson's charms have made it one of this country's fastest-growing cities. It keeps its people serviced across the wide, wide spectrum of daily human wants, with its banks, shops, symphonies, colleges, art galleries, city parks, and more golf courses than you can shake a stick at. By all accounts it's a bountiful source of everything on the human-need checklist, save for just the one thing—the stuff we put in our mouths every few hours to keep us alive. Like many other modern U.S. cities, it might as well be a space station where human sustenance is concerned. Virtually every unit of food consumed there moves into town in a refrigerated module from somewhere far away. Every ounce of the city's drinking, washing, and goldfish-bowl-filling water is pumped from a nonrenewable source—a fossil aquifer that is dropping so fast, sometimes the ground crumbles. In a more recent development, some city water now arrives via a three-hundred-mile-long open canal across the desert from the Colorado River, which—owing to our thirsts—

is a river that no longer reaches the ocean, but peters out in a sand flat near the Mexican border.

If it crosses your mind that water running through hundreds of miles of open ditch in a desert will evaporate and end up full of concentrated salts and muck, then let me just tell you, that kind of negative thinking will never get you elected to public office in the state of Arizona. When this giant new tap turned on, developers drew up plans to roll pink stucco subdivisions across the desert in all directions. The rest of us were supposed to rejoice as the new flow rushed into our pipes, even as the city warned us this water was kind of special. They said it was okay to drink, but don't put it in an aquarium because it would *kill the fish*.

Drink it we did, then, filled our coffee makers too, and mixed our children's juice concentrate with fluid that would gag a guppy. Oh, America the Beautiful, where are our standards? How did Europeans, ancestral cultures to most of us, whose average crowded country would fit inside one of our national parks, somehow hoard the market share of Beautiful? They'll run over a McDonald's with a bulldozer because it threatens the way of life of their fine cheeses. They have international trade hissy fits when we try to slip modified genes into their bread. They get their favorite ham from Parma, Italy, along with a favorite cheese, knowing these foods are linked in an ancient connection the farmers have crafted between the milk and the hogs. Oh. We were thinking *Parmesan* meant, not "coming from Parma," but "coming from a green shaker can." Did they kick us out for bad taste?

No, it was mostly for vagrancy, poverty, or being too religious. We came here for the freedom to make a *Leaves of Grass* kind of culture and hear America singing to a good beat, pierce our navels as needed, and eat whatever we want without some drudge scolding: "You don't know where that's been!" And boy howdy, we do not.

The average food item on a U.S. grocery shelf has traveled farther than most families go on their annual vacations. True fact. Fossil fuels were consumed for the food's transport, refrigeration, and processing, with the obvious environmental consequences. The option of getting our household's food from closer to home, in Tucson, seemed no better to us. The Sonoran desert historically offered to humans baked dirt as a construction material, and for eats, a corn-and-beans diet organized around late summer monsoons, garnished in spring with cactus fruits and wild tubers. The Hohokam and Pima were the last people to live on that land without creating an environmental overdraft. When the Spaniards arrived, they didn't rush to take up the Hohokam diet craze. Instead they set about working up a monumental debt: planting orange trees and alfalfa, digging wells for irrigation, withdrawing millions more gallons from the water table each year than a dozen inches of annual rainfall could ever restore. Arizona is still an agricultural state. Even after the popula-

Oily Food

Americans put almost as much fossil fuel into our refrigerators as our cars. We're consuming about 400 gallons of oil a year per citizen—about 17 percent of our nation's energy use—for agriculture, a close second to our vehicular use. Tractors, combines, harvesters, irrigation, sprayers, tillers, balers, and other equipment all use petroleum. Even bigger gas guzzlers on the farm are not the machines, but so-called inputs. Synthetic fertilizers, pesticides, and herbicides use oil and natural gas as their starting materials, and in their manufacturing. More than a quarter of all farming energy goes into synthetic fertilizers.

But getting the crop from seed to harvest takes only one-fifth of the total oil used for our food. The lion's share is consumed during the trip from the farm to your plate. Each food item in a typical U.S. meal has traveled an average of 1,500 miles. In addition to direct transport, other fuel-thirsty steps include processing (drying, milling, cutting, sorting, baking), packaging, warehousing, and refrigeration. Energy calories consumed by production, packaging, and shipping far outweigh the energy calories we receive from the food.

A quick way to improve food-related fuel economy would be to buy a quart of motor oil and drink it. More palatable options are available. If every U.S. citizen ate just one meal a week (any meal) composed of locally and organically raised meats and produce, we would reduce our country's oil consumption by over 1.1 million barrels of oil *every week.* That's not gallons, but barrels. Small changes in buying habits can make big differences. Becoming a less energy-dependent nation may just need to start with a good breakfast.

STEVEN L. HOPP

tion boom of the mid-nineties, 85 percent of the state's water still went to thirsty crops like cotton, alfalfa, citrus, and pecan trees. Mild winters offer the opportunity to create an artificial endless summer, as long as we can conjure up water and sustain a chemically induced illusion of topsoil.

Living in Arizona on borrowed water made me nervous. We belonged to a far-flung little community of erstwhile Tucson homesteaders, raising chickens in our yards and patches of vegetables for our own use, frequenting farmers' markets to buy from Arizona farmers, trying to reduce the miles-per-gallon quotient of our diets in a gasoholic world. But these gardens of ours had a drinking problem. So did Arizona farms. That's a devil of a choice: Rob Mexico's water or guzzle Saudi Arabia's gas?

Traditionally, employment and family dictate choices about where to live. It's also legitimate to consider weather, schools, and other quality-of-life indices. We added one more wish to our list: more than one out of three of the basic elements necessary for human life. (*Oxygen* Arizona has got.) If we'd had family ties, maybe we'd have felt more entitled to claim a seat at Tucson's lean dining table. But I moved there as a young adult, then added through birth and marriage three more mouths to feed. As a guest, I'd probably overstayed my welcome. So, as the U.S. population made an unprecedented dash for the Sun Belt, one carload of us dog-paddled against the tide, heading for the Promised Land where water falls from the sky and green stuff grows all around. We were about to begin the adventure of realigning our lives with our food chain.

Naturally, our first stop was to buy junk food and fossil fuel.

In the cinder-block convenience mart we foraged the aisles for blue corn chips and Craisins. Our family's natural-foods teenager scooped up a pile of energy bars big enough to pass as a retirement plan for a hamster. Our family's congenitally frugal Mom shelled out two bucks for a fancy green bottle of about a nickel's worth of iced tea. As long as we were all going crazy here, we threw in some 99-cent bottles of what comes free out of drinking fountains in places like Perrier, France. In our present location, 99 cents for good water seemed like a bargain. The goldfish should be so lucky.

As we gathered our loot onto the counter the sky darkened suddenly. After two hundred consecutive cloudless days, you forget what it looks like when a cloud crosses the sun. We all blinked. The cashier frowned toward the plate-glass window.

"*Dang,*" she said, "it's going to rain."

"I hope so," Steven said.

She turned her scowl from the window to Steven. This bleached-blond guardian of gas pumps and snack food was not amused. "It better not, is all I can say."

"But we need it," I pointed out. I am not one to argue with cashiers, but the desert was dying, and this was my very last minute as a Tucsonan. I hated to jinx it with bad precipitation-karma.

"I know that's what they're saying, but I don't care. Tomorrow's my first day off in two weeks, and I want to wash my car."

For three hundred miles we drove that day through desperately parched Sonoran badlands, chewing our salty cashews with a peculiar guilt. We had all shared this wish, in some way or another: that it wouldn't rain on our day off. Thunderheads dissolved ahead of us, as if honoring our compatriot's desire to wash her car as the final benediction pronounced on a dying land. In our desert, we would not see rain again.

⚘

It took us five days to reach the farm. On our first full day there we spent ten hours mowing, clearing brush, and working on the farmhouse. Too tired to cook, we headed into town for supper, opting for a diner of the southern type that puts grits on your plate until noon and biscuits after, whether you ask for them or not. Our waitress was young and chatty, a student at the junior college nearby studying to be a nurse or else, if she doesn't pass the chemistry, a television broadcaster. She said she was looking forward to the weekend, but smiled broadly nevertheless at the clouds gathering over the hills outside. The wooded mountainsides and velvet pastures of southwestern Virginia looked remarkably green to our desert-scorched eyes, but the forests and fields were suffering here too. Drought had plagued most of the southern United States that spring.

A good crack of thunder boomed, and the rain let loose just as the waitress came back to clear our plates. "Listen at that," she clucked. "Don't we need it!"

We do, we agreed. The hayfields aren't half what they should be.

"Let's hope it's a good long one," she said, pausing with our plates balanced on her arm, continuing to watch out the window for a good long minute. "And that it's not so hard that it washes everything out."

It is not my intention here to lionize country wisdom over city ambition. I only submit that the children of farmers are likely to know where food comes from, and that the rest of us might do well to pay attention. For our family, something turned over that evening in the diner: a gas-pump cashier's curse of drought was lifted by a waitress's simple, agricultural craving for rain. I thought to myself: There is hope for us.

⁂

Who is *us,* exactly? I live now in a county whose economic base is farming. A disastrous summer will mean some of our neighbors will lose their farms. Others will have to keep farming *and* go looking for a job at the end of a long commute. We'll feel the effects in school enrollments, local businesses, shifts in land use and tax structure. The health of our streams, soils, and forests is also at stake, as lost farms get sold to developers whose business is to rearrange (drastically) the topsoil and everything on it. When I recognize good agricultural sense, though, I'm not just thinking of my town but also my species. It's not a trivial difference: praying for or against rainfall during a drought. You can argue that wishes don't count, but humans are good at making our dreams manifest and we do, historically speaking, get what we wish for. What are the just deserts for a species too selfish or preoccupied to hope for rain when the land outside is dying? Should we be buried under the topsoil in our own clean cars, to make room for wiser creatures?

We'd surely do better, if only we *knew* any better. In two generations we've transformed ourselves from a rural to an urban nation. North American children begin their school year around Labor Day and finish at the beginning of June with no idea that this arrangement was devised to free up children's labor when it was needed on the farm. Most people of my grandparents' generation had an intuitive sense of agricultural basics: when various fruits and vegetables come into season, which ones keep through the winter, how to preserve the others. On what day autumn's first frost will likely fall on their county, and when to expect the last one in spring. Which crops can be planted before the last frost, and which must wait. Which grains are autumn-planted. What an asparagus patch looks like in August. Most importantly: what animals and vegetables thrive in one's immediate region and how to live well on those, with little else thrown into the mix beyond a bag of flour, a pinch of salt, and a handful of coffee. Few people of my generation, and approximately none of our children, could answer any of those questions, let alone all. This knowledge has vanished from our culture.

We also have largely convinced ourselves it wasn't too important. Consider how Americans might respond to a proposal that agriculture was to become a mandatory subject in all schools, alongside reading and mathematics. A fair number of parents would get hot under the collar to see their kids' attention being pulled away from the essentials of grammar, the all-important trigonometry, to make room for down-on-the-farm stuff. The baby boom psyche embraces a powerful presumption that education is a key to moving *away* from manual labor, and dirt—two undeniable ingredients of farming. It's good enough for us that somebody, somewhere, knows food production well enough to serve the rest of us with all we need to eat, each day of our lives.

If that is true, why isn't it good enough for someone else to know multiplication and the contents of the Bill of Rights? Is the story of bread, from tilled ground to our table, less relevant to our lives than the history of the thirteen colonies? Couldn't one make a case for the relevance of a subject that informs choices we make *daily*—as in, What's for dinner? Isn't ignorance of our food sources causing problems as diverse as overdependence on petroleum, and an epidemic of diet-related diseases?

If this book is not exactly an argument for reinstating food-production classes in schools (and it might be), it does contain a lot of what you might learn there. From our family's gas-station beginnings we have traveled far enough to discover ways of taking charge of one's food, and even knowing where it has been. This is the story of a year in which we made every attempt to feed ourselves animals and vegetables whose provenance we really knew. We tried to wring most of the petroleum out of our food chain, even if that meant giving up some things. Our highest shopping goal was to get our food from so close to home, we'd know the person who grew it. Often that turned out to be *us,* as we learned to produce more of what we needed, starting with dirt, seeds, and enough knowledge to muddle through. Or starting with baby animals and enough sense to refrain from naming them.

This is not a how-to book aimed at getting you cranking out your own food. We ourselves live in a region where every other house has a garden out back, but to many urban people the idea of growing your food must seem as plausible as writing and conducting your own symphonies for your personal listening pleasure. If that is your case, think of the agricultural parts of the story as a music appreciation course for food—acquainting yourself with the composers and conductors can improve the quality of your experience. Knowing the secret natural history of potatoes, melons, or asparagus gives you a leg up on detecting whether those in your market are wholesome kids from a nearby farm, or vagrants who idled away their precious youth in a boxcar. Knowing how foods grow is to know how and when to look for them; such expertise is useful for certain kinds of people, namely, the ones who eat, no matter where they live or grocery shop.

Absence of that knowledge has rendered us a nation of wary label-readers, oddly uneasy in our obligate relationship with the things we eat. We call our food animals by different names after they're dead, presumably sparing ourselves any vision of the beefs and the porks running around on actual hooves. Our words for unhealthy contamination—"soiled" or "dirty"—suggest that if we really knew the number-one ingredient of a garden, we'd all head straight into therapy. I used to take my children's friends out to the garden to warm them up to the idea of eating vegetables, but this strategy sometimes backfired: they'd back away slowly saying, "Oh *man,* those things touched *dirt*!" Adults do the same by pretending it all comes from the clean, well-lighted grocery store. We're like petulant teenagers rejecting our mother. We *know* we came out of her, but *ee-ew.*

We don't know beans about beans. Asparagus, potatoes, turkey drumsticks—you name it, we don't have a clue how the world makes it. I usually think I'm exaggerating the scope of the problem, and then I'll encounter an editor (at a well-known nature magazine) who's nixing the part of my story that refers to pineapples growing from the ground. She insisted they grew on trees. Or, I'll have a conversation like this one:

"What's new on the farm?" asks my friend, a lifelong city dweller who likes for me to keep her posted by phone. She's a gourmet cook, she cares about the world, and has been around a lot longer than I have. This particular conversation was in early spring, so I told her what was up in the garden: peas, potatoes, spinach.

"Wait a minute," she said. "When you say, 'The potatoes are up,' what do you mean?" She paused, formulating her question: "What part of a potato comes *up*?"

"Um, the plant part," I said. "The stems and leaves."

"Wow," she said. "I never knew a potato *had* a plant part."

Many bright people are really in the dark about vegetable life. Biology teachers face kids in classrooms who may not even believe in the metamorphosis of bud to flower to fruit and seed, but rather, some continuum of pansies becoming petunias becoming chrysanthemums; that's the only reality they witness as landscapers come to campuses and city parks and surreptitiously yank out one flower before it fades from its prime, replacing it with another. (My biology-professor brother pointed this out to me.) The same disconnection from natural processes may be at the heart of our country's shift away from believing in evolution. In the past, principles of natural selection and change over time made sense to kids who'd watched it all unfold. Whether or not they knew the terms, farm families understood the processes well enough to imitate them: culling, selecting, and improving their herds and crops. For modern kids who intuitively believe in the spontaneous generation of fruits and vegetables in the produce section, trying to get their minds around the slow speciation of the plant kingdom may be a stretch.

Steven, also a biology professor, grew up in the corn belt of Iowa but has encountered his share of agricultural agnostics in the world. As a graduate student he lived in an urban neighborhood where his little backyard vegetable garden was a howling curiosity for the boys who ran wild in the alley. He befriended these kids, especially Malcolm, known throughout the neighborhood as "Malcolm-get-your-backside-in-here-now-or-you-won't-be-*having*-no-dinner!" Malcolm liked hanging around when Steven was working in the garden, but predictably enough, had a love-hate thing with the idea of the vegetables touching the dirt. The first time he watched Steven pull long, orange carrots out of the ground, he demanded: "How'd you get them *in* there?"

Steven held forth with condensed Intro Botany. Starts with a seed, grows into a plant. Water, sunlight, leaves, roots. "A carrot," Steven concluded, "is actually a root."

"Uh-huh . . . ," said Malcolm doubtfully.

A crowd had gathered now. Steven engaged his audience by asking, "Can you guys think of other foods that might be root vegetables?"

Malcolm checked with his pals, using a lifeline before confidently submitting his final answer: "Spaghetti?"

We can't know what we haven't been taught. Steven couldn't recognize tobacco in vivo before moving in his twenties to southwestern Virginia, where the tobacco leaf might as well be the state flag. One Saturday morning soon after he'd moved, he was standing on a farmer's porch at a country yard sale when a field of giant, pale leaves and tall pink flower spikes caught his eye. He asked the farmer the name of this gorgeous plant. The man grinned hugely and asked, "You're not *from* here, are you, son?"

That farmer is probably still telling this story; Steven is his Malcolm. Every one of us is somebody's Malcolm. Country folks can be as food-chain-challenged as the city mice, in our own ways. Rural southern cooking is famous for processed-ingredient recipes like Coca-Cola cake, and plenty of rural kids harbor a potent dread of compost and earthworms. What we all don't know about farming could keep the farmers laughing until the cows come home. Except that they are barely making a living, while the rest of us play make-believe about the important part being the grocery store.

When we walked as a nation away from the land, our knowledge of food production fell away from us like dirt in a laundry-soap commercial.

Now, it's fair to say, the majority of us don't want to be farmers, see farmers, pay farmers, or hear their complaints. Except as straw-chewing figures in children's books, we don't quite believe in them anymore. When we give it a thought, we mostly consider the food industry to be a *thing* rather than a person. We obligingly give 85 cents of our every food dollar to that thing, too—the processors, marketers, and transporters. And we complain about the high price of organic meats and vegetables that might send back more than three nickels per buck to the farmers: those actual humans putting seeds in the ground, harvesting, attending livestock births, standing in the fields at dawn casting their shadows upon our sustenance. There seems to be some reason we don't want to compensate or think about these hardworking people. In the grocery store checkout corral, we're more likely to learn which TV stars are secretly fornicating than to inquire as to the whereabouts of the people who grew the cucumbers and melons in our carts.

This drift away from our agricultural roots is a natural consequence of migration from the land to the factory, which is as old as the Industrial Revolution. But we got ourselves uprooted entirely by a drastic reconfiguration of U.S. farming, beginning just after World War II. Our munitions plants, challenged to beat their swords into plowshares, retooled to make ammonium nitrate surpluses into chemical fertilizers instead of explosives. The next explosions were yields on midwestern corn and soybean fields. It seemed like a good thing, but some officials saw these new surpluses as reason to dismantle New Deal policies that had helped farmers weather the economic uncertainties notorious to their vocation. Over the next decades, nudged by industry, the government rewrote the rules on commodity subsidies so these funds did not safeguard farmers, but instead guaranteed a supply of cheap corn and soybeans.

These two crops, formerly food for people and animals, became something entirely new: a standardized raw material for a new extractive industry, not so different from logging or mining. Mills and factories were designed for a multibranched production line as complex as the one that turns iron and aluminum ores into the likes of automobiles, paper clips, and antiperspirants. But instead, this new industry made piles of corn and soybeans into high-fructose corn syrup, hydrogenated oils, and thousands of other starch- or oil-based chemicals. Cattle and chickens were brought in off the pasture into intensely crowded and mechanized CAFOs (concentrated animal feeding operations) where corn—which is no part of a cow's natural diet, by the way—could be turned cheaply and quickly into animal flesh. All these different products, in turn, rolled on down the new industrial food pipeline to be processed into the soft drinks, burgers, and other cheap foods on which our nation now largely runs—or sits on its bottom, as the case may be.

This is how 70 percent of all our midwestern agricultural land shifted gradually into single-crop corn or soybean farms, each one of them now, on average, the size of Manhattan. Owing to synthetic fertilizers and pesticides, genetic modification, and a conversion of farming from a naturally based to a highly mechanized production system, U.S. farmers now produce 3,900 calories per U.S. citizen, per day. That is twice what we need, and 700 calories a day more than they grew in 1980. Commodity farmers can only survive by producing their maximum yields, so they do. And here is the shocking plot twist: as the farmers produced those extra calories, the food industry figured out how to get them into the bodies of people who didn't really *want* to eat 700 more calories a day. That is the well-oiled machine we call Late Capitalism.

Most of those calories enter our mouths in forms hardly recognizable as corn and soybeans, or even vegetable in origin: high-fructose corn syrup (HFCS) owns up to its parentage, but lecithin, citric acid, maltodextrin, sorbitol, and xanthan gum, for example, are also manufactured from corn. So are beef, eggs, and poultry, in a different but no less artificial process. Soybeans also become animal flesh, or else a category of ingredient known as "added fats." If every product containing corn or soybeans were removed from your grocery store, it would look more like a hardware store. Alarmingly, the lightbulbs might be naked, since many packaging materials also now contain cornstarch.

With so many extra calories to deliver, the packages have gotten bigger. The shapely eight-ounce Coke bottle of yesteryear became twenty ounces of carbonated high-fructose corn syrup and water; the accompanying meal morphed similarly. So did the American waistline. U.S. consumption of "added fats" has increased by one-third since 1975, and our

HFCS is up by 1000 percent. About a third of all our calories now come from what is known, by community consent, as junk food.

No cashier held a gun to our heads and made us supersize it, true enough. But humans have a built-in weakness for fats and sugar. We evolved in lean environments where it was a big plus for survival to gorge on calorie-dense foods whenever we found them. Whether or not they understand the biology, food marketers know the weakness and have exploited it without mercy. Obesity is generally viewed as a failure of personal resolve, with no acknowledgment of the genuine conspiracy in this historical scheme. People actually did sit in strategy meetings discussing ways to get all those surplus calories into people who neither needed nor wished to consume them. Children have been targeted especially; food companies spend over $10 billion a year selling food brands to kids, and it isn't broccoli they're pushing. Overweight children are a demographic in many ways similar to minors addicted to cigarettes, with one notable exception: their parents are usually their suppliers. We all subsidize the cheap calories with our tax dollars, the strategists make fortunes, and the overweight consumers get blamed for the violation. The perfect crime.

All industrialized countries have experienced some commodification of agriculture and increased consumption of processed foods. But nowhere else on earth has it become normal to layer on the love handles as we do. (Nude beaches are still popular in Europe.) Other well-fed populations have had better luck controlling caloric excess through culture and custom: Italians eat Italian food, the Japanese eat Japanese, and so on, honoring ancient synergies between what their land can give and what their bodies need. Strong food cultures are both aesthetic and functional, keeping the quality and quantity of foods consumed relatively consistent from one generation to the next. And so, while the economies of many Western countries expanded massively in the late twentieth century, their citizens did not.

Here in the U.S. we seem puzzled by these people who refrain from gluttony in the presence of a glut. We've even named a thing we call the French Paradox: How can people have such a grand time eating cheese and fattened goose livers and still stay slim? Having logged some years in France, I have some hunches: they don't suck down giant sodas; they consume many courses in a meal but the portions of the fatty ones tend to be tiny; they smoke like chimneys (though that's changing); and they draw out meals sociably, so it's not just about shoveling it in. The all-you-can-eat buffet is an alien concern to the French, to put it mildly. Owing to certain rules about taste and civility in their heads, their bodies seem to know when enough is enough. When asked, my French friends have confided with varying degrees of tact that the real paradox is how people manage to consume, so very much, the scary food of America.

Why do we? Where are *our* ingrained rules of taste and civility, our ancient treaties between our human cravings and the particular fat of our land? Did they perhaps fly out the window while we were eating in a speeding car?

Food culture in the United States has long been cast as the property of a privileged class. It is nothing of the kind. Culture is the property of a species. Humans don't do everything we crave to do—that is arguably what makes us human. We're genetically predisposed toward certain behaviors that we've collectively decided are unhelpful; adultery and racism are possible examples. With reasonable success, we mitigate those impulses through civil codes, religious rituals, maternal warnings—the whole bag of tricks we call culture. Food cultures concentrate a population's collective wisdom about the plants and animals that grow in a place, and the complex ways of rendering them tasty. These are mores of survival, good health, and control of excess. Living without such a culture would seem dangerous.

And here we are, sure enough in trouble. North America's native cuisine met the same unfortunate fate as its native people, save for a few relics like the Thanksgiving turkey. Certainly, we still have regional specialties, but the Carolina barbecue will almost certainly have California tomatoes in its sauce (maybe also Nebraska-fattened feedlot hogs), and the Louisiana gumbo is just as likely to contain Indonesian farmed shrimp. If either of these shows up on a fast-food menu with lots of added fats or HFCS, we seem unable either to discern or resist the corruption. We have yet to come up with a strong set of generalized norms, passed down through families, for savoring and sensibly consuming what our land and climate give us. We have, instead, a string of fad diets convulsing our bookstores and bellies, one after another, at the scale of the national best seller. Nine out of ten nutritionists (unofficial survey) view this as evidence that we have entirely lost our marbles. A more optimistic view might be this: these sets of mandates captivate us because we're looking hard for a food culture of our own. A profit-driven food industry has exploded and nutritionally bankrupted our caloric supply, and we long for a Food Leviticus to save us from the sinful roil of cheap fats and carbs.

What the fad diets don't offer, though, is any sense of national and biological integrity. A food culture is not something that gets *sold* to people. It arises out of a place, a soil, a climate, a history, a temperament, a collective sense of belonging. Every set of fad-diet rules is essentially framed in the negative, dictating what you must give up. Together they've helped us form powerfully negative associations with the very act of eating. Our most celebrated models of beauty are starved people. But we're still an animal that must eat to live. To paraphrase a famous campaign slogan: it's the biology, stupid. A food culture of anti-eating is worse than useless.

People hold to their food customs because of the *positives:* comfort, nourishment, heavenly aromas. A sturdy food tradition even calls to outsiders; plenty of red-blooded Americans will happily eat Italian, French, Thai, Chinese, you name it. But try the reverse: hand the Atkins menu to a French person, and run for your life.

Will North Americans ever have a food culture to call our own? Can we find or make up a set of rituals, recipes, ethics, and buying habits that will let us love our food and eat it too? Some signs point to "yes." Better food—more local, more healthy, more sensible—is a powerful new topic of the American conversation. It reaches from the epicurean quarters of Slow Food convivia to the matter-of-fact Surgeon General's Office; from Farm Aid concerts to school lunch programs. From the rural routes to the inner cities, we are staring at our plates and wondering where that's *been.* For the first time since our nation's food was ubiquitously local, the point of origin now matters again to some consumers. We're increasingly wary of an industry that puts stuff in our dinner we can't identify as animal, vegetable, mineral, or what. The halcyon postwar promise of "better living through chemistry" has fallen from grace. "No additives" is now often considered a plus rather than the minus that, technically, it is.

We're a nation with an eating disorder, and we know it. The multiple maladies caused by bad eating are taking a dire toll on our health—most tragically for our kids, who are predicted to be this country's first generation to have a *shorter* life expectancy than their parents. That alone is a stunning enough fact to give us pause. So is a government policy that advises us to eat more fruits and vegetables, while doling out subsidies *not* to fruit and vegetable farmers, but to commodity crops destined to become soda pop and cheap burgers. The Farm Bill, as of this writing, could aptly be called the Farm Kill, both for its effects on small farmers and for

Hungry World

All these heirloom eggplants and artisan cheeses from the farmers' market are great for weekend dinner parties, but don't we still need industrial farming to feed the hungry?

In fact, all the world's farms currently produce enough food to make every person on the globe fat. Even though 800 million people are chronically underfed (6 will die of hunger-related causes while you read this article), it's because they lack money and opportunity, not because food is unavailable in their countries. The UN Food and Agriculture Organization (FAO) reports that current food production can sustain world food needs even for the 8 billion people who are projected to inhabit the planet in 2030. This will hold even with anticipated increases in meat consumption, and without adding genetically modified crops.

Is all this the reliable bounty of industrial production? Yes and no—with the "no" being more of a problem in the near future. Industrial farming methods, wherever they are practiced, promote soil erosion, salinization, desertification, and loss of soil fertility. The FAO estimates that over 25 percent of arable land in the world is already compromised by one or more of these problems. The worst-affected areas are those with more arid climates or sloped terrain. Numerous field trials in both the United States and the United Kingdom have shown that organic practices can produce commodity crop yields (corn, soybeans, wheat) comparable to those of industrial farms. By using cover crops or animal manures for fertilizer, these practices improve soil fertility and moisture-holding capacity over seasons, with cumulative benefits. These techniques are particularly advantageous in regions that lack the money and technology for industrial approaches.

what it does to us, the consumers who are financing it. The Green Revolution of the 1970s promised that industrial agriculture would make food cheaper and available to more people. Instead, it has helped more of us become less healthy.

A majority of North Americans do understand, at some level, that our food choices are politically charged, affecting arenas from rural culture to international oil cartels and global climate change. Plenty of consumers are trying to get off the petroleum-driven industrial food wagon: banning fast food from their homes and schools, avoiding the unpronounceable

---

Conventional methods are definitely producing huge quantities of corn, wheat, and soybeans, but not to feed the poor. Most of it becomes animal feed for meat production, or the ingredients of processed foods for wealthier consumers who are already getting plenty of calories. Food sellers prefer to market more food to people who have money, rather than those who have little. World food trade policies most often favor developed countries at the expense of developing countries; distributors, processors, and shippers reap most of the benefits. Even direct food aid for disasters (a small percentage of all the world's hunger) is most profitable for grain companies and shippers. By law, 75 percent of such aid sent from the United States to other nations must be grown, packaged, and shipped by U.S. companies. This practice, called "tied aid," delays shipments of food by as much as six months, increases the costs of the food by over 50 percent, and directs over two-thirds of the aid money to the distributors.

If efficiency is the issue, resources go furthest when people produce their own food, near to where it is consumed. Many hunger-relief organizations provide assistance not in the form of bags of food, but in programs that teach and provide support technology for locally appropriate, sustainable farming. These programs do more than alleviate hunger for a day and send a paycheck to a multinational. They provide a livelihood to the person in need, addressing the real root of hunger, which is not about food production, but about poverty.

For more information, visit www.wn.org, www.journeytoforever.org, or www.heifer.org.

STEVEN L. HOPP

ingredient lists. However, *banning* is negative and therefore fails as a food culture per se.

Something positive is also happening under the surface of our nation's food preference paradigm. It could be called a movement. It includes gardeners who grow some of their own produce—one-quarter of all U.S. households, according to the U.S. Census Bureau. Just as importantly, it's the city dwellers who roll their kids out of bed on Saturday mornings and head down to the farmers' markets to pinch the tomatoes and inhale the spicy-sweet melons—New York, alone, has about a quarter million such shoppers. It involves the farmers' markets themselves, along with a new breed of restaurant owner (and customer) dedicated to buying locally produced food. It has been embraced by farmers who manage to keep family farms by thinking outside the box, learning to grow organic peppers or gourmet mushrooms. It engages schoolchildren and teachers who are bringing food-growing curricula into classrooms and lunchrooms from Berkeley, California, to my own county in southern Appalachia. It includes the kids who get dirty in those outdoor classrooms planting tomatoes and peppers at the end of third grade, then harvesting and cooking their own pizza when they start back into fourth. And it owes a debt to parents who can watch those kids getting dirty, and not make a fuss.

At its heart, a genuine food culture is an affinity between people and the land that feeds them. Step one, probably, is to *live* on the land that feeds them, or at least on the same continent, ideally the same region. Step two is to be able to countenance the ideas of "food" and "dirt" in the same sentence, and three is to start poking into one's supply chain to learn where things are coming from. In the spirit of this adventure, our family set out to find ourselves a real American culture of food, or at least the piece of it that worked for us, and to describe it for anyone who might be looking for something similar. This book tells the story of what we learned, or didn't; what we ate, or couldn't; and how our family was changed by one year of deliberately eating food produced in the same place where we worked, loved our neighbors, drank the water, and breathed the air. It's not at all necessary to live on a food-producing farm to participate in this culture. But it is necessary to know such farms exist, understand something about what they do, and consider oneself basically in their court. This book is about those things.

The story is pegged, as we were, to a one-year cycle of how and when foods become available in a temperate climate. Because food cultures affect everyone living under the same roof, we undertook this project—both the eating and the writing—as a family. Steven's sidebars are, in his words, "fifty-cent buckets of a dollar's worth of goods" on various topics I've mentioned in the narrative. Camille's essays offer a nineteen-year-old's perspective on the local-food project, plus nutritional information, recipes, and meal plans for every season. Lily's contributions were many, including more than fifty dozen eggs and a willingness to swear off Pop-Tarts for the duration, but she was too young to sign a book contract.

Will our single-family decision to step off the nonsustainable food grid give a big black eye to that petroleum-hungry behemoth? Keep reading, but don't hold your breath. We only knew, when we started, that similar choices made by many families at once were already making a difference: organic growers, farmers' markets, and small exurban food producers now comprise the fastest-growing sector of the U.S. food economy. A lot of people at once are waking up to a troublesome truth about cheap fossil fuels: we are going to run out of them. Our jet-age dependence on petroleum to feed our faces is a limited-time-only proposition. Every food calorie we presently eat has used dozens or even hundreds of fossil-fuel calories in its making: grain milling, for example, which turns corn into the ingredients of packaged foods, costs ten calories for every one food calorie produced. That's *before* it gets shipped anywhere. By the time my children are my age, that version of dinnertime will surely be an unthinkable extravagance.

I enjoy denial as much as the next person, but this isn't rocket science: our kids will eventually have to make food differently. They could be assisted by some familiarity with how vegetables grow from seeds, how animals grow on pasture, and how whole ingredients can be made into meals, gee whiz, right in the kitchen. My husband and I decided our children would not grow up without knowing a potato has a plant part. We would take a food sabbatical, getting our hands dirty in some of the actual dying arts of food production. We hoped to prove—at least to ourselves—that a family living on or near green land need not depend for its life on industrial food. We were writing our Dear John letter to a roomie that smells like exhaust fumes and the feedlot.

But sticking it to the Man (whoever he is) may not be the most inspired principle around which to organize one's life. We were also after tangible, healthy pleasures, in the same way that boycotting tobacco, for example, brings other benefits besides the satisfaction of withholding your money from Philip Morris. We hoped a year away from industrial foods would taste so good, we might actually enjoy it. The positives, rather than the negatives, ultimately nudged us to step away from the agribusiness supply line and explore the local food landscape. Doing the right thing, in this case, is not about abstinence-only, throwing out bread, tightening your belt, wearing a fake leather belt, or dragging around feeling righteous and gloomy. Food is the rare moral arena in which the ethical choice is generally the one more likely to make you groan with pleasure. Why resist that?

In Nikos Kazantzakis's novel *Zorba the Greek,* the pallid narrator frets a lot about his weaknesses of the flesh. He lies awake at night worrying about the infinite varieties of lust that call to him from this world; for example, cherries. He's way too fond of cherries. Zorba tells him, Well then, I'm afraid what you must do is stand under the tree, collect a big bowl full, and stuff yourself. Eat cherries like they're going out of season.

This was approximately the basis of our plan: the Zorba diet.

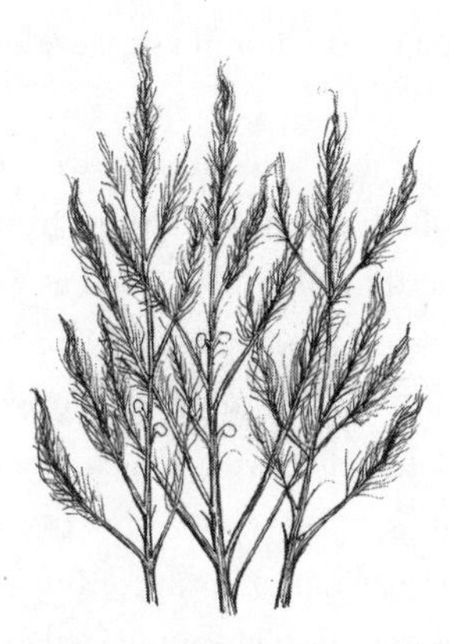

2 · WAITING FOR ASPARAGUS

Late March

A question was nagging at our family now, and it was no longer, "When do we get there?" It was, "When do we *start*?"

We had come to the farmland to eat deliberately. We'd discussed for several years what that would actually mean. We only knew, somewhat abstractly, we were going to spend a year integrating our food choices with our family values, which include both "love your neighbor" and "try not to wreck every blooming thing on the planet while you're here."

We'd given ourselves nearly a year to settle in at the farm and address some priorities imposed by our hundred-year-old farmhouse, such as hundred-year-old plumbing. After some drastic remodeling, we'd moved into a house that still lacked some finishing touches, like doorknobs. And a back door. We nailed plywood over the opening so forest mammals wouldn't wander into the kitchen.

Between home improvement projects, we did find time that first summer to grow a modest garden and can some tomatoes. In October the sober forests around us suddenly revealed their proclivity for cross-dressing. (Trees in Tucson didn't just throw on scarlet and orange like this.) Then came the series of snowfalls that comprised the first inclement winter of the kids' lives. One of our Tucson-bred girls was so dismayed by the cold, she adopted fleece-lined boots as orthodoxy, even indoors; the other was so thrilled with the concept of third grade canceled on account of snow, she kept her sled parked on the porch and developed rituals to enhance the odds.

With our local-food project still ahead of us, we spent time getting to know our farming neighbors and what they grew, but did our grocery shopping in fairly standard fashion. We relied as much as possible on the organic section and skipped the junk, but were getting our food mostly from *elsewhere.* At some point we meant to let go of the food pipeline. Our plan was to spend one whole year in genuine acquaintance with our food sources. If something in our diets came from outside our county or state, we'd need an extraordinary reason for buying it. ("I want it" is not extraordinary.) Others before us have publicized local food experiments: a Vancouver couple had announced the same intention just ahead of us, and were now reported to be eating dandelions. Our friend Gary Nabhan, in Tucson, had written an upbeat book on his local-food adventures, even after he poisoned himself with moldy mesquite flour and ate some road-kill. We were thinking of a different scenario. We hoped to establish that a normal-ish American family could be content on the fruits of our local foodshed.

It seemed unwise to start on January 1. February, when it came, looked just as bleak. When March arrived, the question started to nag: What are we waiting for? We needed an official start date to begin our 365-day experiment. It seemed sensible to start with the growing season, but what did that mean, exactly? When wild onions and creasy greens started to pop up along the roadsides? I drew the line at our family gleaning the ditches in the style of *Les Misérables.* Our neighborhood already saw us as objects of charity, I'm pretty sure. The cabin where we lived before moving into the farmhouse was extremely primitive quarters for a family of four. One summer when Lily was a toddler I'd gone to the hardware store to buy a big bucket in which to bathe her outdoors, because we didn't have a bathtub or large sink. After the helpful hardware guys offered a few things that weren't quite right, I made the mistake of explaining what I meant to use this bucket for. The store went quiet as all pitying eyes fell upon me, the Appalachian mother with the poster child on her hip.

So, no public creasy-greens picking. I decided we should define New Year's Day of our local-food year with something cultivated and wonderful, the much-anticipated first real vegetable of the year. If the Europeans could make a big deal of its arrival, we could too: we were waiting for asparagus.

❦

Two weeks before spring began on the calendar, I was outside with my boots in the mud and my parka pulled over my ears, scrutinizing the asparagus patch. Four summers earlier, when Steven and I decided this farm would someday be our permanent home, we'd worked to create the garden that would feed us, we hoped, into our old age. "Creation" is a large enough word for the sweaty, muscle-building project that took most of our summer and a lot of help from a friend with earth-moving equipment. Our challenge was the same one common to every farm in southern Appalachia: topography. Our farm lies inside a U-shaped mountain ridge. The forested hillsides slope down into a steep valley with a creek running down its center. This is what's known as a "hollow" (or "holler," if you're from here). Out west they'd call it a canyon, but those have fewer trees and a lot more sunshine.

At the mouth of the hollow sits our tin-roofed farmhouse, some cleared fields and orchards, the old chestnut-sided barn and poultry house, and a gravel drive that runs down the hollow to the road. The cabin (now our guesthouse) lies up in the deep woods, as does the origin of our water supply—a spring-fed creek that runs past the house and along the lane, joining a bigger creek at the main road. We have more than a hundred acres here, virtually all of them too steep to cultivate. My grandfather used to say of farms like these, you could lop off the end of a row and let the potatoes roll into a basket. A nice image, but the truth is less fun. We tried cultivating the narrow stretch of nearly flat land along the creek, but the bottomland between our tall mountains gets direct sun only from late morning to mid-afternoon. It wasn't enough to ripen a melon. For years we'd studied the lay of our land for a better plan.

Eventually we'd decided to set our garden into the south-facing mountainside, halfway up the slope behind the farmhouse. After clearing brambles we carved out two long terraces that hug the contour of the hill—less than a quarter of an acre altogether—constituting our only truly level property. Year by year we've enriched the soil with compost and cover crops, and planted the banks between terraces with blueberry bushes, peach and plum trees, hazelnuts, pecans, almonds, and raspberries. So we have come into the job of overseeing a hundred or so acres of woodlands that exhale oxygen and filter water for the common good, and about 4,000 square feet of tillable land that are meant to feed our family. And in one little corner of *that,* on a June day three years earlier, I had staked out my future in asparagus. It took a full day of trenching and planting to establish what I hope will be the last of the long trail of these beds I've left in the wake of my life.

Now, in March, as we waited for a sign to begin living off the land, this completely bare patch of ground was no burning bush of portent. (Though it was blackened with ash—we'd burned the dead stalks of last year's plants to kill asparagus beetles.) Two months from this day, when it would be warm enough to plant corn and beans, the culinary happening of asparagus would be a memory, this patch a waist-high forest of feathery fronds. By summer's end they'd resemble dwarf Christmas trees covered with tiny red balls. Then frost would knock them down. For about forty-eight weeks of the year, an asparagus plant is unrecognizable to anyone except an asparagus grower. Plenty of summer visitors to our garden have stood in the middle of the bed and asked, "What is this stuff, it's beautiful!" We tell them it's the asparagus patch, and they reply, "No, *this,* these feathery little trees?"

An asparagus spear only looks like its picture for one day of its life, usually in April, give or take a month as you travel from the Mason-Dixon line. The shoot emerges from the ground like a snub-nosed green snake headed for sunshine, rising so rapidly you can just about see it grow. If it doesn't get its neck cut off at ground level as it emerges, it will keep growing. Each triangular scale on the spear rolls out into a branch, until the snake becomes a four-foot tree with delicate needles. Contrary to lore, fat spears are no more tender or mature than thin ones; each shoot begins life with its own particular girth. In the hours after emergence it lengthens, but does not appreciably fatten.

To step into another raging asparagus controversy, white spears are botanically no different from their green colleagues. White shoots have been deprived of sunlight by a heavy mulch pulled up over the plant's crown. European growers go to this trouble for consumers who prefer the stalks before they've had their first blush of photosynthesis. Most Americans prefer the more developed taste of green. (Uncharacteristically, we're opting for the better nutritional deal here also.) The same plant could produce white or green spears in alternate years, depending on how it is treated. If the spears are allowed to proceed beyond their first exploratory six inches, they'll green out and grow tall and feathery like the houseplant known as asparagus fern, which is the next of kin.

Older, healthier asparagus plants produce chunkier, more multiple shoots. Underneath lies an octopus-shaped affair of chubby roots (called a crown) that stores enough starch through the winter to arrange the phallic send-up when winter starts to break. The effect is rather sexy, if you're the type to see things that way. Europeans of the Renaissance swore by it as an aphrodisiac, and the church banned it from nunneries.

The earliest recipes for this vegetable are about 2,500 years old, written in ancient Greek and Egyptian hieroglyphics, suggesting the Mediterranean as the plant's homeland. The Caesars took their asparagus passion to extravagant lengths, chartering ships to scour the empire for the best spears and bring them to Rome. Asparagus even inspired the earliest frozen-food industry, in the first century, when Roman charioteers would hustle fresh asparagus from the Tiber River Valley up into the Alps and keep it buried there in snow for six months, all so it could be served with a big ta-daa at the autumnal Feast of Epicurus. So we are not the first to go to ridiculous lengths to eat foods out of season.

Northern Europeans didn't catch on to asparagus until much later, but by the time they came to the New World, they couldn't leave it behind. It's a long-lived plant whose seeds are spread by birds from gardens to hedgerows, so we have wild populations of it growing in every temperate part of North America where enough rain falls to keep it alive. It likes light soils where the top few inches of the ground freeze in winter. It's especially common along roadsides and railroad right-of-ways that are kept clear of overlying vegetation. Wild asparagus is not always tastiest but offers the advantage of being free. My father used to love bringing home bundles of it in early spring when house calls took him out on the country roads where it grew. The biggest problem is finding it, among tall weeds, in the first day after emergence when it has to be cut. Dad always made it a point to notice tall stands of wild asparagus later in the summer wherever they waved in the breeze. He would stop his car, get out, and mark the location of the patch with orange flagging tape he carried for this purpose. If the highway department or winter weather didn't take down his flags, we'd have well-marked asparagus checkpoints all over the county the next spring. We kids loved the idea of eating anything stolen, especially with lots of butter.

In my adult life I have dug asparagus beds into the property of every house I've owned, and some I rented—even tiny urban lots and student ghettoes—always leaving behind a vegetable legacy waving in the wake of my Johnny-Asparagus-seed life. I suppose in those unsettled years I was aspiring to a stability I couldn't yet purchase. A well-managed asparagus bed can keep producing for twenty or thirty years, but it's a ludicrous commitment to dig one into the yard of a student rental. It's hard work to dig the trench, fill it with compost, and tuck in a row of asparagus crowns ordered from a seed company. Then you wait *three years* for a harvest. A too-young plant gets discouraged when you whack off its every attempt to send up new shoots in the spring, abuse that will make the plant sink into vegetable despair and die.

After the plant has had two full summers to bulk up, then you can begin cutting off its early efforts—but only for two weeks in the first year of harvest. Even with fully mature plants, the harvester must eventually back off from this war between producer and consumer, and let the plant win. After about eight weeks of daily cutting, the asparagus farmer puts away the knife, finally letting the spears pass beyond edibility into the lanky plants they long to be. For most crop species, the season ends when all the vegetable units have been picked and the mother plant dies or gets plowed under. Asparagus is different: its season ends by declaration, purely out of regard for the plant. The key to the next spring's action is the starch it has stored underground, which only happens if the plant has enough of a summer life to beef up its bank account. Of all our familiar vegetables, the season for local, fresh asparagus is the very shortest, for this reason.

Don't expect baby asparagus tips any time other than March, April, or

May, unless you live in New Zealand or South America. Some California farmers have worked out a way to cut a brief second harvest in late fall, but this is exceptional. For most of us, if we see asparagus in any month far removed from April, we're looking at some hard traveling. At our house we only eat asparagus for the weeks it's in season, but during those weeks we eat it *a lot*—the spears must be cut every day. About the same time the asparagus plant is getting weary of our management plan, we're starting to feel the same way. It works out.

From the outlaw harvests of my childhood, I've measured my years by asparagus. I sweated to dig it into countless yards I was destined to leave behind, for no better reason than that I believe in vegetables in general, and this one in particular. Gardeners are widely known and mocked for this sort of fanaticism. But other people fast or walk long pilgrimages to honor the spirit of what they believe makes our world whole and lovely. If we gardeners can, in the same spirit, put our heels to the shovel, kneel before a trench holding tender roots, and then wait three years for an edible incarnation of the spring equinox, who's to make the call between ridiculous and reverent?

The asparagus plant's life history sets it apart, giving it a special edge as the year's first major edible. It's known botanically as a perennial, with a life span of many years. The rest of our plant foods are almost always the leaves, flowers, fruits, or seeds of plants that begin life in spring as seedlings and perish just a few months later when they're frozen by autumn, or eaten, whichever comes first. (The exceptions are the fruits we call "fruits," which grow on berry bushes or trees, and root crops, which operate a bit differently; more about these later.) Annuals tend to grow more quickly than perennials and have been cultivated as food crops for thousands of years. The grass family (whose seed heads are our grains) is especially speedy, with corn the clear winner in the carbon-fixing efficiency race. But asparagus wins the vegetable prize for living longer than one year. That's why it is the very first one to leap up in springtime, offering edible biomass when other vegetables are still at the seedling stage; it had a head start.

The plant's edible portion, however, is direly short-lived. The moment the asparagus neck goes under the knife, an internal starting gun fires "*Go!*" and it begins to decompose, metabolizing its own sugars and trying—because it knows no other plan—to keep growing. It's best eaten the day it is cut, period. When transported, even as refrigerated cargo, the plant's tight bud scales loosen and start to reveal the embryonic arms that were meant to become branches. The fresh stems have the tight, shiny sex appeal of dressed-up matrons on the dance floor of a Latin social club, but they lose their shine and crispness so quickly when the song is over. The sweetness goes starchy.

We don't even know all the things that go wrong in the swan song of a vegetable, since flavor and nutritional value both result from complex interactions of living phytochemical systems. Early in the twentieth century, Japanese food scientist Kikunae Ikeda first documented that asparagus had a flavor that lay outside the range of the four well-known tastes of sweet, sour, bitter, and salty. Its distinctive tang derives from glutamic acid, which Dr. Ikeda named "the fifth taste," or umami. This was, for once, the genuine discovery of a taste sensation. (Later came the invention of an artificial umami flavoring known as monosodium glutamate.) But the flavor chemicals quickly lose their subtlety. Asparagus that's not fresh tastes simple or even bitter, especially when overcooked.

Pushing a refrigerated green vegetable from one end of the earth to another is, let's face it, a bizarre use of fuel. But there's a simpler reason to pass up off-season asparagus: it's inferior. Respecting the dignity of a spectacular food means enjoying it at its best. Europeans celebrate the short season of abundant asparagus as a form of holiday. In the Netherlands the first cutting coincides with Father's Day, on which restaurants may feature all-asparagus menus and hand out neckties decorated with asparagus spears. The French make a similar party out of the release of each year's Beaujolais; the Italians crawl over their woods like harvester ants in the autumn mushroom season, and go gaga over the summer's first tomato.

Waiting for foods to come into season means tasting them when they're good, but waiting is also part of most value equations. Treating foods this way can help move "eating" in the consumer's mind from the

Routine Maintenance Department over to the Division of Recreation. It's hard to reduce our modern complex of food choices to unifying principles, but this is one that generally works: eating home-cooked meals from whole, in-season ingredients obtained from the most local source available is eating well, in every sense. Good for the habitat, good for the body.

A handful of creative chefs have been working for years to establish this incipient notion of a positive American food culture—a cuisine based on our own ingredients. Notable pioneers are Alice Waters of Chez Panisse in San Francisco, and Rick Bayless of Chicago's Frontera Grill, along with cookbook maven Deborah Madison. However, to the extent that it's even understood, this cuisine is widely assumed to be the property of the elite. Granted, in restaurants it can sometimes be pricey, but the do-it-yourself version is not. I am not sure how so many Americans came to believe only our wealthy are capable of honoring a food aesthetic. Anyone who thinks so should have a gander at the kitchens of working-class immigrants from India, Mexico, anywhere really. Cooking at home is cheaper than buying packaged foods or restaurant meals of comparable quality. Cooking *good* food is mostly a matter of having the palate and the skill.

The main barrier standing between ourselves and a local-food culture is not price, but attitude. The most difficult requirements are patience and a pinch of restraint—virtues that are hardly the property of the wealthy. These virtues seem to find precious little shelter, in fact, in any modern quarter of this nation founded by Puritans. Furthermore, we apply them selectively: browbeating our teenagers with the message that they should wait for sex, for example. Only if they wait to experience intercourse under the ideal circumstances (the story goes), will they know its true value. "*Blah blah blah,*" hears the teenager: words issuing from a mouth that can't even wait for the right time to eat tomatoes, but instead consumes tasteless ones all winter to satisfy a craving for everything *now*. We're raising our children on the definition of promiscuity if we feed them a casual, indiscriminate mingling of foods from every season plucked from the supermarket, ignoring how our sustenance is cheapened by wholesale desires.

Waiting for the quality experience seems to be the constitutional article that has slipped from American food custom. If we mean to reclaim it, asparagus seems like a place to start. And if the object of our delayed gratification is a suspected aphrodisiac? That's the sublime paradox of a food culture: restraint equals indulgence.

⚘

On a Sunday in early April we sat at the kitchen table putting together our grocery list for the coming week. The mood was uncharacteristically grave. Normally we all just penciled our necessities onto a notepad stuck onto the fridge. Before shopping, we'd consolidate our foraging plan. The problem now was that we wanted to be a different kind of animal—one that doesn't jump the fence for every little thing. We kept postponing our start date until the garden looked more hospitable, but if we meant to do this for a whole year, we would have to eat in April sooner or later. We had harvested and eaten asparagus now, twice. That was our starting gun: ready, set . . . ready?

Like so many big ideas, this one was easier to present to the board of directors than the stockholders. Our family now convened around the oak table in our kitchen; the milk-glass farmhouse light above us cast a dramatic glow. The grandfather clock ticked audibly in the next room. We'd fixed up our old house in the architectural style known as recycling: we'd gleaned old light fixtures, hardware, even sinks and a bathtub from torn-down buildings; our refrigerator is a spruced-up little 1932 Kelvinator. It all gives our kitchen a comfortable lived-in charm, but at the moment it felt to me like a set where I was auditioning for a part in either *Little House on the Prairie* or *Mommie Dearest*.

They all sat facing me. Steven: my faithful helpmeet, now quite happy to let me play the heavy. And whose idea this whole thing was in the first place, I'm pretty sure. Camille: our redheaded teenager, who in defiance of all stereotypes has the most even temperament in our family. From birth, this child has calmly studied and solved every problem in her path, never asking for special help from the Universe or her parents. At eighteen she now functioned in our household as a full adult, cooking and planning meals often, and was also a dancer who fueled her calorie-intensive passions with devotedly healthy rations. If this project was going to impose a burden, she would feel it. And finally, Lily: earnest, dark-pigtailed persuader and politician of our family who could, as my grandfather would say, charm the socks off a snake. I had a hunch she didn't really know what was coming. Otherwise she'd already be lobbying the loopholes.

Six eyes, all beloved to me, stared unblinking as I crossed the exotics off our shopping list, one by one. All other pastures suddenly looked a whole lot greener than ours. All snack foods come from the land of Oz, it seems, even the healthy ones. Cucumbers, in April? Nope. Those would need passports to reach us right now, or at least a California license. Ditto for those make-believe baby carrots that are actually adult carrots whittled down with a lathe. And all prewashed salad greens emanate from California. Even salad dressing was problematic because of all the ingredients—over a dozen different foods logging their own mileage to get to a salad dressing factory, and then to us. As fuel economy goes, I suppose the refrigerated tropicals like bananas and pineapples are the Humvees of the food world, but multi-ingredient concoctions are sneaky sports cars. I drew a pencil line through one item after another. "Salad dressing is easy to make," I said. The vinegar and oil in our pantry were not local, of course, but with a small effort, thirty seconds spent shaking things together in a jar, we could improve the gas mileage of our vinaigrette. In the herb garden we already had garlic chives and oregano, the hardiest of the spicy Mediterranean perennials, braving the frosts of late winter.

We were getting plenty of local eggs too, so in a reckless burst of confidence I promised to make mayonnaise. It's supposed to be pretty easy. I had a recipe I'd been saving since my high school French class, waiting for the right time to try it, because of one irresistible step that translates as follows: "Whip heartily for two minutes while holding only pleasant thoughts in mind."

Back to the grocery list, trying for that positive mindset: a few more items fell without significant protest. Then I came to block letters in Camille's hand, underlined: FRESH FRUIT, PLEASE???

We were about to cross the Rubicon.

I shifted tactics. Instead of listing what we can't have, I said, we should outline what we knew we *could* get locally. Vegetables and meat—which constitute the bulk of our family's diet—would be available in some form throughout the coming year. We had met or knew of farmers in our county who sold pasture-fed chickens, turkeys, beef, lamb, and pork. Many more were producing vegetables. Like so many other towns, large and small, ours holds a farmers' market where local growers set up booths twice a week from mid-April to October. Soon our garden would also be feeding us. Our starting point would be this: we would take a loyalty oath to our own county's meat and produce, forsaking all others, however sexy the veggies and flesh of California might be.

What else does a family need? Honey would do instead of sugar, in a county where beekeepers are as thick as thieves. Eggs, too, were an easy local catch. Highly processed convenience foods we try to avoid, so those would not be a problem categorically. The other food groups we use in quantity are grains, dairy, and olive oil. We knew of some good dairies in our state, but olives don't grow in this climate. No reasonable substitute exists, and no other oil is produced here. Likewise, we knew of a local mill that ground corn, wheat, and other flours, but its wheat was outsourced from other states. If we purchased only these two foods from partly or wholly nonlocal sources—grains and olive oil—we would be making a sea change in our household economy, keeping an overwhelming majority of our food transactions local. We would try to buy our grains in the least processed, easiest-to-transport form available (bulk flour and some North American rice) so those food dollars would go mostly to farmers.

I put down the list, tried not to chew my pencil, and consciously shut out the image of my children going hungry . . . Lily begging leftovers from somebody's lunchbox at school.

Let me be clear about one thing: I have no interest in playing poor. I've logged some years in frugal material circumstances, first because I was born into a fairly modest rural social order, and later due to years of lousy paychecks. I understand Spam as a reasonable protein source. Both Steven and I have done our time on student stipends, government cheese, and the young-professional years of beans and rice. A huge turning point for me was a day in my mid-thirties when I walked into the supermarket and realized I could buy any ten things there I wanted. Not the lobsters in the aquarium, okay, but not just dented cans in the bargain bin, either. I appreciate the privilege of food choices.

So why give them back voluntarily? It is both extraordinary and unsympathetic in our culture to refrain from having everything one can afford. Yet people do, mostly because they are allergic, or religious. We looked around the table at one another, knowing we had our reasons too. Strange, though, how much it felt like stepping into a spaceship and slamming shut the hatch.

"It won't be that bad," I said. "We're coming into spring."

It wasn't spring yet, however. We were in for some lean months before the midsummer bounty started flooding us with the real rewards of local flavor and color. But April is a forward-looking time on the farm, full of work and promise. It seemed best to jump in now. Sink or swim.

Hedging, we decided to allow ourselves one luxury item each in limited quantities, on the condition we'd learn how to purchase it through a channel most beneficial to the grower and the land where it grows. Steven's choice was a no-brainer: coffee. If he had to choose between coffee and our family, it might be a tough call. Camille's indulgence of choice was dried fruit; Lily's was hot chocolate. We could get all those from fair trade organizations that work with growers in Africa, Asia, and South America. I would rely on the same sources for spices that don't grow locally; a person can live without turmeric, cinnamon, and cloves, I've heard, but I am not convinced. Furthermore, dry goods like these, used by most households in relatively tiny quantities, don't register for much on the world's gas-guzzling meter.

With that, our hopeful agreement in place with its bylaws and backstops, we went back one last time to our grocery list. Almost everything left fell into the grains category: bread flour and rolled oats are big-ticket items, since Steven makes most of our bread, and oatmeal is our cool-weather breakfast of choice. We usually buy almonds and raisins to put in our oatmeal, but I crossed those off, hoping to find local substitutes. Then we came back around to the sticky one. FRESH FRUIT, PLEASE???

At the moment, fruits were only getting ripe in places where people were wearing bikinis. Correlation does not imply causation: putting on our swimsuits would not make it happen here. "Strawberries will be coming in soon," I said, recognizing this as possibly the first in a long line of pep talks to come.

The question remained, What about now?

"Look," I said, "the farmers' market opens this Saturday. We'll go see what's there." Around the table went the *Oh sure, Mom* face that mothers everywhere know and do not love.

⚜

Saturday dawned dark, windy, and fiercely cold. The day's forecast was for *snow.* Spring had been slapped down by what they call around here "dogwood winter," a hard freeze that catches the dogwoods in bloom—and *you* thinking you were about to throw your sweaters into the cedar chest. April fool.

The cold snap was worrisome for our local orchards, since apple and peach trees had broken dormancy and blossomed out during the last two sunny weeks. They could lose the year's productivity to this one cold spell. If anybody was going to be selling fruit down at the farmers' market today, in the middle of blasted dogwood winter, I'd be a monkey's uncle. Nevertheless, we bundled up and headed on down. We have friends who sell at the market, some of whom we hadn't seen in a while. On a day like this they'd need our moral support.

It was a grim sight that met us in the parking lot. Some of the vendors huddled under awnings that snapped and flapped like the sails of sinking ships in a storm. Others had folded up their tents and stood over their boxes with arms crossed and their backs to the mean wind. Only eight vendors had turned out today, surely the bravest agricultural souls in the county, and not another customer in sight. What would they have this early, anyway—the last of last year's shriveled potatoes?

Hounded by the dogs of *Oh sure, Mom,* I made up my mind to buy something from everyone here, just to encourage them to come back next week. My farm advocacy work for the day.

We got out of the car, pulled our hoods over our ears, and started our tour of duty. Every vendor had something better than shriveled potatoes. Charlie, a wiry old man who is the self-appointed comedian of our market, was short on cheer under the circumstances but did have green onions. We'd run out of our storage onions from last year's garden, and missed them. At least half our family's favorite dishes begin with a drizzle of oil in the skillet, a handful of chopped onions and garlic tossed in. We bought six fat bundles of Charlie's onions. This early in the season their white bulbs were only the size of my thumb, but when chopped with their green tops they would make spicy soups and salads.

From Mike and Paul, at the next two booths, we bought turkey sausage and lamb. At the next, the piles of baby lettuce looked to me like money in the bank, and I bagged them. Fruitless though our lives might be, we would have great salads this week, with chunks of sausage, hard-

How to Find a Farmer

Whether you're a rural or urban consumer, it's easier than ever to find local or regionally grown food. Following the passage of the Farmer-to-Consumer Direct Marketing Act of 1976, active U.S. farmers' markets have grown from about 350 to well over 3,500 today, or an average of 75 per state. Most urban areas host farmers' markets from spring until fall; some are open all year. Market rules usually guarantee that the products are fresh and local.

Farmers' markets are also a good place to ask about direct sales from farmer to consumer. Options include roadside stands, U-pick operations, artisanal products, buyers' clubs, and community-supported agriculture (CSA). In a typical CSA, subscribers pay a producer in early spring and then receive a weekly share of the produce all season long.

Don't be afraid to ask producers what else they might have available. Someone selling eggs on Saturday morning probably has eggs the rest of the week. Farmers can also tell you which local stores may sell their eggs, meat, or produce. Many grocery and health-food stores now stock local foods, as more consumers ask. Locally owned stores are better bets, since chains rely on regional distribution. And don't overlook small ethnic or specialty grocers.

You can find your nearest farmers' markets and local producers on the USDA Web site: www.ams.usda.gov/farmersmarkets. Also check: www.localharvest.org and www.csacenter.org.

STEVEN L. HOPP

boiled eggs, and experimental vinaigrettes. Next down the line we found black walnuts, painstakingly shelled out by hand. Walnut is a common wild tree here, but almost nobody goes to the trouble to shell them—nowhere but at the farmers' market would you find local nuts like these. The vendor offered us a sample, and we were surprised by the resinous sweetness. They would be good in our oatmeal and a spectacular addition to Steven's whole-grain bread.

Each of our purchases so far was in the one- to three-dollar range except the nuts, which were seven a pound—but a pound goes a long way. I frankly felt guilty getting so much good fresh stuff for so little money, from people who obviously took pains to bring it here. I pushed on to the end, where Lula sold assorted jams and honey. We were well fixed for these already, given to us by friends or made ourselves last fall. Lula's three children shivered on the ground, bundled in blankets. I scanned the table harder, unwilling to walk away from those kids without plunking down some bucks.

That's where I spotted the rhubarb. Big crimson bundles of it, all full of itself there on the table, loaded with vitamin C and tarty sweetness and just about screaming, "Hey, look at me, I'm fruit!" I bought all she had, three bundles at three dollars apiece: my splurge of the day.

Rhubarb isn't technically fruit, it's an overgrown leaf petiole, but it's a fine April stand-in. Later at home when we looked in Alice Waters's *Chez Panisse Fruit* for some good recipes, we found Alice agreed with us on this point. "Rhubarb," she writes, "is the vegetable bridge between the tree fruits of winter and summer." That poetic injunction sent us diving into the chest freezer, retrieving the last package of our frozen Yellow Transparent apple slices from last summer. For dinner guests we threw together an apple-rhubarb cobbler to ring out the old year and ring in the new. Rhubarb, the April fruit. I'm a monkey's uncle.

If not for our family's local-food pledge to roust us out of our routine, I'm sure we would not have bothered going down to the market on that miserable morning. Most of us are creatures so comforted by habit, it can take something on the order of religion to invoke new, more conscious behaviors—however glad we may be afterward that we went to the trouble. Tradition, vows, something *like* religion was working for us now, in our search for a new way to eat. It had felt arbitrary when we sat around the table with our shopping list, making our rules. It felt almost silly to us, in fact, as it may now seem to you. Why impose restrictions on ourselves? Who cares?

The fact is, though, millions of families have food pledges hanging over their kitchens—subtle rules about going to extra trouble, cutting the pasta by hand, rolling the sushi, making with care instead of buying on the cheap. Though they also may be busy with jobs and modern life, people the world over still take time to follow foodways that bring their families happiness and health. My family happens to live in a country where the main foodway has a yellow line painted down the middle. If we needed rules we'd have to make our own, going on faith that it might bring us something worthwhile.

On Saturday morning at the market as we ducked into the wind and started back toward our car, I clutched my bags with a heady sense of accomplishment. We'd found a lot more than we'd hoped for. We chatted a little more with our farmer friends who were closing up shop behind us, ready to head home too. Back to warm kitchens, keeping our fingers crossed in dogwood winter for the fruits of the coming year.

The Truth About Asparagus

BY CAMILLE

When I was little, asparagus season wasn't my favorite time of year. The scene was the same every spring: my parents would ceremoniously bring the first batch of asparagus to the table, delicately cooked and arranged on a platter, and I would pull up my nose.

"Just try one, Camille, you might like it this year," Mom would say, enthusiastically serving me a single green shoot.

Yeah, right! I would think. Ever so gingerly, I would spear the menacing vegetable with my fork and bring the tip of it to my tongue. That's as far as I would get before dramatically wincing and flicking the asparagus down in disgust.

"Okay, good," she would respond with a sparkle in her eye, "more for us then!"

I decided that even if I grew up to love asparagus, I would always tell my Mom I hated it. I didn't want her to be right about my personal preferences. My stubborn plot fell by the wayside, however, several springs later when I bit into a tender asparagus shoot and realized how delicious it was. I decided a lifetime of lies and deception wouldn't be worth it.

Little kids seem to choose arbitrarily which foods they absolutely won't eat (for me it was asparagus; for my sister it's peppers). Most of us eventually get eased or taunted out of thumb-sucking, but it's easy to carry childish food decisions into adulthood. I feel lucky that I was pushed out of that box. Not beaten with a stick, but exposed continuously by grown-ups who were obviously enjoying something I was too immature to like. Now my family laughs about my days of asparagus hatred, especially since I've become such a veggie hog, they have to move the dish out of my reach so there might be a chance of having some leftovers. I think my body won out over my brain, recognizing vegetables as being crucial to good health. Asparagus is one of the best natural sources of folic acid, vitamins A, C, and K, some B vitamins, potassium, phosphorus, and glutathione, which is a potent antioxidant and anticarcinogen. Nutritional benefits can motivate some to incorporate vegetables into their diets, but for most people taste is the key. I'm glad my parents did their eat-your-vegetables act in a household where fresh vegetables were the norm. If they had tried this with the flavorless, transported kind widely available in this country, I might still be the wincing asparagus drama queen.

When asparagus is rolling in from the garden every day, the simplest way to prepare it is a quick sauté over high heat. This technique caramelizes the sugars and brings out the very best of the asparagus flavor. Using butter in the skillet (alone or in combination with olive oil) provides just enough water to steam the spears as they're also being sautéed. Gently shake them in the pan, cooking just until they turn bright green with slightly blackened undersides. Grilling asparagus in a grill basket produces similar tasty results. Steaming is fine too, for a minimalist treatment. Freshly harvested asparagus is so tender, we never bother with peeling it.

Naturally, in April we can't live on asparagus alone. In late winter and spring the local food enthusiast will have to rely on some foods stored from last year—we use our stored root vegetables, winter squash, dried tomatoes, and some frozen packages of last year's garden goods to round out the spring greens. Some herbs, like cilantro and oregano, may start coming in now along with baby greens, early broccoli, peas, and onions. A typical week's meal plan might look like the following. Always bear in mind, with this and other meal plans, that leftovers and creative recycling are options. Every chicken dinner at our house, for example, ends with the carcass going into the stock pot (basically a big kettle of water) to boil for several hours. When the stock is thick and gold-colored and the meat falls off the bones, they're both ready to freeze or just refrigerate for a couple of days for an easy chicken salad and/or soup.

LATE WINTER MEAL PLAN

Sunday ~ Herbed roast chicken with potatoes, beets, and carrots

Monday ~ Chef salad with boiled eggs or sausage, green onions, and dried tomatoes, with bread

Tuesday ~ Chicken soup with carrots, kale, and rice (or fresh bread)

Wednesday ~ Twice-baked potatoes stuffed with sautéed spring onions, broccoli, and cheese

Thursday ~ Asparagus omelette or frittata and baby-greens salad

Friday ~ Pizza with dried tomatoes, olives, feta, and sautéed onions

Saturday ~ Lamb chops, mesclun salad, asparagus, and rhubarb-apple crisp

3 · SPRINGING FORWARD

April is the cruelest month, T. S. Eliot wrote, by which I think he meant (among other things) that springtime makes people crazy. We expect too much, the world burgeons with promises it can't keep, all passion is really a setup, and we're doomed to get our hearts broken yet again. I agree, and would further add: Who cares? Every spring I go there anyway, around the bend, unconditionally. I'm a soul on ice flung out on a rock in the sun, where the needles that pierced me begin to melt all as one.

On the new edge of springtime when I stand on the front porch shading my eyes from the weak morning light, sniffing out a tinge of green on the hill and the scent of yawning earthworms, oh, boy, *then*! I roll like a bear out of hibernation. The maple buds glow pink, the forsythia breaks into its bright yellow aria. These are the days when we can't keep ourselves indoors around here, any more than we believe what our eyes keep telling us about the surrounding land, i.e., that it is still a giant mud puddle, now lacking its protective covering of ice. So it comes to pass that one pair of boots after another run outdoors and come back mud-caked—more shoes than we even knew we had in the house, proliferating like wild portobellos in a composty heap by the front door. So what? Noah's kids would have felt like this when the flood had *almost* dried up: muddy boots be hanged. Come the end of the dark days, I am more than joyful. I'm nuts.

Our household was a week into high spring fever when Lily and I decided it was safe to carry out some of the seedlings we'd started indoors on homemade shelves under fluorescent bulbs. The idea of eating from our home ground for a year had moved us to start a grocery store from seed. We'd been tucking tomatoes into seed flats since January, proceeding on to the leafy greens and broccoli, the eggplants, peppers, okra, and some seed catalog mysteries we just had to try: rock melons, balloon flowers! By mid-March our seed-starting shelves were overwhelmed.

Then began the lover's game we play with that irresistible rascal partner, March weather. He lulls us into trust one day with smiles and sunshine and daytime highs in the sixties, only to smack us down that very night with a hard freeze. On our farm we have a small unheated greenhouse that serves as a halfway house, a battered-seedling shelter if you will, where the little greenlings can enjoy the sun but are buffered from cold nights by five degrees or so. Usually that's enough of a safety margin. But then will come a drear night when the radio intones, *Lows tonight in the teens,* and we run to carry everything back inside, dashing in the back door, setting flats all over the table and counters until our kitchen looks like the gullet and tonsils of a Chia Pet whale.

This is what's cruel about springtime: no matter how it treats you, you can't stop loving it. If the calendar says it's the first day of spring, *it is.* Lily and I had been lured up the garden path, literally, carrying flats of broccoli, spinach, and cilantro seedlings to the greenhouse on the bank just uphill from the house.

The greatest rewards of living in an old farmhouse are the stories and the gardens, if they're still intact in any form. We are lucky enough to have both. The banks all around us are crowded with flowering shrubs and hummocks of perennial bulbs that never fail to please and startle us, like old friends leaping from behind the furniture to yell, "Surprise!" These flowers are gifts from a previous century, a previous dweller here—a tale, told in flowers, of one farm wife's fondness for beauty and this place. In a few more months we'd be drunk on the scent of Lizzie Webb's mock oranges and lilacs, but the show begins modestly in April with her tiny Lenten roses, white-petaled snowdrops, and the wildish little daffodils called jonquils that have naturalized all over the grassy slopes. As Lily and I walked single file up the path to the greenhouse, I noticed these were up, poking their snub, yellow-tipped noses through a fringe of leaves.

"Oh, Mama," Lily cried, "look what's about to bloom—the *tranquils.*"

There went the last of the needles of ice around my heart, and I understood I'd be doomed to calling the jonquils *tranquils* for the rest of my days. Lily is my youngest. Maybe you know how these things go. In our family, those pink birds with the long necks are called *flingmos* because of how their real name was cutely jumbled by my brother's youngest child—and that was, yikes, twenty years ago.

That's how springtime found us: grinning from ear to ear, hauling out our seedlings, just as the rest of our neighborhood began to haul out the plastic lawn flingmos and little Dutch children kissing and those spooky plywood silhouettes of cowboys leaning against trees. Lawn decoration is high art in the South, make no mistake about it. Down here in Dixie, people do not just fling out a couple or three strands of twinkly lights in December and call it a year, no *ma'am.* In our town it's common practice for folks to jazz up their yards even for the minor decorative occasions of Valentine's Day and the Fourth of July, and on Easter to follow an unsettling tradition of hanging stuffed bunny rabbits from the crab apple trees (by the neck, until dead, to all appearances). If they're serious about it, they'll surround the cottontailed unfortunate with a ferociously cheerful crop of dangling plastic eggs.

I grew up in this territory, and my recollection from childhood is that every community had maybe just *one* person with a dolled-up yard—such as my paternal grandmother, who was known to be gifted in the decor department—but for the average citizen it was enough to plop a tire out there and plant petunias in it. Or marigolds in a defunct porcelain toilet, in neighborhoods with a different sense of decorum. But those days are gone, my friend, and never more will it be so easy to keep up with the Joneses.

Our recent construction work on the farmhouse had turned the front yard into what you might euphemistically call a blank slate. (Less euphemistically: mud.) With my hands full of vegetable seedlings and the wild rumpus of spring in my head, I considered planting lettuces and red Russian kale in place of a lawn, maybe in the visage of the Mona Lisa. Or letters that spelled out something, like "MOW ME." But in the end, after considering the fact that nobody can see our house from the road, I desisted in fancying myself the Michelangelo of Kale. These greens would go into the vegetable bed, though I'll tell you I am not above tucking a tomato or two into a handsome perennial border. Vegetables are gorgeous, especially spring greens, arriving brightly as they do after a long winter of visually humble grains and stored root crops.

Bronze Arrowhead lettuces, Speckled Trout romaine, red kale—this is the rainbow of my April garden, and you'll find similar offerings then at a farmers' market or greengrocer. It's the reason I start our vegetables from seed, rather than planting out whatever the local nursery has to offer: variety, the splendor of vegetables. I have seen women looking at jewelry ads with a misty eye and one hand resting on the heart, and I only know what they're feeling because that's how I read the seed catalogs in January. In my mind the garden grows and grows, as I affix a sticky note to every page where there's something I need. I swoon over names like Moon and Stars watermelon, Cajun Jewel okra, Gold of Bacau pole bean, Sweet Chocolate pepper, Collective Farm Woman melon, Georgian Crystal garlic, mother-of-thyme. Steven walks by, eyes the toupee of yellow sticky notes bristling from the top of the catalog, and helpfully asks, "Why don't you just mark the one you *don't* want to order?"

Heirloom vegetables are irresistible, not just for the poetry in their names but because these titles stand for real stories. Vegetables acquire histories when they are saved as seeds for many generations, carefully maintained and passed by hand from one gardener to another. Heirlooms are open-pollinated—as opposed to hybrids, which are the onetime product of a forced cross between dissimilar varieties of a plant. These crosses do rely on the sex organs of the plant to get pollen into ovaries, so they're still limited to members of the same species: tall corn with early corn, for example, or prolific cucumbers with nonprickly ones, in blends that combine the ideal traits of both parents for one-time-only offspring. These whiz-kid hybrid seeds have slowly colonized and then dominated our catalogs and our croplands. Because of their unnatural parentage they offer special vigor, but the next generation from these crosses will be of an unpredictable and mostly undesirable character. Thus, hybrid seeds have to be purchased again each year from the companies that create them.

Genetic modification (GM) takes the control even one step further from the farmer. Seed companies have made and sold hybrids since the 1920s (starting with the Hybrid Corn Company, now a subsidiary of DuPont), but GM is a newer process involving direct manipulation of genes in the laboratory. Freed from the limits of natural sex, the gene engineer may combine traits of creatures that aren't on speaking terms in the natural world: animal or bacterial genes spliced into the chromosomes of plants, for example, and vice versa. The ultimate unnatural product of genetic engineering is a "terminator gene" that causes a crop to commit genetic suicide after one generation, just in case some maverick farmer might want to save seed from his expensive, patented crop, instead of purchasing it again from the company that makes it.

By contrast to both GM and hybridization, open-pollinated heirlooms are created the same way natural selection does it: by saving and reproducing specimens that show the best characteristics of their generation, thus gradually increasing those traits in the population. Once bred to a given quality, these varieties yield the same characteristics again when their seeds are saved and grown, year after year. Like sunshine, heirloom seeds are of little interest to capitalism if they can't be patented or owned. They have, however, earned a cult following among people who grow or buy and eat them. Gardeners collect them like family jewels, and Whole Foods Market can't refrain from poetry in its advertisement of heirlooms, claiming that the tomatoes in particular make a theatrical entrance in the summertime, "stealing the summer produce scene. Their charm is truly irresistible. Just the sound of the word 'heirloom' brings on a warm, snuggly, bespectacled grandmother knitting socks and baking pies kind of feeling."

They've hired some whiz-bang writers down at Whole Foods, for sure, but the hyperbolic claims are based on a genuine difference. Even a child who dislikes tomatoes could likely tell the difference between a watery mass-market tomato and a grandmotherly (if not pie-baking) heirloom. Vegetables achieve historical status only if they deserve it. Farmers are a class of people not noted for sentimentality or piddling around. Seeds get saved down the generations for a reason, or for many, and in the case of vegetables one reason is always flavor. Heirlooms are the tangiest or sweetest tomatoes, the most fragrant melons, the eggplants without a trace of bitterness.

Most standard vegetable varieties sold in stores have been bred for uniform appearance, mechanized harvest, convenience of packing (e.g. square tomatoes), and a tolerance for hard travel. None of these can be mistaken, in practice, for actual flavor. Homegrown tomatoes are famously superior to their supermarket counterparts, but the disparity is just as great (in my experience) for melons, potatoes, asparagus, sweet corn, broccoli, carrots, certain onions, and the Japanese edible soybeans called edamame. I have looked for something to cull from my must-grow list on the basis of its being reasonably similar to the supermarket version. I have yet to find that vegetable.

How did supermarket vegetables lose their palatability, with so many people right there watching? The Case of the Murdered Flavor was a contract killing, as it turns out, and long-distance travel lies at the heart of the plot. The odd notion of transporting fragile produce dates back to the early twentieth century when a few entrepreneurs tried shipping lettuce and artichokes, iced down in boxcars, from California eastward over the mountains as a midwinter novelty. Some wealthy folks were charmed by the idea of serving out-of-season (and absurdly expensive) produce items to their dinner guests. It remained little more than an expensive party trick until mid-century, when most fruits and vegetables consumed in North America were still being produced on nearby farms.

Then fashion and marketing got involved. The interstate highway system became a heavily subsidized national priority, long-haul trucks were equipped with refrigeration, and the cost of gasoline was nominal. The state of California aggressively marketed itself as an off-season food producer, and the American middle class opened its maw. In just a few decades the out-of-season vegetable moved from novelty status to such an ordinary item, most North Americans now don't know what out-of-season means.

While marketers worked out the logistics of moving every known vegetable from every corner of the planet to somewhere else, agribusiness learned to breed varieties that held up in a boxcar, truck, or ship's cargo hold. *Indestructible* vegetables, that is to say: creations that still looked decent after a road trip. Vegetable farmers had little choice but to grow what the market demanded. In the latter half of the twentieth century they gradually dropped from their repertoire thousands of flavorful varieties traditionally grown for the table, concentrating instead on the handful of new varieties purchased by transporters, restaurant chains, and processed-food manufacturers. Modern U.S. consumers now get to taste less than 1 percent of the vegetable varieties that were grown here a century ago. Those old-timers now lurk only in backyard gardens and on farms that specialize in direct sales—if they survive at all. Many heirlooms have been lost entirely.

The same trend holds in other countries, wherever the influence of industrial-scale agriculture holds sway. In Peru, the original home of potatoes, Andean farmers once grew some four thousand potato varieties, each with its own name, flavor, and use, ranging in size from tiny to gigantic and covering the color spectrum from indigo-purple to red, orange, yellow, and white. Now, even in the regions of Peru least affected by the modern market, only a few dozen potato varieties are widely grown. Other indigenous crops elsewhere in the world have followed the same path, with the narrowing down of corn and amaranth varieties in Central America, squashes in North America, apples in Europe, and grains in the Middle East. And it's not just plant varieties but whole species that are being lost. As recently as ten years ago farmers in India still grew countless indigenous oil crops, including sesame, linseed, and mustards; in 1998 all the small mills that processed these oils were ordered closed, the same year a ban on imported soy oil was lifted. A million villages lost their mills, ten million farmers lost their living, and GM soy found a vast new market.

According to Indian crop ecologist Vandana Shiva, humans have eaten some 80,000 plant species in our history. After recent precipitous changes, three-quarters of all human food now comes from just eight species, with the field quickly narrowing down to genetically modified corn, soy, and canola. If woodpeckers and pandas enjoy celebrity status on the endangered-species list (dubious though such fame may be), food crops are the forgotten commoners. We're losing them as fast as we're losing rain forests. An enormous factor in this loss has been the new idea of plant varieties as patentable properties, rather than God's gifts to humanity or whatever the arrangement was previously felt to be, for all of prior history. God lost that one in 1970, with the Plant Variety Protection Act. Anything owned by humans, of course, can be taken away from others; the removal of crop control from farmers to agribusiness has been power-

The Strange Case of Percy Schmeiser

In 1999, a quiet middle-aged farmer from Bruno, Saskatchewan, was sued by the largest biotech seed producer in the world. Monsanto Inc. claimed that Percy Schmeiser had damaged them, to the tune of $145,000, by having their patented gene in some of the canola plants on his 1,030 acres. The assertion was not that Percy had actually planted the seed, or even that he obtained the seed illegally. Rather, the argument was that the plants on Percy's land contained genes that belonged to Monsanto. The gene, patented in Canada in the early 1990s, gives genetically modified (GM) canola plants the fortitude to withstand spraying by glyphosate herbicides such as Roundup, sold by Monsanto.

Canola, a cultivated variety of rapeseed, is one of over three thousand species in the mustard family. Pollen from mustards is transferred either by insects, or by wind, up to one-third of a mile. Does the patented gene travel in the pollen? Yes. Are the seeds viable? Yes, and can remain dormant up to ten years. If seeds remain in the soil from previous years, it's illegal to harvest them. Further, if any of the seeds from a field contain the patented genes, it is illegal to save them for use. Percy had been saving his canola seeds for fifty years. Monsanto was suing for possession of intellectual property that had drifted onto his plants. The laws protect possession of the gene itself, irrespective of its conveyance. Because of pollen drift and seed contamination, the Monsanto genes are ubiquitous in Canadian canola.

Percy lost his court battles: he was found guilty in the Federal Court of Canada, the conviction upheld in the court of appeals. The Canadian Supreme Court narrowly upheld the decision (5–4), but with no compensation to Monsanto. This stunning case has drawn substantial attention to the problems associated with letting GM genies out of their bottle. Organic canola farmers in Saskatchewan have now sued Monsanto and another company, Aventis, for making it impossi-

ful and swift. Six companies—Monsanto, Syngenta, DuPont, Mitsui, Aventis, and Dow—now control 98 percent of the world's seed sales. These companies invest heavily in research whose purpose is to increase food production capacity only in ways that can be controlled strictly. Terminator technology is only one (extreme) example. The most common genetic modifications now contained in most U.S. corn, soy, cotton, and canola do one of two things: (1) put a bacterial gene into the plant that kills caterpillars, or (2) alter the crop's physiology so it withstands the her-

ble for Canadian farmers to grow organic canola. The National Farmers Union of Canada has called for a moratorium on all GM foods. The issue has spilled over the borders as well. Fifteen countries have banned import of GM canola, and Australia has banned all Canadian canola due to the unavoidable contamination made obvious by Monsanto's lawsuit. Farmers are concerned about liability, and consumers are concerned about choice. Twenty-four U.S. states have proposed or passed various legislation to block or limit particular GM products, attach responsibility for GM drift to seed producers, defend a farmer's right to save seeds, and require seed and food product labels to indicate GM ingredients (or allow "GM-free" labeling).

The U.S. federal government (corporate-friendly as ever) has stepped in to circumvent these proconsumer measures. In 2006 Congress passed the National Uniformity for Food Act, which would eliminate more than two hundred state-initiated food safety and labeling laws that differ from federal ones. Thus, the weakest consumer protections would prevail (but they're *uniformly* weak!). Here's a clue about who really benefits from this bill: it's endorsed by the American Frozen Food Institute, ConAgra, Cargill, Dean Foods, Hormel, and the National Cattlemen's Beef Association. It's opposed by the Consumers Union, the Sierra Club, the Union of Concerned Scientists, the Center for Food Safety, and thirty-nine state attorney generals. Keeping GM's "intellectual" paws out of our bodies, and our fields, is up to consumers who demand full disclosure on what's in our food.

For more information, visit www.biotech-info.net or www.organicconsumers.org.

STEVEN L. HOPP

bicide Roundup, so that chemicals can be sprayed over the crop. (The crop stays alive, the weeds die.) If you guessed Monsanto controls sales of both the resistant seed and the Roundup, give yourself a star. If you think you'd never eat such stuff, you're probably wrong. GM plants are virtually everywhere in the U.S. food chain, but don't have to be labeled, and aren't. Industry lobbyists intend to keep it that way.

Monsanto sells many package deals of codependent seeds and chemicals, including so-called traitor technologies in which a crop's disease resistance relies on many engineered genes resting in its tissues—genes that can only be turned on, as each disease arises, by the right chemical purchased from Monsanto.

It's hardly possible to exaggerate the cynicism of this industry. In internal reports, Monsanto notes "growers who save seed from one year to the next" as significant competitors, and allocates a $10 million budget for investigating and prosecuting seed savers. Agribusinesses can patent plant varieties for the purpose of removing them from production (Seminis dropped 25 percent of its total product line in one recent year, as a "cost-cutting measure"), leaving farmers with fewer options each year. The same is true for home gardeners, who rarely suspect when placing seed orders from Johnny's, Territorial, Nichols, Stokes, and dozens of other catalogs that they're likely buying from Monsanto. In its 2005 annual report, Monsanto describes its creation of American Seeds Inc. as a licensing channel that "allows us to marry our technology with the high-touch, local face of regional seed companies." The marriage got a whopping dowry that year when Monsanto acquired Seminis, a company that already controlled about 40 percent of the U.S. vegetable seed market. Garden seed inventories show that while about 5,000 nonhybrid vegetable varieties were available from catalogs in 1981, the number in 1998 was down to 600.

Jack Harlan, a twentieth-century plant geneticist and author of the classic *Crops and Man,* wrote about the loss of genetic diversity in no uncertain terms: "These resources stand between us and catastrophic starvation on a scale we cannot imagine. . . . The line between abundance and disaster is becoming thinner and thinner."

The "resources" Harlan refers to are old varieties, heirlooms and land races—the thousands of locally adapted varieties of every crop plant important to humans (mainly but not limited to wheat, rice, corn, and potatoes), which historically have been cultivated in the region where each crop was domesticated from its wild progenitor. Peru had its multitude of potatoes, Mexico its countless kinds of corn, in the Middle East an infinity of wheats, each subtly different from the others, finely adapted to its region's various microclimates, pests and diseases, and the needs of the humans who grew it. These land races contain a broad genetic heritage that prepares them to coevolve with the challenges of their environments.

Disease pathogens and their crop hosts, like all other predators and prey, are in a constant evolutionary dance with each other, changing and improving without cease as one evolves a slight edge over its opponent, only to have the opponent respond to this challenge by developing its own edge. Evolutionary ecologists call this the Red Queen principle (named in 1973 by Leigh Van Valen), after the Red Queen in *Through the Looking Glass,* who observed to Alice: "In this place it takes all the running you can do to keep in the same place." Both predator and prey must continually change or go extinct. Thus the rabbit and fox both get faster over the generations, as their most successful offspring pass on more genes for speediness. Humans develop new and stronger medicines against our bacterial predators, while the bacteria continue to evolve antibiotic-resistant strains of themselves. (The people who don't believe in evolution, incidentally, are just as susceptible as the rest of us to this observable occurrence of evolution. Ignorance of the law is no excuse.)

Plant diseases can attack their host plants in slightly new ways each season, encouraged by changes in prevailing conditions of climate. This is where genetic variability becomes important. Genetic engineering cannot predict or address such broad-spectrum challenges. Under highly varied environmental conditions, the resilience of open-pollinated land races can be compared approximately with the robust health of a mixed-breed dog versus the finicky condition of a pooch with a highly inbred pedigree. The mongrel may not perform as predictably under perfectly controlled conditions, but it has the combined smarts and longevity of all the sires that ever jumped over the fence. Some of its many different genes are likely to come in handy, in a pinch.

The loss of that mongrel vigor puts food systems at risk. Crop failure is a possibility all farmers understand, and one reason why the traditional farmstead raised many products, both animal and vegetable, unlike the monocultures now blanketing our continent's midsection. History has regularly proven it drastically unwise for a population to depend on just a few varieties for the majority of its sustenance. The Irish once depended on a single potato, until the potato famine rewrote history and truncated many family trees. We now depend similarly on a few corn and soybean strains for the majority of calories (both animal and vegetable) eaten by U.S. citizens. Our addiction to just two crops has made us the fattest people who've ever lived, dining just a few pathogens away from famine.

⚜

Woe is us, we overfed, undernourished U.S. citizens—we are eating poorly for so very many reasons. A profit-driven, mechanized food industry has narrowed down our variety and overproduced corn and soybeans. But we let other vegetables drop from the menu without putting up much of a fight. In our modern Café Dysfunctional, "eat your vegetables" has become a battle cry of mothers against presumed unwilling subjects. In my observed experience, boys in high school cafeterias treat salad exactly as if it were a feminine hygiene product, and almost nobody touches the green beans. Broccoli was famously condemned in the 1990s from the highest office in the land. What's a mother to do? Apparently, she's to shrug and hand the kids a gigantic cup of carbonated corn syrup. Corn is a vegetable, right? Good, because on average we're consuming 54.8 gallons of soft drinks, per person, per year.

Mom is losing, no doubt, because our vegetables have come to lack two features of interest: nutrition and flavor. Storage and transport take predictable tolls on the volatile plant compounds that subtly add up to taste and food value. Breeding to increase shelf life also has tended to decrease palatability. Bizarre as it seems, we've accepted a tradeoff that amounts to: "Give me every vegetable in every season, even if it tastes like a cardboard picture of its former self." You'd think we cared more about the *idea* of what we're eating than about what we're eating. But then, if you examine the history of women's footwear, you'd think we cared more about the idea of showing off our feet than about, oh, for example, walking. Humans can be fairly ridiculous animals.

I wouldn't dare predict what will happen next with women's footwear, but I did learn recently that the last couturier in China who made shoes for bound feet is about to go out of business (his last customers are all in their nineties), and in a similar outburst of good sense, the fashion in vegetables may come back around to edibility. Heirlooms now sometimes appear by name on restaurant menus, and are becoming an affordable mainstay of farmers' markets. Flavor in food is a novelty that seems to keep customers coming back.

Partly to supply this demand, and partly because some people have cared all along, national and international networks exist solely to allow farmers and gardeners to exchange and save each other's heirloom seeds. The Seed Savers' Exchange, headquartered on a farm in Decorah, Iowa, was founded by Diane and Kent Whealy after Diane's grandfather left her the seeds of a pink tomato that his parents brought from Bavaria in the 1870s. Seeds are living units, not museum pieces; in jars on a shelf their viability declines with age. Diane and Kent thought it seemed wise to move seed collections into real gardens. Their idea has grown into a network of 8,000 members who grow, save, and exchange more than 11,000 varieties from their own gardening heritage, forming an extensive living collection. The *Seed Savers' Yearbook* makes available to its members the seeds of about twice as many vegetable varieties as are offered by all U.S. and Canadian mail-order seed catalogs combined. Native Seeds/SEARCH is a similar network focused on Native American crops; the North American Fruit Explorers promotes heirloom fruit and nut tree collections. Thanks to these and other devotees, the diversity of food crops is now on the rise in the United States.

The world's largest and best known save-the-endangered-foods organization is Slow Food International. Founded in Italy in 1986, the organization states that its aim is "to protect the pleasures of the table from the homogenization of modern fast food and life." The group has 83,000 members in Italy, Germany, Switzerland, the United States, France, Japan, and Great Britain. The organization promotes gastronomic culture, conserves agricultural biodiversity and cultural identities tied to food production, and protects traditional foods that are at risk of extinction. Its Ark of Taste initiatives catalog and publicize forgotten foods—a Greek fava bean grown only on the island of Santorini, for instance, or the last indigenous breed of Irish cattle. Less than ten years after its launch, the Italian Ark has swelled to hold some five hundred products. A commission in the United States now catalogs uniquely American vegetable and animal varieties and products that are in danger of extinction, making the Ark of Taste a worldwide project.

You can't save the whales by eating whales, but paradoxically, you *can* help save rare, domesticated foods by eating them. They're kept alive by gardeners who have a taste for them, and farmers who know they'll be able to sell them. The consumer becomes a link in this conservation chain by seeking out the places where heirloom vegetables are sold, taking them home, whacking them up with knives, and learning to incorporate their exceptional tastes into personal and family expectations. Many foods placed on the Ark of Taste have made dramatic recoveries, thanks to the seed savers and epicurean desperadoes who defy the agents of gene control, tasting the forbidden fruits, and planting more.

If I could only save one of my seed packets from the deluge, the heirloom vegetable I'd probably grab is five-color silverbeet. It is not silver (*silverbeet* is Australian for Swiss chard), but has broad stems and leaf ribs vividly colored red, yellow, orange, white, or pink. Each plant has one stem color, but all five colors persist in a balanced mix in this beloved variety. It was the first seed variety I learned to save, and if in my dotage I end up in an old-folks' home where they let me grow one vegetable in the yard, it will be this one. It starts early, produces for months, looks like a bouquet when you cut it, and is happily eaten by my kids. They swear the different colors taste different, and in younger days were known to have blindfolded color-taste contests. (What kids will do, when deprived of ready access to M&Ms.) One of the first recipes we invented as a family, which we call "eggs in a nest," was inspired by the eggs from Lily's first flock of hens and five-color chard from the garden.

Children are, of course, presumed to hate greens, so assiduously that a cartoon character with spinach-driven strength was invented to inspire them. I suspect it's about the preparation. Even poor Popeye only gets miserably soppy-looking stuff out of a can. (Maybe sucking it in through his pipe gives it extra flavor.) The rule of greens is that they should be *green,* cooked with such a light touch that the leaf turns the color of the light that means "go!" and keeps some personality. Overcooking turns it nearly black. To any child who harbors a suspicion of black foods, I would have to say, with the possible exception of licorice, I'm with you.

Leafy greens are nature's spring tonic, coming on strong in local markets in April and May, and then waning quickly when weather gets hot. Chard will actually put up a fight against summer temperatures, but lettuce gives up as early as late May in the south. As all good things must come to an end, the leafy-greens season closes when the plant gets a cue from the thermometer—85 degrees seems to do it for most varieties. Then they go through what amounts to plant puberty: shooting up, transforming practically overnight from short and squat to tall and graceful, and of course it is all about sex. The botanical term is *bolting.* It ends with a cluster of blossoms forming atop the tall stems; for lettuces, these flowers are tiny yellow versions of their cousin, the dandelion. And like any adolescent, the bolting lettuce plant has volatile chemicals coursing through its body; in the case of lettuce, the plant is manufacturing a burst of sesquiterpene lactones, the compounds that make a broken lettuce stem ooze milky white sap, and which render it suddenly so potently, spit-it-out bitter. When lettuce season is over, it's *over.*

These compounds are a family trait of the lettuce clan, accounting for the spicy tang of endives, arugula, and radicchio, while the pale icebergs have had most of these chemicals bred out of them (along with most of their nutrients). Once it starts to go tall and leggy, though, even an innocent iceberg becomes untouchable. This chemical process is a vestige of a plant's life in the wild, an adaptation for protecting itself from getting munched at that important moment when sexual reproduction is about to occur.

But on a brisk April day when the tranquils are up and Jack Frost might still come out of retirement on short notice, hot weather is a dream. This is the emerald season of spinach, kale, endive, and baby lettuces. The chard comes up as red and orange as last fall's leaves went out. We lumber out of hibernation and stuff our mouths with leaves, like deer, or tree sloths. Like the earth-enraptured primates we once were, and could learn to be all over again. In April I'm happiest with mud on the knees of my jeans, sitting down to the year's most intoxicating lunch: a plate of greens both crisp and still sun-warmed from the garden, with a handful of walnuts and some crumbly goat cheese. This is the opening act of real live food.

By the time the lettuce starts to go flowery and embittered, who cares? We'll have fresh broccoli by then. When you see stuffed bunnies dangling from the crab apple trees, the good-time months have started to roll.

First, Eat Your Greens

BY CAMILLE

I've grown up in a world that seems to have a pill for almost everything. College kids pop caffeine pills to stay up all night writing papers, while our parents are at home popping sleeping pills to prevent unwelcome all-nighters. We can take pills for headaches, stomachaches, sinus pressure, and cold symptoms, so we can still go to work sick. If there's no time to eat right, we have nutrition pills, too. Just pop some vitamins and you're good to go, right?

Both vitamin pills and vegetables are loaded with essential nutrients, but not in the same combinations. Spinach is a good source of both vitamin C and iron. As it happens, vitamin C boosts iron absorption, allowing the body to take in more of it than if the mineral were introduced alone. When I first started studying nutrition, I became fascinated with these coincidences, realizing of course they're not coincidences. Human bodies and their complex digestive chemistry evolved over millennia in response to all the different foods—mostly plants—they raised or gathered from the land surrounding them. They may have died young from snakebite or blunt trauma, but they did not have diet-related illnesses like heart disease and Type II diabetes that are prevalent in our society now, even in some young adults and children.

Our bodies aren't adapted to absorb big loads of nutrients all at once (many supplements surpass RDA values by 200 percent or more), but tiny quantities of them in combinations—exactly as they occur in plants. Eating a wide variety of different plant chemicals is a very good idea, according to research from the American Society for Nutritional Sciences. You don't have to be a chemist, but color vision helps. By eating plant foods in all different colors you'll get carotenoids to protect body tissues from cancer (yellow, orange, and red veggies); phytosterols to block cholesterol absorption and inhibit tumor growth (green and yellow plants and seeds); and phenols for age-defying antioxidants (blue and purple fruits). Thousands of the phytochemicals we eat haven't even been studied or named yet, because there are so many, with such varied roles, finely tuned as fuel for our living bodies. A head of broccoli contains more than a thousand.

Multivitamins are obviously a clunky substitute for the countless subtle combinations of phytochemicals and enzymes that whole foods contain. One way to think of these pills might be as emergency medication for lifestyle-induced malnutrition. I'm coming of age in a society where the majority of adults are medically compromised by that particular disease. Not some, but *most;* that's a scary reality for a young person. It's helpful to have some idea how to take preventive action. My friends sometimes laugh at the weird food combinations that get involved in my everyday quest to squeeze more veggies into a meal, while I'm rushing to class. (Peanut butter and spinach sandwiches?) But we all are interested in staying healthy, however we can.

Leafy greens, like all plants, advertise their nutritional value through color: dark green or red leaves with a zesty tang bring more antioxidants to your table. But most any of them will give you folic acid (*folic* equals "foliage"), a crucial nutrient for pregnant women that's also needed by everyone for producing hemoglobin. From Popeye to Thumper the rabbit, the message that "you have to finish your greens" runs deep in kid culture, for good reason. Parents won't have to work so hard at bribing their kids with desserts if they don't serve slimy greens. When fresh and not overcooked, spinach, chard, kale, bok choy, and other greens are some of my favorite things.

Here are some recipes that bring out the best in dark, leafy greens. These are staple meals for our family in the season when greens are coming up in our garden by the bushel.

❋ EGGS IN A NEST

(This recipe makes dinner for a family of four, but can easily be cut in half.)

2 cups uncooked brown rice

Cook rice with 4 cups water in a covered pot while other ingredients are being prepared.

Olive oil—a few tablespoons

1 medium onion, chopped, and garlic to taste

Sauté onions and garlic in olive oil in a wide skillet until lightly golden.

Carrots, chopped

½ cup dried tomatoes

Add and sauté for a few more minutes, adding just enough water to rehydrate the tomatoes.

1 *really large* bunch of chard, coarsely chopped

Mix with other vegetables and cover pan for a few minutes. Uncover, stir well, then use the back of a spoon to make depressions in the cooked leaves, circling the pan like numbers on a clock.

8 eggs

Break an egg into each depression, being careful to keep yolks whole. Cover pan again and allow eggs to poach for 3 to 5 minutes. Remove from heat and serve over rice.

❋ SPINACH LASAGNA

1 pound whole-grain lasagna noodles

Prepare according to package directions.

4 cups chopped spinach

Steam for 2–3 minutes, let excess water drain.

16 ounces tomato sauce

2 cups fresh ricotta

2 cups mozzarella

Spread a thin layer of tomato sauce on the bottom of a large casserole. Cover surface with a layer of noodles, ½ of the ricotta, ½ of the spinach, ⅓ of the remaining sauce, and ⅓ of the mozzarella. Lay down another layer of noodles, the rest of the ricotta, the rest of the spinach, ⅓ of the sauce, and ⅓ of the mozzarella. Spread a final layer of noodles, the remainder of the sauce and mozzarella; bake uncovered at 350° for 40 minutes.

Download these and all *Animal, Vegetable, Miracle* recipes at www.AnimalVegetableMiracle.com

GREENS SEASON MEAL PLAN

Sunday ~ Greek roasted chicken and potatoes, chard-leaf dolmades with béchamel sauce

Monday ~ Eggs in a Nest

Tuesday ~ Chicken salad (from Sunday's leftover chicken) on a bed of baby greens

Wednesday ~ Pasta tossed with salmon, sautéed fresh chard, and dried tomatoes

Thursday ~ Dinner salad with boiled eggs, broccoli, nuts, and feta; fresh bread

Friday ~ Pizza with chopped sautéed spinach, mushrooms, and cheese

Saturday ~ Spinach lasagna

4 · STALKING THE VEGETANNUAL

If potatoes can surprise some part of their audience by growing leaves, it may not have occurred to everyone that lettuce has a flower part. It does, they all do. Virtually all nonanimal foods we eat come from flowering plants. Exceptions are mushrooms, seaweeds, and pine nuts. If other exotic edibles exist that you call food, I salute you.

Flowering plants, known botanically as angiosperms, evolved from ancestors similar to our modern-day conifers. The flower is a handy reproductive organ that came into its own during the Cretaceous era, right around the time when dinosaurs were for whatever reason getting downsized. In the millions of years since then, flowering plants have established themselves as the most conspicuously successful terrestrial life forms ever, having moved into every kind of habitat, in infinite variations. Flowering plants are key players in all the world's ecotypes: the deciduous forests, the rain forests, the grasslands. They are the desert cacti and the tundra scrub. They're small and they're large, they fill swamps and tolerate drought, they have settled into most every niche in every kind of place. It only stands to reason that we would eat them.

Flowering plants come in packages as different as an oak tree and a violet, but they all have a basic life history in common. They sprout and leaf out; they bloom and have sex by somehow rubbing one flower's boy stuff against another's girl parts. Since they can't engage in hot pursuit, they lure a third party, such as bees, into the sexual act—or else (depending on species) wait for the wind. From that union comes the blessed event, babies made, in the form of seeds cradled inside some form of fruit. Finally, sooner or later—because after *that,* what's the point anymore?—they die. Among the plants known as annuals, this life history is accomplished all in a single growing season, commonly starting with spring and ending with frost. The plant waits out the winter in the form of a seed, safely protected from weather, biding its time until conditions are right for starting over again. The vegetables we eat may be leaves, buds, fruits, or seeds, but each comes to us from some point along this same continuum, the code all annual plants must live by. No variations are allowed. They can't set fruit, for example, before they bloom. As obvious as this may seem, it's easy enough to forget in a supermarket culture where the plant stages constantly present themselves in random order.

To recover an intuitive sense of what will be in season throughout the year, picture a season of foods unfolding as if from one single plant. Take a minute to study this creation—an imaginary plant that bears over the course of one growing season a cornucopia of all the different vegetable products we can harvest. We'll call it a vegetannual. Picture its life passing before your eyes like a time-lapse film: first, in the cool early spring, shoots poke up out of the ground. Small leaves appear, then bigger leaves. As the plant grows up into the sunshine and the days grow longer, flower buds will appear, followed by small green fruits. Under midsummer's warm sun, the fruits grow larger, riper, and more colorful. As days shorten into the autumn, these mature into hard-shelled fruits with appreciable seeds inside. Finally, as the days grow cool, the vegetannual may hoard the sugars its leaves have made, pulling them down into a storage unit of some kind: a tuber, bulb, or root.

So goes the year. First the leaves: spinach, kale, lettuce, and chard (here, that's April and May). Then more mature heads of leaves and flower heads: cabbage, romaine, broccoli, and cauliflower (May–June). Then tender young fruit-set: snow peas, baby squash, cucumbers (June), followed by green beans, green peppers, and small tomatoes (July). Then more mature, colorfully ripened fruits: beefsteak tomatoes, eggplants, red and yellow peppers (late July–August). Then the large, hard-shelled fruits with developed seeds inside: cantaloupes, honeydews, watermelons, pumpkins, winter squash (August–September). Last come the root crops, and so ends the produce parade.

Plainly these don't all come from the same plant, but each comes from a *plant,* that's the point—a plant predestined to begin its life in the spring and die in the fall. (A few, like onions and carrots, are attempting to be biennials, but we'll ignore that for now.) Each plant part we eat must come in its turn—leaves, buds, flowers, green fruits, ripe fruits, hard fruits—because that is the necessary order of things for an annual plant. For the life of them, they can't do it differently.

Some minor deviations and a bit of overlap are allowed, but in general, picturing an imaginary vegetannual plant is a pretty reliable guide to what will be in season, wherever you live. If you find yourself eating a watermelon in April, you can count back three months and imagine a place warm enough in January for this plant to have launched its destiny. Mexico maybe, or southern California. Chile is also a possibility. If you're inclined to think this way, consider what it took to transport a finicky fruit the size of a human toddler to your door, from that locale.

Our gardening forebears meant watermelon to be the juicy, barefoot taste of a hot summer's end, just as a pumpkin is the trademark fruit of late October. Most of us accept the latter, and limit our jack-o'-lantern activities to the proper botanical season. Waiting for a watermelon is harder. It's tempting to reach for melons, red peppers, tomatoes, and other late-summer delights before the summer even arrives. But it's actually possible to wait, celebrating each season when it comes, not fretting about its being absent at all other times because something else good is at hand.

If many of us would view this style of eating as deprivation, that's only because we've grown accustomed to the botanically outrageous condition of having everything, always. This may be the closest thing we have right now to a distinctive national cuisine. Well-heeled North American epicures are likely to gather around a table where whole continents collide discreetly on a white tablecloth: New Zealand lamb with Italian porcinis, Peruvian asparagus, and a hearty French Bordeaux. The date on the calendar is utterly irrelevant.

I've enjoyed my share of such meals, but I'm beginning at least to notice when I'm consuming the United Nations of edible plants and animals all in one seating. (Or the WTO, is more like it.) On a winter's day not long ago I was served a sumptuous meal like this, finished off with a dessert of raspberries. Because they only grow in temperate zones, not the tropics, these would have come from somewhere deep in the Southern Hemisphere. I was amazed that such small, eminently bruisable fruits could survive a zillion-mile trip looking so good (I myself look pretty wrecked after a mere red-eye from California), and I mumbled some reserved awe over that fact. I think my hostess was amused by my country-mouse naïveté. "This is New York," she assured me. "We can get anything we want, any day of the year."

So it is. And I don't wish to be ungracious, but we get it at a price. Most of that is not measured in money, but in untallied debts that will be paid by our children in the currency of extinctions, economic unravelings, and global climate change. I do know it's impolite to raise such objections at the dinner table. Seven raspberries are not (I'll try to explain

The Global Equation

By purchasing local vegetables instead of South American ones, for example, aren't we hurting farmers in developing countries? If you're picturing Farmer Juan and his family gratefully wiping sweat from their brows when you buy that Ecuadoran banana, picture this instead: the CEO of Dole Inc. in his air-conditioned office in Westlake Village, California. He's worth $1.4 billion; Juan gets about $6 a day. Much money is made in the global reshuffling of food, but the main beneficiaries are processors, brokers, shippers, supermarkets, and oil companies.

Developed nations promote domestic overproduction of commodity crops that are sold on the international market at well below market price, undermining the fragile economies of developing countries. Often this has the effect of driving small farmers into urban areas for jobs, decreasing the agricultural output of a country, and forcing the population to purchase those same commodities from abroad. Those who do stay in farm work are likely to end up not as farm *owners,* but as labor on plantations owned by multinationals. They may find themselves working in direct conflict with local subsistence. Thus, when Americans buy soy products from Brazil, for example, we're likely supporting an inter-

someday to my grandkids) the end of the world. I ate them and said "Thank you."

Human manners are wildly inconsistent; plenty of people before me have said so. But this one takes the cake: the manner in which we're allowed to steal from future generations, while commanding them not to do that to us, and rolling our eyes at anyone who is tediously PC enough to point this out. The conspicuous consumption of limited resources has yet to be accepted widely as a spiritual error, or even bad manners.

Our culture is not unacquainted with the idea of food as a spiritually loaded commodity. We're just particular about which spiritual arguments we'll accept as valid for declining certain foods. Generally unacceptable reasons: environmental destruction, energy waste, the poisoning of workers. Acceptable: it's prohibited by a holy text. Set down a platter of country ham in front of a rabbi, an imam, and a Buddhist monk, and you may have just conjured three different visions of damnation. Guests with high blood pressure may add a fourth. Is it such a stretch, then, to make moral

national company that has burned countless acres of Amazon rain forest to grow soy for export, destroying indigenous populations. Global trade deals negotiated by the World Trade Organization and World Bank allow corporations to shop for food from countries with the poorest environmental, safety, and labor conditions. While passing bargains on to consumers, this pits farmers in one country against those in another, in a downward wage spiral. Product quality is somewhat irrelevant.

Most people no longer believe that buying sneakers made in Asian sweatshops is a kindness to those child laborers. Farming is similar. In every country on earth, the most humane scenario for farmers is likely to be feeding those who live nearby—if international markets would allow them to do it. Food transport has become a bizarre and profitable economic equation that's no longer really about feeding anyone: in our own nation we export 1.1 million tons of potatoes, while we also import 1.4 million tons. If you care about farmers, let the potatoes stay home. For more information visit: www.viacampesina.org.

STEVEN L. HOPP

choices about food based on the global consequences of its production and transport? In a country where 5 percent of the world's population glugs down a quarter of all the fuel, also belching out that much of the world's waste and pollution, we've apparently made big choices about consumption. They could be up for review.

The business of importing foods across great distances is not, by its nature, a boon to Third World farmers, but it's very good business for oil companies. Transporting a single calorie of a perishable fresh fruit from California to New York takes about 87 calories worth of fuel. That's as efficient as driving from Philadelphia to Annapolis, and back, in order to walk three miles on a treadmill in a Maryland gym. There may be people who'd do it. Pardon me while I ask someone else to draft my energy budget.

In many social circles it's ordinary for hosts to accommodate vegetarian guests, even if they're carnivores themselves. Maybe the world would likewise become more hospitable to diners who are queasy about fuel-guzzling foods, if that preference had a name. Petrolophobes? Seasonaltarians? Local eaters, Homeys? Lately I've begun seeing the term *locavores,* and I like it: both scientifically and socially descriptive, with just the right hint of "Livin' *la vida loca.*"

Slow Food International has done a good job of putting a smile on this eating style, rather than a pious frown, even while sticking to the quixotic agenda of fighting overcentralized agribusiness. The engaging strategy of the Slowies (their logo is a snail) is to celebrate what we have, standing up for the pleasures that seasonal eating can bring. They have their work cut out for them, as the American brain trust seems mostly blank on that subject. Consider the frustration of the man who wrote in this complaint to a food columnist: having studied the new food pyramid brought to us by the U.S. Dietary Guidelines folks (impossible to decipher but bless them, they do keep trying), he had his marching orders for "2 cups of fruit, 2½ cups of vegetables a day." So he marched down to his grocery and bought (honest to Pete) eighty-three plums, pears, peaches, and apples. Outraged, he reported that virtually the entire lot was "rotten, mealy, tasteless, juiceless, or hard as a rock and refusing to ripen."

Given the date of the column, this had occurred in February or March.

The gentleman lived in Frostburg, Maryland, where they would still have been deeply involved in a thing called winter. I'm sure he didn't really think tasty tree-ripened plums, peaches, and apples were hanging outside ripe for the picking in the orchards around . . . um, *Frost*-burg. Probably he didn't think "orchard" at all—how many of us do, in the same sentence with "fruit"? Our dietary guidelines come to us without a roadmap.

Concentrating on local foods means thinking of fruit invariably as the product of an orchard, and a winter squash as the fruit of an early-winter farm. It's a strategy that will keep grocery money in the neighborhood, where it gets recycled into your own school system and local businesses. The green spaces surrounding your town stay green, and farmers who live nearby get to grow more food next year, for you. But before any of that, it's a win-win strategy for anyone with taste buds. It begins with rethinking a position that is only superficially about deprivation. Citizens of frosty worlds unite, and think about marching past the off-season fruits: you have nothing to lose but mealy, juiceless, rock-hard and refusing to ripen.

5 · MOLLY MOOCHING

April

In the year 1901, Sanford Webb ran a dozen head of cattle on his new farm and watched to see where they settled down every night. The place they chose, he reasoned, would be the most sheltered spot in the hollow. That was where he built his house, with clapboard sides, a steep tin roof, and a broad front porch made of river rock. The milled door frames and stair rails he ordered from Sears, Roebuck. He built the house for his new bride, Lizzie, and the children they would raise here—eleven in all—during the half-century to come.

In the 1980s those children put the home place up for sale. They weren't keen to do it, but had established farmsteads of their own by the time their parents passed away. All were elderly now, and none was in a position to move back to the family farm and fix up the home place. They decided to let it go out of the family.

Steven walked into this picture, and a deal with fate was sealed. He didn't know that. He was looking for a bargain fixer-upper he could afford on his modest academic salary, a quiet place to live where he could listen to the birds and maybe grow some watermelons. He was a bachelor, hardly looking for a new family at that moment, but the surviving Webbs—including the youngest sisters, who lived adjacent, now in their seventies—observed that he needed some looking after. Dinner invitations ensued. When Steven eventually brought a wife and kids to the farm, we were gathered into that circle. The Webbs unfailingly invite us to their family reunions. Along with the pleasures of friendship and help with anything from binding a quilt to canning, we've been granted a full century's worth of stories attached to this farm.

The place is locally famous, it turns out. Sanford Webb was a visionary and a tinkerer who worked as a civil engineer for the railroad but also was the first in the neighborhood—or even this end of the state—to innovate such things as household electricity, a grain mill turned by an internal combustion engine, and indoor food refrigeration. The latter he fashioned by allowing a portion of the farm's cold, rushing creek to run through a metal trough inside the house. (We still use a version of this in our kitchen for no-electricity refrigeration.) Creativity ran in the family. In the upstairs bedroom the older Webb boys once surprised their mother by smuggling up, one part at a time, everything necessary to build, crank up, and start a Model-T Ford. These inventive brothers later founded a regional commercial airline, Piedmont Air, and paid their kid sister Neta a dime a day to come down and sweep off the runway before each landing.

Sanford was also forward-thinking in the ways of horticulture. He worked on the side as a salesman for Stark's Nursery at a time when the normal way to acquire fruit trees was to borrow a scion from a friend. Mr. Webb proposed to his neighbors the idea of buying named varieties of fruit trees, already grafted onto root stock, that would bear predictably and true. Stayman's Winesaps, Gravensteins, and Yellow Transparents began to bloom and bear in our region. For every sixteen trees Mr. Webb sold, he received one to plant himself. The lilacs, mock oranges, and roses of Sharon he brought home for Lizzie still bloom around our house. So does a small, frost-hardy citrus tree called a trifoliate orange, a curiosity that has nearly gone extinct in the era of grocery-store oranges. (We know of only one nursery that still sells them.)

The man was passionate about fruit trees. Throughout our hollow, great old pear trees now stand a hundred feet tall, mostly swallowed by a forest so deep they don't get enough sun to bear fruit. But occasionally when I'm walking up the road I'll be startled by the drop (and smash) of a ripe pear fallen from a great height. The old apple orchard we've cleared and pruned, and it bears for us. We keep the grass mowed between the symmetrically spaced trunks, and on warm April days when the trees are in bloom, it's an almost unbearably romantic place for a picnic. As the white petals rain down like weightless, balmy snow, Lily dances in circles and says things like, "Quick, somebody needs to have a wedding!" I recently called the Stark's Nursery company in Missouri to let them know an orchard of their seventy-five-year-old antique apple varieties still stood and bore in Virginia. They were fairly enthused to hear it.

Nearly every corner of this farm, in Webb family lore, has a name attached to it, or a story, or both. There is Pear Orchard Hill, Dewberry Hill, the Milk Gap, where the cows used to cross the road coming home to the barn. We still call it the Milk Gap, though no cows use it now. A farm has its practical geography; when you tell someone to go close the gate, she needs to know whether you mean the Milk Gap gate or the barn lot gate. We have the Garden Road, the Woods Road, the Paw-Paw Cemetery, and the New Orchard. And once a year, for a few days, we have a spot that rises from obscurity to prominence: Old Charley's Lot.

Old Charley was a billy goat that belonged to the Webbs some seventy years ago. In the customary manner of billy goats, he stank. For that reason they kept him penned on a hillside nearly half a mile up the hollow from the house: that was Old Charley's lot, figuratively and literally. Most of the time now it's just a steep, unvisited place on the back side of our farm, but for one week of the year (for reasons unrelated to the goat, as far as we know), it's the pot of gold at the end of our rainbow. What grows there sells for upward of twenty dollars a pound in city markets: the most prized delicacy that ever comes to our table.

⚘

The mountainsides of our farm stand thickly wooded with poplar, beech, and oak. Two generations ago they were clear pastures, grazed by livestock or plowed for crops. This used to be a tobacco farm. It astonishes us when our neighbors look at our tall woods and say: "That's where we grew our corn. Our tobacco patch was on top, for the better sun." Such steep hillsides were worked with mules and a lot of hand labor (a tractor would roll like a boulder), which is exactly why so much of the farmland around here has now grown up into medium-sized stands of trees. The deciduous eastern woodlands of North America can bear human alteration with a surprising grace. Barring the devastations of mining or the soil disaster of clear-cut logging, this terrain can often recover its wildness within the span of a human lifetime.

Farming a hillside with mules had its own kind of grace, I am sure, but it's mostly a bygone option. The scope of farming in southern Appalachia has now retreated to those parts of the terrain that are tractor-friendly, which is to say, the small pieces of relatively flat bottomland that lie between the steep slopes. Given that restriction, only one crop fit the bill here for the past half-century, and that was tobacco; virtually no other legal commodity commands such a high price per acre that farmers could stay in business with such small arable fields. That, plus the right climate, made Kentucky and southwestern Virginia the world's supplier of burley tobacco.

That plant works well here for cultural reasons also: it's the most labor-intensive commodity crop still grown in the United States, traditionally cultivated by an extended family or cooperative communities. Delicate tobacco seedlings have to be started in sheltered beds, then set by hand into the field and kept weed-free. Once mature, the whole plant is cut, speared with a sharp stick, and the entire crop painstakingly hung to dry in voluminous, high-roofed, well-ventilated barns. Once the fragile leaves are air-cured to dark brown, they must be stripped by hand from the stalk, baled, and taken to the auction house.

On the flat, wide farms of Iowa one person with a tractor can grow enough corn to feed more than a hundred people. But in the tucked-away valleys of Appalachia it takes many hands to make just one living, and only if they work at growing a high-priced product. The same small acreage planted in corn would hardly bring in enough income to pay the property tax. For this reason, while the small family farm has transformed elsewhere, it has survived as a way of life in the burley belt. Tobacco's economy makes an indelible imprint on the look of a place—the capacious architecture of its barns, the small size of its farms—and on how a county behaves, inducing people to know and depend on one another. It makes for the kind of place where, when you walk across the stage at high school graduation, every person in the audience knows your name and how much work you've been known to do in a day. Tobacco even sets the date of graduation, since the end and beginning of the school year must accommodate spring setting time and fall cutting.

This was the context of my childhood: I grew up in a tobacco county. Nobody in my family smoked except for my grandmother, who had one cigarette per afternoon, whether she needed it or not, until the day in her ninth decade when she undertook to quit. But we knew what tobacco meant to our lives. It paid our schoolteachers and blacktopped our roads. It was the sweet scent of the barn loft where I hid out and read books on summer afternoons; it was the brown powder that clung to our jeans after an afternoon of playing in old outbuildings. It was the reason my first date had to end early on a Friday night: he had to get up early on Saturday to work the tobacco. For my classmates who went to college, it was tobacco that sent them. Me too, since my family could not have stayed solvent without other family economies that relied on tobacco.

From that society I sallied out into a world where, to my surprise, *farmer* was widely presumed synonymous with *hee-haw,* and tobacco was the new smallpox. I remember standing in someone's kitchen once at a college party—one of those intensely conversational gatherings of the utterly enlightened—listening while everyone present agreed on the obvious truth about tobacco: it should be eliminated from this planet and all others. I blurted out, foolishly, "But what about the tobacco farmers?"

You'd have thought I'd spoken up for child porn. Somebody asked, "Why should I care about tobacco farmers?"

I'm still struggling to answer that. Yes, I do know people who've died wishing they'd never seen a cigarette. Yes, it's a plant that causes cancer after a long line of people (postfarmer) have specifically altered and abused it. And yes, it takes chemicals to keep the blue mold off the crop. *And* it sends people to college. It makes house payments, buys shoes, and pays doctor bills. It allows people to live with their families and shake hands with their neighbors in one of the greenest, kindest places in all this world. Tobacco is slowly going extinct as a U.S. crop, and that is probably a sign of good civic sense, but it's also a cultural death when all those who grew it must pack up, go find an apartment somewhere, and work in a factory. What is family farming worth?

Most tobacco farmers wish they could grow something else. As of now, most will have to. Federal price supports, which have safeguarded the tobacco livelihood since the Depression, officially ended in 2005. Extension services and agriculture schools throughout the region have anticipated that deadline for more than twenty years, hoping to come up with a high-value crop to replace tobacco. No clear winner has yet emerged. When I was in high school, the family of one of my best friends tried growing bell peppers, the latest big idea of the era. They lost the entire year's income when the promised markets failed to materialize. I still get a knot in my stomach remembering the day their field of beautiful peppers, representing months of the family's labor and their year's livelihood, had to be plowed back into the dirt, in the end worth more as compost than as anything else. If people out in the world were irate about the human damage of tobacco, why wouldn't they care enough—and pay enough—to cover the costs of growing vegetables? I can date from that moment my awareness of how badly our food production system is deranged, and how direly it is stacked against the farmer.

The search for a good substitute crop is still on, but now that the modest price supports have ended, farmers in tobacco country have only a year or two more to figure out how to stay on their land. Vegetables are a high-value crop, especially if they're organic, but only in areas that have decent markets for them and a good infrastructure for delivering perishable goods to these markets. The world's most beautiful tomato, if it can't get into a shopper's basket in less than five days, is worth exactly nothing. Markets and infrastructure depend on consumers who will at least occasionally choose locally grown foods, and pay more than rock-bottom prices.

In my county, two of the best tobacco-transition experiments to date are organic vegetables and sustainable lumber. A program in our area offers farmers expert advice on creating management plans for the wooded hillsides that typically occupy so much of the acreage of local farms. Mature trees can be harvested carefully from these woodlots in a way that leaves the forests healthy and sustainably productive. The logs are milled into lumber, kiln-dried, and sold to regional buyers seeking alternatives to rain-forest teak or clear-cut redwood. When we needed new oak flooring in our farmhouse, we were able to purchase it from a friend's woodlot nearby. No farmer earns a whole livelihood from this, but the family farm has a tradition of cobbling together solvency from many crops.

Experimental programs like these, though new and small, are notable for the way they turn a certain economic paradigm on its head. U.S. po-

Is Bigger Really Better?

Which are more economically productive, small family farms or big industrial farms? Most people assume they know the answer, and make a corollary assumption: that small farmers are basically asking to go bankrupt, they're inefficient even though their operations are probably more environmentally responsible, sustainable, diverse, and better connected to their communities than the big farms are. But isn't it really just about the profits?

If so, small farms win on that score too, hands down. According to USDA records from the 1990s, farms less than four acres in size had an average net income of $1,400 per acre. The per-acre profit declines steadily as farm size grows, to less than $40 an acre for farms above a thousand acres. Smaller farms maximize productivity in three ways: by using each square foot of land more intensively, by growing a more diverse selection of products suitable to local food preferences, and by selling more directly to consumers, reaping more of the net earnings. Small-farm profits are more likely to be sustained over time, too, since these farmers tend to be better stewards of the land, using fewer chemical inputs, causing less soil erosion, maintaining more wildlife habitat.

If smaller is economically better, why are the little guys going out of business? Aside from their being more labor-intensive, marketing is the main problem. Supermarkets prefer not to bother with boxes of vegetables if they can buy truckloads. Small operators have to be both grower and marketer, selling their products one bushel at a time. They're doing everything right, they just need customers.

Food preference surveys show that a majority of food shoppers are willing to pay more for food grown locally on small family farms. The next step, following up that preference with real buying habits, could make or break the American tradition of farming. For more information, visit www.nffc.net.

STEVEN L. HOPP

litical debate insistently poses economic success and environmental health as enemies, permanently at odds. Loggers or owls? People or green spaces? The presumed antagonism between "Man" and "Nature" is deeply rooted in our politics, culture, bedtime stories (Red Riding Hood's grandma, or the wolf?), and maybe even our genes. But farming at its best optimizes both economic and environmental health at the same time. A strategy that maximizes either one at the cost of the other is a fair working definition of *bad* farming. The recent popularity of agriculture that damages soil fertility still does not change the truth: what every farmer's family needs is sustainability, the capacity to coax productiveness out of the same plot of ground year after year. Successful partnerships between people and their habitats were once the hallmark of a healthy culture. After a profoundly land-altering hiatus, the idea may be regaining its former shine.

⁂

The people of southern Appalachia have a long folk tradition of using our woodlands creatively and knowing them intimately. The most caricatured livelihood, of course, is the moonshine still hidden deep in the hollow, but that is not so much about the woodlands as the farms; whiskey was once the most practical way to store, transport, and add value to the small corn crops that were grown here.

These hills have other secrets. One of them is a small, feisty cousin of garlic known as the ramp. Appalachian mothers used to regard these little bulbs as a precious spring tonic—one that schoolboys took willingly because it rendered them so odoriferous, they'd likely get barred from the schoolhouse for several days. For reasons not entirely clear to the outsider, ramps are still prized by those who know where to find them in the earliest days of spring. The emergence of ramps elicits joyous, stinky ramp festivals throughout the region.

Ginseng, another Appalachian botanical curiosity, is hunted and dug up for its roots, which sell for enormous prices to consumers on the other side of the world. "Sang" hunters know where to look, and tend to keep their secrets. So do Molly Mooch hunters, and if you think you've never heard of such, you just don't know the code. A Molly Mooch is a morel.

The mushroom genus *Morchella* contains some of the most highly prized of all edible wild fungi. Morels fruit in the spring, and just to keep you on your toes regarding the wild mushroom situation, they contain toxic hemolysins that destroy red blood cells—chemicals that are rendered harmless during cooking. (Just don't eat them raw.) Their close relatives, the *Gyromitra,* are pure poison, but the edible morels look different enough from anything else that they're safe to collect, even for novice mushroomers like me. Their distinctive tall caps are cupped and wrinkled in a giraffish pattern unique to their kind. Here in the eastern woodlands we have the black, common, tulip, and white morels, and one unfortunate little cousin called (I am so sorry) the Dog Pecker. All are edible except for the last one. They're similar enough in ecology and fruiting time that we've sometimes gathered many types on the same day, from the same wooded areas. I've heard them called Molly Mooch, sponge mushrooms, haystacks, dryland fish, and snakeheads. What everyone agrees on is that they're delicious.

Wild mushrooms are among the few foods North Americans still eat that must be hunted and gathered. Some fungi are farmed, but exotics like the morel defy all attempts at domestication. Maybe that's part of what we love about them. "With their woodsy, earthy, complex flavors and aromas, and their rich, primeval colors and forms," writes Alice Waters, wild mushrooms bring to our kitchens "a reminder that all the places we inhabit were once wildernesses." They are also incredibly hard to find, very good at looking exactly like a little pile of curled, dead brown leaves on the forest floor. In my early days of Molly Mooching, I could stand with my boots touching one without spotting it until it was pointed out to me. They're both particular and mysterious about where they grow: in old apple orchards, some people vow, while others insist it's only around the roots of tulip poplars or dying elms. Whatever the secret, the Molly Mooches do know it, because they tend to show up in the same spot year after year.

On our farm we could have walked the woods for the rest of our lives without finding one, because they don't grow near our roads or trails, they've never shown up in our old apple orchard, and they're shy of all other places we normally frequent. Where they do grow is in Old Charley's

Lot. We know that only because our friends who grew up on this farm showed us where to look. This is the kind of knowledge that gets lost if people have to leave their land. Farmers aren't just picturesque technicians. They are memory banks, human symbionts with their ground.

My family is now charged with keeping the secret history of a goat, a place, and a mushroom. Just as our local-food pledge had pushed us toward the farmers' market on the previous Saturday, it pushed us out the back door on the following cold, rainy Monday. Morels emerge here on the first warm day after a good, soaking mid-April rain. It's easy to get preoccupied with life and miss that window, or to coast past it on the lazy comfort of a full larder. This April our larder was notably empty, partly I suppose for just this reason—to force us to pay attention to things like the morels. Steven came home from his teaching duties, donned jeans and boots, and headed up toward Old Charley's Lot with a mesh bag in hand. Mushroom ethics mandate the mesh collecting bag, so the spores can scatter as you carry home your loot.

No loot was carried that day. We really knew it was still too cold. We'd had snow the previous day. But we were more than usually motivated, so on Tuesday in only slightly nicer weather, Steven headed out again. I was mending a broken leg that spring and could not yet navigate the steep, slick mountainside, so I was consigned to the wifely role of waiting for Man the Hunter to return. After the first hour I moved on to the wifely custom of worrying he'd fallen into a sinkhole. But no, he eventually returned from the woods, empty-handed but intact. He was just being thorough.

On Wednesday he went out again, and came back through the kitchen door with a conspicuous air of conquest. Triumphantly he held up his mesh bag: a few dozen fawn-colored, earthy, perfect morels. It wasn't a huge catch, but it was big enough. By the weekend there would be more, enough to share with our neighbors. I grinned, and went to the refrigerator. A little while earlier I'd gone up to the garden and returned with my own prize lying across my forearm like two dozen long-stemmed roses: our most spectacular asparagus harvest ever.

We put our Mollies in a bowl of salt water to soak briefly prior to cooking. I'm not sure why, but our mushroom-hunting friends say to do this with morels, and I am not one to argue with wild mushroomers who claim the distinction of being still alive. I sat down at the kitchen table with Deborah Madison's gorgeous cookbook *Local Flavors,* which works from the premise that any week of the year can render up, from very near your home, the best meal of your life. Deborah's word is good. We cooked up her "Bread pudding with asparagus and wild mushrooms" for a fantastic Wednesday supper, seduced by the fragrance even before we took it out of the oven. Had I been worried that cutting the industrial umbilicus would leave us to starve? Give me this deprivation, any old day of the week.

⚘

On Saturday the weather was still cold and windy. I pulled my seed potatoes out of storage to check on them. Not a pretty picture: sick to death of the paper bags in which they'd been stored since last fall, they were sending long, white, exploratory sprouts into the darkness of the bottom drawer of the refrigerator.

We decided for their sakes that the wind had dried the ground enough for us to till the potato patch with the tractor. A few weeks ago we'd tried that too early, and the too-wet ground behind the tractor rolled over in long curls of thick, unworkable clay clods. Today the soil was still a bit too clumpy to be called perfect, but "perfect" is not the currency of farming. I followed behind the tiller breaking up clods with a hefty Italian grape hoe, the single piece of equipment I rely on most for physical fitness and sometimes therapy. We hoed out three deep rows, each about seventy feet long, in which to drop our seed potatoes. If that seems like a lot for one family, it's not. We do give some away, and save some for next year's seed, but mostly we eat them: new potatoes all summer, fingerlings in the fall, big indigo blues and Yukon Gold bakers all winter. In my view, homeland security derives from having enough potatoes.

On the same long day we dropped peas into furrows, seeded carrots, and set out more of the broccoli we'd planted in succession since mid-March. My baby onion plants (two hundred of them) were ready, so I tucked the string-bean-sized seedlings into rows along the cold, damp edge of the upper field: Stockton Reds, Yellow Sweets, Torpedos, and a small, flat Italian favorite called "Borretana cipollini." I was anticipating our family's needs, knowing I would not be purchasing vegetables from the grocery store next winter. Two onions per week seemed reasonable.

Onion plants can take a light frost, so they don't have to wait until the full safety of late spring. Their extreme sensitivity is to day length: "short day" onions like Vidalias, planted in autumn in the deep South, are triggered to fatten into bulbs when day length reaches about ten hours, in May or so. By contrast, "long day" onions are planted in spring in the north, as early as the ground thaws so they'll have enough growth under their belts to make decent-sized bulbs when triggered to do so by the fifteen-hour days of a northern-latitude high summer. If you didn't realize onion farmers had to be this scientific about what varieties they plant, that's just the start. They're also required by law to live within a seventeen-county area centered on Vidalia, Georgia, in order to sell you a Vidalia onion, or in the Walla-Walla, Washington, region to print "Walla Walla Sweets" on the bag. French wine growers are not the only farmers who can market the subtleties of soil and climate, the things that translate into the regionally specific flavor they call *terroir*. The flavor of an onion, like that of a wine grape, is influenced by climate, soil chemistry, even soil microbes. It's surely true of other vegetables, or would be, if we knew enough of our own local flavors to recognize them.

The earliest plantings are onions, potatoes, peas, and the cole crops (broccoli, cauliflower, kale, cabbage, and Brussels sprouts are all the same genus and species). All these do well in cool weather and don't mind a few weeks of frost, or even snow on their heads. Broccoli buds will start to pop up above their leaves in the last cool days of May. In the same weather, snap peas twine up their trellises lightning fast and set their pods. Along with rosy new potatoes and green onions pulled early from bed, these will be the first garden proceeds, with asparagus, cold-hardy spinach, and other salad greens. The garden-bereft don't have to live without these pleasures. In most of the country, farmers' markets get going in April or early May. Especially in the Northeast, market gardeners are also savvy about stretching the season with cold frames, so these treasures can fill their stalls very early, in limited quantities that will go to the early risers.

If you picture the imaginary vegetannual, you'll see these are the earliest tendrils of annual growth—leaves, shoots, and buds—filled out with early production coaxed from overwintered roots. Grocery-store-habituated shoppers may only have eyes for the Fourth-of-July-fireworks kind of garden bounty: big, blowsy tomatoes, eggplants, and summer squash. But many weeks before that big bang, subtle pleasures begin pushing up through the chill, come and gone again if you weren't ready. I'd offer the same advice I include in my directions for finding our little town: Don't blink. You'll miss us.

Getting It While You Can

BY CAMILLE

When our family first started our local food project, I was daunted. How ironic, I thought: while most parents are harping on their kids to eat more fruits and veggies, my sister and I are being told to give up the juicy pleasure of a fresh peach or pear all winter. I tried to picture how I would get through the months when there are no apples, no plums, and the strawberries of spring seem light-years away. This may sound dramatic, but fruit is my favorite food.

I was forced to get creative. The first step, shopping, is actually easier. When you peruse the farmers' market for fresh produce, the options are clear. You don't miss what's *not* there, either, like Skittles placed at a third-grader's eye level in the checkout. No wailing kids or annoying tabloids (omigod . . . is Brangelina really over?!). Just wonderful, fresh things to eat. As the seasons change, different fruits and vegetables come and go, so as a shopper you learn a get-it-while-you-can mentality.

The first strawberries showed up at our farmers' market in late spring, on a day when I'd stopped in alone on my way home from a morning class. When I saw giant boxes of strawberries piled on the tailgate of a farmer's truck, I didn't waste ten seconds asking myself the questions I would mull over in a conventional grocery: "Hmm, do I really want berries today? Are these overpriced? Are they going to mold the minute I get them home?" I power-walked past other meandering shoppers and bought a bucket load. I drove home daydreaming about the creations I could cook up with my loot.

The key to consuming enough produce and reaping maximum nutritional benefits is planning meals around whatever you have. This presents opportunities to get inventive in the kitchen and try new things, like stuffed zucchini. How many spinach dishes can you have in one week without getting sick of it? When working with fresh ingredients, the answer is, *a lot*!

The variety comes automatically, as you eat loads of leafy greens in April, but find them petering out soon, with broccoli becoming the dark green vegetable of choice. You won't have time to get too tired of any one food, and your nutrient needs will be met. There's no need to measure out every tedious one-quarter cup of raisins and half cup of chopped peppers.

Two things that are impossible to get tired of are asparagus and morels, because neither one stays around long enough. If you have them on the same day in April, you'll forget all about peaches and can make this dish from *Local Flavors,* by Deborah Madison.

❋ ASPARAGUS AND MOREL BREAD PUDDING

3 cups milk
1 cup chopped spring onions with green shoots

Add onions to milk in saucepan and bring to a boil; remove from heat and set aside to steep.

1 loaf stale or toasted multigrain bread, broken into crouton-sized crumbs

Pour milk over crumbs and allow bread to soak.

1 pound asparagus

Chop into ½-inch pieces and simmer in boiling water until bright green.

2 tablespoons butter
1 pound morels (or other wild mushrooms)
Salt and pepper to taste

Melt butter in skillet, cook mushrooms until tender, add salt and pepper, and set aside.

4 eggs
⅓ cup chopped parsley
3 tablespoons oregano
3 cups grated Swiss cheese

Break eggs and beat until smooth, add herbs and plenty of salt and pepper, add bread crumbs with remaining milk, asparagus, mushrooms with their juices, and ⅔ of the cheese. Mix thoroughly and pour into a greased 8 by 12-inch baking dish; sprinkle remaining cheese on top and bake at 350° for about 45 minutes (until puffy and golden).

Download this and all *Animal, Vegetable, Miracle* recipes at www.AnimalVegetableMiracle.com

6 · THE BIRDS AND THE BEES

One of my favorite short stories is Eudora Welty's "Why I Live at the P.O." It's a dead-on comic satire of a certain spirit of family life, and it feeds my private fantasy that I too might someday take up residence in the USPO. If other people don't share this ambition, they just haven't been blessed as I have. Latest in the line of my estimable mail associates was Postmistress Anne, manager of all things postal in our little town—these things taking place within a building about the size of a two-car garage. When we moved, I rented one of their largest boxes for my writer mail, apologizing in advance for the load of stuff I'd be causing them to handle. Anne and her colleagues insisted the pleasure was all theirs. We're lucky we still have a P.O. in our little town, they explained. The government keeps track of what's moving through, and if the number is too low they'll close the branch. "I'll bet you get lots of interesting things in the mail," they surmised.

I thought to myself: You have no idea.

So I turned over three pieces of ID to prove I was citizen enough to rent a postal box in the Commonwealth of Virginia, and since then I have wondered if they've ever had second thoughts. Such surprising gifts come to me through the U.S. mail: a "Can-Jo" (rhymes with "banjo," with a body made from a Mountain Dew can) hand-crafted by a man who felt I needed to have one. (So did, he felt, President Carter.) Class projects. Paintings of imaginary people. More books than probably burned in the

Alexandria library. Anne sounded unfazed on the morning she called to say, "You'd better come get this one, it's making a pretty good racket."

Maybe in places like Hollywood, California, postal clerks would be uneasy about weighing packages and selling stamps while twenty-eight baby chickens peeped loudly into their right ear and four crates of angry insects buzzed in their left. Not here. They all just grinned when Lily and I came in. The insects weren't ours, but Anne invited Lily to check them out anyway. "Come on back behind the counter, hon, take a look at these bees. They've got honey dripping out already."

Bees? Ho-hum, just an ordinary day at our P.O. I adjusted my notion of myself as a special-needs postal customer.

Lily bent over the bee cages, peering at the trembling masses of worker bees humming against the wire mesh sides of the boxes. The sticky substance dripping out was actually sugar water sent along to sustain the bees through their journey. Down below the buzzing clots of worker bees sat the queen bees with their enormous hind ends, each carefully encapsulated in her own special chamber. These big-bootied ladies were replacement queens ordered by local beekeepers from a bee supply company, to jump-start hives whose previous queens were dead or otherwise inadequate.

But Lily quickly turned to the box with *her* name on it: a small cardboard mailing crate with dime-sized holes on all sides and twenty-eight loud voices inside: the noise-density quotient of one kindergarten packed into a shoebox. Lily picked it up and started crooning like a new mother. This was the beginning of a flock she'd been planning for many months and will be tending, I presume, until we see her off to college. Because we knew the chicks were coming this morning, I had allowed her to stay home from school to wait for the call. She wasn't sure the principal would consider this an excused absence. I assured her it was a responsibility large enough to justify a few hours of missed class. I hadn't even known we'd be having lessons in the birds *and* the bees.

Once she'd brought them home, taken her twenty-eight chicks out of that tiny box, and started each one on its path to a new life under her care, Lily was ready to get back to third grade. When we signed her in at the principal's office, the secretary needed a reason for Lily's tardiness. Lily threw back her shoulders and announced, "I had to start my own chicken business this morning."

The secretary said without blinking, "Oh, okay, farming," and entered the code for "Excused, Agriculture." Just another day at our elementary school, where education comes in many boxes.

⚘

I already knew what we'd be in for when we got the baby chicks home. I'd been through the same drill a week earlier with my own poultry project: fifteen baby turkeys. I'd lifted each one out of the box and they hit the ground running, ready to explore the newspaper-lined crate I'd set up in the garage. Right away they set about pecking at every newsprint comma and period they could find. These peeps were hungry, which meant they were born two days earlier. Poultry hatchlings don't need to eat or drink for the first forty-eight hours of life, as they are born with a margin of safety called the yolk sac—the yolk of the egg absorbed into the chick's belly just before hatching. This adaptation comes in handy for birds like chickens and turkeys that have to get up and walk right away, following Mom around to look for something edible. (Other baby birds live in a nest for the first weeks, waiting for a parent to bring takeout.) Newborn poultry can safely be put in a box right after hatching and shipped anyplace they'll reach in two days. Some animal-rights groups have tried to make an issue of it, but mail-order chicks from reputable hatcheries have virtually a 100 percent survival rate.

Until I opened up the box and let in the sunlight, my poultry babies must have presumed they'd spent their last two days of incubation in an upgraded, community egg. Now they were out, with yolk-sac tummies crying, *Time's up!* I scattered a handful of feed around the bottom of their crate. Some of the less gifted pushed the food aside so they could keep pecking at the attractive newsprint dots. Oh, well, we don't grow them for their brains. I filled a shallow water container and showed them how to drink, which they aren't born knowing how to do. They are born, in fact, knowing a good deal of the nothing a turkey brain will ever really grasp, but at this stage their witlessness was lovable. I picked up each one and dipped its tiny beak into the water. Soon they caught on and it was the rage, this water drinking, as all the poults tried dipping and stretching like yodelers, now urgently pecking at any shiny thing, including my wristwatch.

From time to time one of the babies would be overtaken by the urge for a power nap. Staggering like a drunk under the warm glow of the brooder lamp, it would shut its eyes and keel over, feet and tiny winglets sprawled out flat. More siblings keeled onto the pile, while others climbed over the fuzzy tumble in a frantic race to nowhere.

It's a good thing they don't stay this adorable forever. I'd raised turkeys before, with the cuteness factor being a huge worry at the start. When they imprinted on me as Mama and rushed happily to greet me whenever I appeared, I just felt that much more like Cruella De Vil. Inevitably, though, all adorable toddlers turn into something else. These babies would lose their fluff to a stiff adult plumage, and by Thanksgiving they'd just be beasts—in the case of the toms, testosterone-driven beasts that swagger and charge blindly at anything that might be a live female turkey (i.e., anything that moves). As time took its course, turkey nature itself would nudge us toward the task of moving them from barnyard to the deep freeze. This one little shoebox of fluff, plus grain, grass, and time, would add up to some two hundred pounds of our year's food supply.

I can't claim I felt emotionally neutral as I took these creatures in my hands, my fingers registering downy softness and a vulnerable heartbeat. I felt maternal, while at the same time looking straight down the pipe toward the purpose of this enterprise. These babies were not pets. I know this is a controversial point, but in our family we'd decided if we meant to eat *anything,* meat included, we'd be more responsible tenants of our food chain if we could participate in the steps that bring it to the table. We already knew a lot of dying went into our living: the animals, the plants in our garden, the beetles we pull off our bean vines and crunch underfoot, the weeds we rip from the potato hills. Plants have the karmic advantage of creating their own food out of pure air and sunlight, whereas we animals, lacking green chlorophyll in our skin, must eat some formerly living things every single day. You can leave the killing to others and pretend it never happened, or you can look it in the eye and know it. I would never presume to make that call for anyone else, but for ourselves we'd settled on a strategy of giving our food a good life until it was good on the table. Our turkeys would be pampered as children, and then allowed a freedom on open pasture that's unknown to conventionally raised poultry. Thanksgiving was still far away. And some of these birds would survive the holidays, if all went according to plan. Our goal was to establish a breeding flock. These were some special turkeys.

Of the 400 million turkeys Americans consume each year, more than 99 percent of them are a single breed: the Broad-Breasted White, a quick-fattening monster bred specifically for the industrial-scale setting. These are the big lugs so famously dumb, they can drown by looking up at the rain. (Friends of mine swear they have seen this happen.) If a Broad-Breasted White should escape slaughter, it likely wouldn't live to be a year old: they get so heavy, their legs collapse. In mature form they're incapable of flying, foraging, or mating. That's right, reproduction. Genes that make turkeys behave like animals are useless to a creature packed wing-to-wing with thousands of others, and might cause it to get uppity or suicidal, so those genes have been bred out of the pool. Docile lethargy works better, and helps them pack on the pounds. To some extent, this trend holds for all animals bred for confinement. For turkeys, the scheme that gave them an extremely breast-heavy body and ultra-rapid growth has also left them with a combination of deformity and idiocy that renders them unable to have turkey sex. Poor turkeys.

So how do we get more of them? Well you might ask. The sperm must be artificially extracted from live male turkeys *by a person,* a professional turkey sperm-wrangler if you will, and artificially introduced to the hens, and that is all I'm going to say about that. If you think they send the toms off to a men's room with little paper cups and *Playhen Magazine,* that's not how it goes. I will add only this: if you are the sort of parent who threatens your teenagers with a future of unsavory jobs when they ditch school, here's one more career you might want to add to the list.

When our family considered raising turkeys ourselves, we knew we weren't going to go *there.* I was intrigued by what I knew of older breeds, especially after Slow Food USA launched a campaign to reacquaint American palates with the flavors of heritage turkeys. I wondered if the pale, grain-fattened turkeys I'd always bought at the supermarket were counterparts to the insipid vegetable-formerly-known-as-tomato. All the special qualities of heirloom vegetables are found in heirloom breeds of domestic animals too: superior disease resistance, legendary flavors, and scarcity, as modern breeds take over the market. Hundreds of old-time varieties of hoof stock and poultry, it turns out, are on the brink of extinction.

The Price of Life

Industrial animal food production has one goal: to convert creatures into meat. These intensively managed factory farms are called concentrated animal feeding operations (CAFOs). The animals are chosen for rapid growth, ability to withstand confinement (some literally don't have room to turn around), and resistance to the pathogens that grow in these conditions. Advocates say it's an efficient way to produce cheap, good-quality meat for consumers.

Opponents raise three basic complaints: first, the treatment of animals. CAFOs house them as tightly as possible where they never see grass or sunlight. If you can envision one thousand chickens in your bathroom, in cages stacked to the ceiling, you're honestly getting the picture. (Actually, a six-foot-by-eight-foot room could house 1,152.)

A second complaint is pollution. So many animals in a small space put huge volumes of excrement into that small space, creating obvious waste storage and water quality problems. CAFO animals in the United States produce about six times the volume of fecal matter of all humans on our planet. Animals on pasture, by contrast, enrich the soil.

A third issue is health. Confined animals are physically stressed, and are routinely given antibiotics in their feed to ward off disease. Nearly three-quarters of all antibiotics in the United States are used in CAFOs. Even so, the Consumers Union reported that over 70 percent of supermarket chickens harbored campylobacter and/or salmonella bacteria. The antibiotic-resistant strains of bacteria that grow in these conditions are a significant new threat to humans.

Currently, 98 percent of chickens in the United States are produced by large corporations. If you have an opportunity to buy some of that other 2 percent, a truly free-range chicken from a local farmer, it will cost a little more. So what's the going price these days for compassion, clean water, and the public health?

STEVEN L. HOPP

The American Livestock Breeds Conservancy keeps track of rare varieties of turkeys, chickens, ducks, sheep, goats, pigs, and cattle that were well known to farmers a century ago, but whose numbers have declined to insignificance in the modern market. In addition to broader genetic diversity and disease resistance, heritage breeds tend to retain more of their wild ancestors' sense about foraging, predator avoidance, and reproduction—traits that suit them for life in the pasture and barnyard rather than a crowded, windowless metal house. Many heritage breeds are adapted to specific climates. Above all, they're superior in the arena for which these creatures exist in the first place: as food.

Heritage livestock favorites are as colorfully named as heirloom vegetables. You can have your Tennessee Fainting goats, your Florida Cracker cattle, your Jersey Giant chickens, your Gloucester Old Spots hogs. Among draft animals, let us not forget the American Mammoth Jack Ass. The American Livestock Breeds Conservancy publishes directories of these animals and their whereabouts, allowing member farmers to communicate and exchange bloodlines.

We decided to join the small club of people who are maintaining breeding flocks of heritage turkeys—birds whose endearing traits include the capacity to do their own breeding, all by themselves. Eight rare heritage turkey breeds still exist: Jersey Buff, Black Spanish, Beltsville Small White, Standard Bronze, Narragansett, Royal Palm, Midget White, and Bourbon Red. We picked the last one. They are handsome and famously tasty, but for me it was also a matter of rooting for the home team. This breed comes from Bourbon County, Kentucky, a stone's throw from where I grew up. I imagined my paternal grandmother playing in the yard of the farm where these birds were originally bred—an actual possibility. Fewer than two thousand Bourbon Reds now remain in breeding flocks. It struck me as a patriotic calling that I should help spare this American breed from extinction.

Slow Food has employed the paradox of saving rare breeds by getting more people to eat them, and that's exactly what happened in its 2003 Ark of Taste turkey project. So many people signed up in the spring for heirloom Thanksgiving turkeys instead of the standard Butterball, an unprecedented number of U.S. farmers were called upon to raise them. The demand has continued. So we jumped on that wagon, hoping to have our rare birds and eat them too.

⚜

Lily's chickens, however, were a different story: her own. The day of their promised arrival had been circled on her calendar for many months: *April 23, my babies due!* Some parents would worry about a daughter taking on maternal responsibility so early in life, but Lily was already experienced. She started keeping her first small laying flock as a first-grader, back in Tucson where the coop had to be fortified against coyotes and bobcats. The part of our move to Virginia that Lily most dreaded, in fact, was saying good-bye to her girls. (The friends who adopted them are kind enough to keep us posted on their health, welfare, and egg production.) We prepared her for the move by promising she could start all over again once we got to the farm. It would be a better place for chickens with abundant green pasture for a real free-range flock, not just a handful of penned layers. "You could even sell some of the eggs," I'd added casually.

Say no more. She was off to her room to do some calculations. Lily is the sole member of our family with gifts in the entrepreneurial direction. Soon she was back with a notebook under her arm. "It's okay to move," she said. "I'll have an egg business."

A few days later she brought up the subject again, wanting to be reassured that our Virginia hens would just be for eggs, not for meat. Lily knew what farming was about, and while she'd had no problem eating our early turkey experiments, chickens held a different place in her emotional landscape. How can I convey her fondness for chickens? Other little girls have ballerinas or Barbie posters on their bedroom walls; my daughter has a calendar titled "The Fairest Fowl." One of the earliest lessons in poultry husbandry we had to teach her was "Why we don't kiss chickens on the mouth." On the sad day one of her hens died, she wept loudly for an entire afternoon. I made the mistake of pointing out that it was *just a chicken*.

"You don't understand, Mama," she said, red-eyed. "I love my chickens as much as I love *you*."

Well, shut me up. She realized she'd hurt my feelings, because she crept out of her room an hour later to revise the evaluation. "I didn't really mean that, Mama," she sniffled. "I'm sorry. If I love my chickens six, I love you seven."

Oh, good. I'm not asking who's a ten.

So I knew, in our discussions of poultry commerce, I needed to be reassuring. "They'll be your chickens," I told her. "You're the boss. What you sell is your decision."

As weeks passed and her future on the farm began to take shape in her mind, Lily asked if she'd also be able to have a horse. Her interest in equines surpasses the standard little-girl passion of collecting plastic ones with purple manes and tail; she'd lobbied for riding lessons before she could ride a bike. I'd long assumed a horse was on our horizon. I just hoped it could wait until Lily was tall enough to saddle it herself.

In the time-honored tradition of parents, I stalled. "With your egg business, you can raise money for a horse yourself," I told her. "I'll even match your funds—we'll get a horse when you have half the money to buy one."

When I was a kid, I would have accepted these incalculable vagaries without a second thought, understanding that maybe a horse was out there for me but I'd just have to wait and see. The entrepreneurial gene apparently skips generations. Lily got out her notebook and started asking questions.

"How much does a horse cost?"

"Oh, it depends," I hedged.

"Just a regular mare, or a gelding," she insisted. When it comes to mares and geldings, she knows the score. I'd recently overheard her explaining this to some of her friends. "A stallion is a boy that's really fierce and bossy," she told them. "But they can give them an operation that makes them gentle and nice and helpful. You know. *Like our daddies.*"

Okay, then, this girl knew what she was looking for in horseflesh. What does an animal like that cost, she inquired? "Oh, about a thousand dollars," I said, wildly overestimating, pretty sure this huge number would end the conversation.

Her eyes grew round.

"Yep," I said. "You'll have to earn half. Five hundred."

She eyed me for a minute. "How much can I sell a dozen eggs for?"

"Nice brown organic eggs? Probably two-fifty a dozen. But remember, you have to pay for feed. Your profit might be about a dollar a dozen."

She disappeared into her room with the notebook. She was only a second-grader then, as yet unacquainted with long division. I could only assume she was counting off dollar bills on the calendar to get to five hundred. In a while she popped out with another question.

"How much can you sell chicken meat for?"

"Oh," I said, trying to strike a morally neutral tone in my role as financial adviser, "organic chicken sells for a good bit. Maybe three dollars a pound. A good-size roasting bird might net you ten dollars, after you subtract your feed costs."

She vanished again, for a very long time. I could almost hear the spiritual wrestling match, poultry vs. equines, fur and feathers flying. Many hours later, at dinner, she announced: "Eggs and meat. We'll only kill the mean ones."

⚘

I know I'm not the first mother to make an idle promise I'd come to regret. My mother-in-law has told me that Steven, at age seven, dashed through her kitchen and shouted on the way through, "Mom, if I win a monkey in a contest, can I keep it?" Oh, sure honey, Joann said, stirring the pasta. She had seven children and, I can only imagine, learned to tune out a lot of noise. But Steven won the monkey. And yes, they kept it.

In my case, what I'd posed as a stalling tactic turned out to be a powerful nudge, moving Lily from the state of loving something *as much as her mother* (or six-sevenths as much) to a less sentimental position, to put it mildly. I watched with interest as she processed and stuck to her choices. I really had no idea where this would end.

Chicks must be started no later than April if they're to start laying before cold weather. We moved to the farm in June, too late. From friends we acquired a few mature hens to keep us in eggs, and satisfy Lily's minimum daily requirement of chicken love. But the farm-fresh egg business had to wait. Finally, toward the end of our first winter here, we'd gotten out the hatchery catalog and curled up on the couch to talk about a spring poultry order. Lily shivered with excitement as we discussed the pros and cons of countless different varieties. As seed catalogs are to me, so are the hatchery catalogs for my daughter. Better than emeralds and diamonds, these Rocks, Wyandottes, and Orpingtons. She turned the pages in a trance.

"First of all, some Araucanas," she decided. "Because they lay pretty green eggs. My customers will like those."

I agreed, impressed with her instincts for customer service.

"And for the main laying flock I want about ten hens," she said. "We'll keep one of the roosters so we can have chicks the next spring."

She read listings for the heavy breeds, studying which ones were strong winter layers, which were good mothers (some breeds have motherhood entirely bred out of them and won't deign to sit on their own eggs). She settled on a distinguished red-and-black breed called Partridge Rocks. We ordered sixteen of these, straight run (unsexed), of which about half would grow up to be females. Lily knows you can't have too many roosters in a flock—she had mentioned we would "keep" one of the males, implying the rest would be dispatched. I didn't comment. But it seemed we were now about seven roosters closer to a horse. I hoped they would all be very, very mean.

She paused over a section of the catalog titled "Broilers, Roasters and Fryers."

"Look at these," she said, showing me a picture of an athletic-looking fowl, all breast and drumstick. "Compact bodies and broad, deep breasts . . . ," she read aloud. "These super meat qualities have made the Dark Cornish a truly gourmet item."

"You're sure you want to raise meat birds too?" I asked. "Only if you want to, honey." I was starting to crumble. "You'll get your horse someday, no matter what."

"It's okay," she said. "I won't name them. I'll have my old pet hens to love."

"Of course," I said. "Pets are pets. Food is food."

Out on the near horizon, Lily's future horse pawed the ground and whinnied.

Eating My Sister's Chickens

BY CAMILLE

During my first year of college, one of my frequent conversations went like this:

"Camille, you're a vegetarian, right?"

"Well, no."

"No? You really seem like the type."

"Well, I only eat free-range meat."

"Free *who?*"

I guess I do seem like the type. Personal health and the environment are important to me, and my vocational path even hints at vegetarianism—I teach yoga, and may study nutrition in graduate school. The meat-eating question is one I've considered from a lot of angles, but that's not easy to explain in thirty seconds. A lunch line is probably not the best place to do it, either. For one thing, all meat is not created equal. Cows and chickens that spent their lives in feedlots, fattening up on foods they did not evolve to eat, plus antibiotics, produce different meat from their counterparts that lived outdoors in fresh air, eating grass. That's one nutritional consideration to bear in mind while weighing the pros and cons of vegetarianism.

There are others, too. Vegetarians and vegans should consider taking iron supplements because the amount of this nutrient found in plant sources is minuscule compared with the amount found in meat. Of course, eating plenty of iron-containing dark leafy greens, legumes, and whole grains is a good plan. Along with a host of other essential nutrients, they do offer a good bit of iron, but in some cases it may not be enough to keep the body producing hemoglobin. Vitamin B12 is also tricky; in its natural form it's found only in animal products. There are traces of it in fermented soy and seaweed, but the Vegetarian Society warns that the form of B12 in plant sources is likely unavailable to human digestion. This means that vegans—people who eat no meat, dairy, or eggs—need to rely on supplements or foods fortified with B12 to prevent this dangerous deficiency. Vegan diets also tend to be skimpy in the calcium department, so supplements there can be helpful as well.

Humans are naturally adapted to an omnivorous diet: we have canine teeth for tearing meat and plenty of enzymes in our guts to digest the proteins and fats found in animals. Ancestral societies in every part of the world have historically relied on some animal products for sustenance. Even the ancient Hindu populations of India were not complete vegetarians—though they did not know this. Traditional harvesting techniques always left a substantial amount of insect parts, mostly termite larvae and eggs, in their grain supply. When vegan Hindu populations began moving to England, where food sanitation regulations are stricter, they began to suffer from a high incidence of anemia. Just a tiny amount of meat (even bug parts!) in the diet makes a big difference. Of course, abstaining from meat for relatively short periods for spiritual reasons is a practice common to many societies. During these times, traditionally, we're meant to be less active and more contemplative, reducing our need for the nutrients supplied by animal products.

Generally speaking, people who are not strict vegetarians will find more options in their local-food scene. Pasture-based meat and eggs are produced nearly everywhere in the country, unlike soybean curd and other products that may anchor a vegan diet. Chicken, lamb, and other meats from small farms are available throughout the year. And while animal fats—even for meat eaters—are considered nutrition ogres, they are actually much healthier than the hydrogenated oils that replace them in many processed foods. (Trans fat, a laboratory creation, has no nutritional function in our bodies except to float around producing free radicals that can damage tissue.) Good-quality animal fats contain vitamins A, D, B6, and B12, and some essential minerals. Free-range meat and eggs have a healthy rather than unhealthy cholesterol content, because of what the animal ate during its happy little life.

The following is a chicken recipe we invented that reminds us of Tucson. We use free-range chicken, and fresh vegetables in summer, but in early spring we rely on our frozen zucchini and corn from last summer. The one essential fresh ingredient—cilantro—begins to show up early in farmers' markets. If you don't like cilantro, leave it out, of course, but the dish will lose its southwestern flavor.

❋ CHICKEN RECUERDOS DE TUCSON

1 whole cut-up chicken, or thighs and legs
Olive oil (for sauté)
1 medium onion, sliced
2–3 cloves garlic, minced

Brown chicken in a large kettle. Remove chicken, add oil, and gently sauté the onion and garlic.

1 teaspoon cumin seed
Green chiles to taste, chopped
2 red or green peppers
1 large or 2 medium zucchini or other squash, thickly sliced

Add to kettle and sauté, add small amount of broth if necessary.

1 cup tomatoes (fresh, frozen, canned, or ½ cup dehydrated, depending on season)
2 cups corn kernels
2 teaspoons oregano
1 teaspoon basil
2 cups chicken broth or water

Add to kettle along with browned chicken, add water or broth (more if using dried vegetables), cover, and simmer for 30 to 40 minutes, until chicken is done to bone. Garnish with fresh cilantro.

Download this and all other *Animal, Vegetable, Miracle* recipes at www.AnimalVegetableMiracle.com

7 · GRATITUDE

May

On Mother's Day, in keeping with local tradition, we gave out tomato plants. Elsewhere this may be the genteel fête of hothouse orchids, but here the holiday's most important botanical connection is with tomatoes. Killing spring frosts may safely be presumed behind us, and it's time to get those plants into the garden.

We grow ours from seed, so it's not just the nursery-standard Big Boys for us; we raise more than a dozen different heirloom varieties. For our next-door neighbor we picked out a narrow-leaved early bearer from the former Soviet Union with the romantic name of "Silvery Fir Tree." Carrying the leggy, green-smelling plant, our family walked down the gravel driveway to her house at the bottom of our hollow.

"Oh, well, goodness," she said, taking the plant from us and admiring it. "Well, look at that."

Every region has its own language. In ours, it's a strict rule that you *never* say "Thank you" for a plant. I don't know why. I was corrected many times on this point, even scolded earnestly, before I learned. People have shushed me as I started to utter the words; they put their hands over their ears. "*Why* can't I say thank you?" I've asked. It's hard. Southern manners are so thoroughly bred into my brain, accepting a gift without a thank-you feels like walking away from changing a tire without washing my hands.

"Just don't," people insist. If you do say it, they vow, the plant will wither up straightaway and die. They have lots of stories to back this up. They do not wish to discuss whether plants have ears, or what. Just *don't*.

So we knew what our neighbor was trying not to say. We refrained from saying "You're welcome," had a nice Sunday afternoon visit, and managed not to jinx this plant—it grew well. Of all the tomato plants that ultimately thrived in her garden, she told us the Silvery Fir Tree was the first to bear.

⚜

On the week of May 9 we set out our own tomatoes, fourteen varieties in all: first, for early yields, Silvery Fir Tree and Siberian Early, two Russian types that get down to work with proletarian resolve, bred as they were for short summers. For a more languid work ethic but juicy midseason flavor we grow Brandywines, Cherokee Purples, orange Jaune Flammès, and Green Zebra, which is lemony and bright green striped when fully ripe. For spaghetti sauces and canning, Martino's Roma; Principe Borghese is an Italian bred specifically for sun-drying. Everything we grow has its reason, usually practical but sometimes eccentric, like the Dolly Partons given us by an elderly seed-saving friend. ("What do the tomatoes look like?" I asked. She cupped her hands around two enormous imaginary orbs and mugged, "Do you have to ask?") Most unusual, probably, is an old variety called Long Keeper. The fruits never fully ripen on the vine, but when harvested and wrapped in newspaper before frost, they slowly ripen by December.

That's just the tomatoes. Also in the second week of May we set out pepper seedlings and direct-seeded the corn, edamame, beets, and okra. Squash and cucumber plants went into hills under long tents of row-cover fabric to protect them from cool nights. We weeded the onions, pea vines, and potatoes; we planted seeds of chard, bush beans, and sunflowers, made bamboo tepees for the pole beans, and weathered some spring thunderstorms. That's one good week in food-growing country.

By mid-month, once warmth was assured, we and all our neighbors set out our sweet potato vines (there was a small melee down at the Southern States co-op when the management underordered sweet-potato sets). We also put out winter squash, pumpkins, basil seedlings, eggplants, and melons, including cantaloupes, honeydews, rock melons, perfume melons, and four kinds of watermelons. Right behind planting come the weeding, mulching, vigilance for bugs and birds, worry over too much rain or not enough. It so resembles the never-ending work and attention of parenting, it seems right that it all should begin on Mother's Day.

For people who grow food, late spring is the time when we pay for the relative quiet of January, praying for enough hours of daylight to get everything done. Many who farm for a living also have nine-to-five jobs off the farm and *still* get it done. In May we push deadlines, crunch our other work, borrow time, and still end up parking the tractor with its headlamp beams pointed down the row to finish getting the last plants heeled into place. All through May we worked in rain or under threat of it, playing chicken with lightning storms. We worked in mud so thick it made our boots as heavy as elephant feet. On work and school days we started predawn to get an early hour in, then in the late afternoon picked up again where we'd left off. On weekends we started at daybreak and finished after dusk, aching and hungry from the work of making food. Labors like this help a person appreciate why good food costs what it does. It ought to cost more.

In the midst of our busy spring, one of us had a birthday. Not just a run-of-the-mill birthday I could happily ignore, but an imposing one, involving an even fraction of one hundred. We cooked up a party plan, setting the date for Memorial Day so out-of-town guests could stay for the long weekend. We sent invitations and set about preparing for a throng of guests, whom we would certainly want to feed. Our normal impulse would have been to stock up on standard-issue, jet-propelled edibles. But we were deep enough into our local-food sabbatical by now, that didn't seem entirely normal.

Something had changed for us, a rearrangement of mindset and the contents of our refrigerator. Our family had certainly had our moments of longing for the illicit: shrimp, fresh peaches, and gummy worms, respectively. Our convictions about this project had been mostly theoretical to begin with. But gradually they were becoming fixed tastes that we now found we couldn't comfortably violate for our guests, any more than a

Hindu might order up fast-food burgers just because she had a crowd to feed.

It put us in a bit of a pickle, though, to contemplate feeding a huge crowd on the products of our county this month. If my mother had borne me in some harvest-festival month like October, it would have been easy. But she (like most sensible mammals, come to think of it) had all her children in the springtime, a fact I'd never minded until now. Feeding just my own household on the slim pickings of our local farms had been a challenge in April. The scene was perking up in May, but only slightly. Our spring had been unusually wet and cool, so the late-spring crops were slow coming in. We called a friend who cooks for a living, who came over to discuss the game plan.

Apparently, the customary starting point for caterers in a place that lacks its own food culture is for the client to choose a food theme that is somebody else's land-based food culture. Then all you have to do is import the ingredients from somebody else's land. Mediterranean? A banquet of tomato-basil-mozzarella salads, eggplant caponata, and butternut ravioli—that's a crowd pleaser. And out of the question. No tomatoes or eggplants yet existed in our landscape. Our earliest of early tomatoes was just now at the blossom stage. Mexican? Enchiladas and chipotle rice? Great, except no peppers or tomatillos were going to shine around here. Siberian Tundra was maybe the cuisine we were after. We began to grow glum, thinking of borscht.

Not to worry, said Kay. A good food artist knows her sources. She would call the farmers she knew and see what they had. Starting with ingredients, we'd build our menu from there. As unusual as this might seem, it is surely the world's most normal way of organizing parties—the grape revels of Italy and France in September, the Appalachian ramp hoedowns in April, harvest festivals wherever and whenever a growing season ends. That's why Canadian Thanksgiving comes six weeks before ours: so does Canadian winter. We were determined to have a feast, but if we meant to ignore the land's timetable of generosity and organize it instead around the likes of birthdays, a good travel weekend, and the schedules of our musician friends, that was *our* problem.

Kay called back with a report on our county's late May pantry. There would be asparagus, of course, plus lots of baby lettuces and spinach by then. Free-range eggs are available here year-round. Our friend Kirsty had free-range chicken, and the Klings, just a few miles from us, had grass-fed lamb. The Petersons had strawberries, Charlie had rhubarb, another family was making goat cheese. White's Mill, five miles from our house, had flour. If we couldn't pull together a feast out of that, I wasn't worth the Betty Crocker Homemaker of Tomorrow Award I won in 1972. (Kind of by accident, but that is another story.)

The menu wrote itself: Lamb kabobs on the grill, chicken pizza with goat cheese, asparagus frittata, an enormous salad of spring greens, and a strawberry-rhubarb crisp. To fill out the menu for vegan friends we added summer rolls with bean sprouts, carrots, green onions, and a spicy dipping sauce. We had carrots in the garden I had nursed over the winter for an extra-early crop, and Camille ordinarily grew bean sprouts by the quart in our kitchen windowsill; she would ramp up her production to a couple of gallons. We might feed our multitudes after all.

As the RSVPs rolled in, we called farmers to plead for more strawberries, more chickens. They kindly obliged. The week of the party, I cut from our garden the first three giant heads of Early Comet broccoli—plants we'd started indoors in February and set out into nearly frozen soil in March. Without knowing it, I'd begun preparing for this party months ago. I liked seeing now how that whole process, beginning with seeds, ending with dinner, fixed me to some deeper than usual sense of hospitality. Anyone who knows the pleasure of cooking elaborately for loved ones understands this. Genesis and connection with annual cycles: by means of these, a birthday could be more than a slap on the back and jokes about memory loss.

On Tuesday, four days pre-party, Camille and I hoed weeds from around corn seedlings and planted ten hills of melons for some distant, future party: maybe we'd have corn and cantaloupe by Lily's birthday in July. By dusk the wind was biting our ears and the temperature was falling fast. We hoped the weather would turn kinder by this weekend. We expected well over a hundred people—about thirty spending the weekend. Rain would wreck any chance for outdoor dancing, and camping in the yard would be grim. We scowled at the clouds, remembering (ruefully) the cashier who'd jinxed the rain in Tucson. We weren't in drought here, so we decided we could hope with impunity. And then take what came.

On Wednesday we checked the bean sprouts Camille had started in two glass gallon jars. Their progress was unimpressive; if they intended to fill out a hundred fat, translucent summer rolls in three days, they had some work to do. We tried putting them in a sunnier window, but the day was cloudy. Suddenly inspired, we plugged in a heating pad and wrapped it around the jars. Just an hour did the trick. I'm sure we violated some principle of Deep Ecology, but with just a quick jolt from the electric grid our sprouts were on their way, splitting open their seeds and pushing fat green tails into the world.

On Thursday I went to the garden for carrots, hoping for enough. With carrots you never know what you've got until you grab them by the green hair and tug them up. These turned out to be gorgeous, golden orange, thicker than thumbs, longer than my hand. Shaved into slivers with green onions and our indolent sprouts, two dozen carrots would be plenty. I could only hope the lambs and chickens were cooperating as well. I stood for a minute clutching my carrots, looking out over our pasture to Walker Mountain on the horizon. The view from our garden is spectacular. I thought about people I knew who right at that moment might be plucking chickens, picking strawberries and lettuce, just for us. I felt grateful to the people involved, and the animals also. I don't say this facetiously. I sent my thanks across the county, like any sensible person saying grace before a meal.

Guests began to trickle in on Friday: extended family from Kentucky, old college friends from South Carolina, our musician friends John, Carrie, and Robert. I was bowled over by the simultaneous presence of so many people I care about, from as far away as Tucson and as near as next door. We made all the beds and couches, and pitched tents. We walked in the garden and visited. All those under age twelve welded into a pack and ran around like wild things. I overheard a small platoon leader in the garden command: "You, whatever your name is, go down that way and I'll hide and we'll scare the girls." I only made two rules: Don't injure each other, and don't flatten the crops. With the exception of one scraped finger and the tiniest mishap with a Dolly Parton, they obliged.

We set up a sound system on the back patio, dragged bales of straw into benches, and eyed the sky, which threatened rain all day Saturday but by late afternoon had not delivered. We carried a horse trough out of the barn and filled it with ice to chill our Virginia Chambourcin and Misty River wines, and beer from a nearby microbrewery. The lamb kabobs on the grill made all our mouths water for an hour while Kay and her helpers worked their mojo in our kitchen. The food, when it came out, was applauded: the summer rolls were saucy, the lamb succulent, the frittata puffy and light. The strawberry-rhubarb crisp vanished into thin air. Here's what we *didn't* have: the shrimp arranged in a ring like pink poker chips; those rock-hard broccoli wedges and lathed carrots surrounding the ubiquitous white dip; the pile of pineapple and melon chunks on a platter. Nobody seemed too disappointed.

Some of us were in fact sticking our fingers into the rhubarb-crisp pans to lick up crumbs when the music started. The three-year-olds were the first ones out on the flagstone dance floor, of course, followed closely by my seventy-five-year-old parents, the teenagers and the elders and the middle-aged, recklessly dancing across age categories. And it still didn't rain. Nobody fell in the creek, nobody went hungry, and nobody's husband refused to dance. When the night chilled us we built a huge bonfire, and nobody fell into that either. Midnight found me belting out backup harmonies with my cousin Linda to "You Can't Always Get What You Want" by the Rolling Stones. The over-fifty crowd stayed on its feet until two in the morning. You get what you need.

❦

I'd asked for no presents. The stuff-acquisition curve of my life has long since peaked and lately turned into a campaign against accumulation, with everyday skirmishes on the kitchen table. Not just mail and school papers, either, I mean *stuff* on that table. (Shoes, auto parts, live arthropods in small wire cages.) "No presents," I said. "Really." But here in Dixie we will no more show up to a party empty-handed than barebottomed, because that's how we were raised. A covered dish is standard, but was unnecessary in this case. To make everyone comfortable we had to suggest an alternative.

Camille made the call, and it was inspired: a plant. The tiniest posy, anything would serve. And truthfully, while we'd put prodigious efforts into our vegetable garden and orchards, our front yard lay sorry and neglected. Anything people might bring to set into that ground would improve it. Thus began the plan for my half-century Birthday Garden: higgledy-piggledy, florescent and spontaneous, like friendship itself.

This is what my friends brought: dug from their own backyards, a division of a fifty-year-old peony, irises, a wisteria vine, spicy sweetshrub, beebalm, hostas, datura, lilies, and a flowering vine whose name none of us knows. My parents brought an Aristocrat pear, a variety bred by an old friend from our hometown. A geographer friend brought Portuguese collards, another indulged my fondness for red-hot chile peppers. Rosemary and sage, blueberry and raspberry, fountain grass, blue sweetgrass, sunshine-colored roses, blue-and-white columbines, scarlet poppies, butterfly bush and "Sunset" echinacea—the color scheme of my garden is "Crayola." Our neighbor, to whom we'd taken the tomato plant, dug some divisions of her prettiest lemon lilies. "Oh, well, goodness," I said as I received each of these botanical gifts. "Well, look at that."

I thanked my parents for having me, thanked the farmers for the food, thanked family and friends for the music, the dancing, the miles traveled, and the stunning good luck of having them all in my life. But I did not say "Thank you" for a plant. My garden lives.

Happy Returns

BY CAMILLE

Birthdays have always been a big deal in my family. Mom would cook exactly what we wanted for dinner and bake any kind of cake for Lily and me on our special days. Our only limit was the one-year, one-person rule: each year I could invite as many people to my party as the age I was turning. So, three kids at my three-year-old party, ten at my tenth. As you can probably guess, things got increasingly wild. At one point I remember asking, "Mama, when I turn thirty can I really have thirty people at my party?" I was shocked when she said, "Sure, absolutely." Wow, I thought, excited already at the thought of that huge celebration. Now I just think it's funny that I assumed I'd still be living with my parents on my thirtieth birthday.

Last spring when my mom turned fifty she broke her own rule, big-time. None of us minded. What could be better than a long weekend of live music, good food, and friends and family from all over? We spent weeks setting up for 150 guests and trying to come up with a perfect Plan B (which we never really did) if it rained. I was inspired to see so many people in the fifty-plus age group up and dancing well past midnight, grandparents included. I hope I'll still be dancing—and breaking my own rules—when I'm that age.

Two of our friend Kay's best recipes from that party are good ways to feature the early spring vegetables, and the only fruits you're likely to find in May: strawberries and rhubarb.

❋ ASIAN VEGETABLE ROLLS

2 oz. thin rice noodles

Drop noodles into boiling water, remove from heat, and let stand for 8–10 minutes, stirring occasionally. Drain, rinse with cold water, and drain again.

1 cup bean sprouts
10 soft lettuce leaves
1 cup carrots, finely shredded
2 to 3 green onions, finely chopped
½ cup mint leaves
½ cup cilantro leaves
8 rice paper wrappers (about 8 inches square)

Lay out noodles and vegetables in an assembly line. Heat a pan of water until it's almost too hot to handle. Soak one rice paper wrapper in the hot water for 15–20 seconds, then take it out and lay it flat. Flatten out one lettuce leaf on top (this helps prevent other fillings from poking through the wrapper). Next, place a finger-sized bunch of noodles close to one side of the paper and roll that side over the noodles. Continue this same pattern for the vegetable fillings, laying each ingredient parallel to the noodles and rolling the paper over. After the mint and cilantro leaves have gone in, fold the ends of the wrapper in, then fold the remaining side over them to secure. Set roll on a platter, seam side down. Keep rolls moist until served, and separated so they don't stick together (the wrappers will rip).

Serve whole or cut in half, with your choice of spicy dipping sauce. One simple option is to add a few tablespoons of rice vinegar and sesame oil to a half cup of soy sauce.

❋ STRAWBERRY RHUBARB CRISP

3 cups strawberries, halved
3 cups rhubarb, chopped
½ cup honey

Mix together thoroughly and place in an 8-by-8-inch ungreased pan.

½ cup flour
½ cup rolled oats
½ cup brown sugar (or a bit more, to taste)
¾ teaspoon cinnamon

½ teaspoon allspice
⅓ cup butter

Mix until crumbly, sprinkle over fruit mixture, and bake at 350° for 40 to 50 minutes, until golden.

Download these and all other *Animal, Vegetable, Miracle* recipes at www.AnimalVegetableMiracle.com

MAY OPENS UP MANY MORE POSSIBILITIES IN OUR WEEKLY MEAL PLAN:

Sunday ~ Grilled chicken, fresh bread, and a giant salad of fresh greens, carrots, and peas
Monday ~ Asparagus and morel bread pudding
Tuesday ~ Asian summer rolls with spicy peanut sauce, served with rice
Wednesday ~ Vegetarian tacos with refried beans, pea shoots, lettuce, spring onions, and cheese
Thursday ~ Cheese ravioli tossed with stir-fried spring vegetables, oregano, and olive oil
Friday ~ Chicken pizza with olives and feta
Saturday ~ Frittata packed with cheese and vegetables, salad, strawberry-rhubarb crisp

8 · GROWING TRUST

Mid-June

Twice a year, on opposite points around the calendar's circle, the perpetual motion of our garden life goes quiet. One is obvious: midwinter, when fields lie under snow. Our animals need extra care then, but any notion of tomato is history. The other vegetal lull is in June, around Midsummer's Day. Seeds are in the ground if they're going to be. Corn and beans are up, cukes and tomatoes are blossoming. Broccoli and asparagus are harvested; peas are winding down.

It isn't that we walk out into the field on June 10 and say, "Wow, nothing to do anymore." There will always be more weeds. Everything could be mulched better, fed more compost, protected better against groundhogs. A thriving field of vegetables is as needy as a child, and similarly, the custodian's job isn't done till the goods have matured and moved out. But you can briefly tiptoe away from the sleeping baby. It's going to wake up wailing, but if you need the rest, you get while the getting is good.

We had planned our escape for late June, the one time between May planting and September harvest when it seemed feasible to take a short vacation from our farm. If we'd been marketing to customers or retailers, this would still be a breakneck time of getting orders lined up and successive plantings laid out. Our farming friends all agree this is the most trying challenge of the job: lost mobility. It's nearly impossible to leave fields and animals for just one day, let alone a week. Even raising food on our relatively modest scale required that we put in overtime to buy a respite. We mulched everything heavily to keep root systems moist, discourage weeds, and prevent late blight.

My favored mulching method is to cover the ground between rows of plants with a year's worth of our saved newspapers; the paper and soy-based ink will decompose by autumn. Then we cover all that newsprint—comics, ax murderers, presidents, and all—with a deep layer of old straw. It is grand to walk down the rows dumping armloads of moldy grass glop onto the faces of your less favorite heads of state: a year in review, already starting to compost.

Believe it or not, weeds will still come up through all this, but it takes a while. With neighbors on call to refill the poultry waterers, open and close coops, and keep an eye on the green things, we figured on escaping for a week and a half. Our plan was to head north in a big loop through New England, up to Montreal, and back through Ohio, staying with friends and relatives all along the way.

We nearly had the car packed when it started raining cherries. We'd been watching our huge cherry trees, which every June bear fruit enough to eat and freeze for pies and sorbets all year long. This, plus our own sparse peaches, plums, Asian pears, and a local orchard's autumn apples, were the only local sources of tree fruits we knew about, and we didn't want to miss any of them. Our diet had turned our attention keenly to fruit, above all else. Like the Frostburg man with his eighty-three mealy peaches and so forth, we wanted our USDA requirement. This winter we'd be looking at applesauce and whatever other frozen fruit we could put by ahead of time, at whatever moment it came into season. Tree-ripened fruit, for the local gourmand, is definitely worth scheduling your vacation around.

Last year, my journal said, the cherries had ripened around June 8. This year summer was off to such a cool, slow start, we stood under the tree and tried to generate heat with our heart's desire. Then it happened: on June 15, one day before our planned departure, the hard red spheres turned to glossy black, all at once. The birds showed up in noisy gangs, and up we went to join them. Standing on ladders and the roof of the truck, we picked all afternoon into dusk, till we were finding the fruits with our fingers instead of our eyes.

Like the narrator of Kazantzakis's *Zorba the Greek,* I have a resolute weakness for cherries. Annually I take Zorba's advice on the cure; so far it hasn't worked. All of us were smitten, filling gallon buckets, biting cherries alive from their stems. This was our first taste of firm, sweet fruit flesh in months, since the early April day when we'd taken our vows and foresworn all exotics. Fruit is what we'd been hankering for, the only deprivation that kept needling us. Now we ate our fill, delivered some to neighbors, and put two gallons in the freezer, rejoicing. Our fructose celibacy was over.

The next day our hands were still stained red as Lady Macbeth's, but it was time to go, or we'd never get our trip. We packed up some gifts for the many friends whose hospitality and guest beds lay ahead of us: cherries of course, bottles of local wine, and a precious few early tomatoes we had managed to pamper to ripeness—by June 12, a record for our neighborhood. We threw bags of our salad greens and snap peas into a cooler with some cheese and homemade bread for munching along the way. If we waited around, some other task would start to fall on our heads; we could practically hear the weeds clawing at the president's face. We hit the gas and sped away.

⚘

Farming is not for everybody; increasingly, it's hardly for anybody. Over the last decade our country has lost an average of 300 farms a week. Large or small, each of those was the life's work of a real person or family, people who built their lives around a promise and watched it break. The loss of a farm is a darkness leading to some of life's bitterest ends. Keeping one, on the other hand, may mean also working in a factory at the end of a long daily drive, behind and ahead of the everyday work of farming.

Wherever farms are still living, it's due to some combination of luck, courage, and adaptability. In my home state, Kentucky, our agriculture is known for two nonedible commodities: tobacco and racehorses. The latter is a highly capitalized industry that spreads little of its wealth into the small family farm; the former was the small farm's bottom dollar, until the bottom dropped out. In my lifetime Kentucky farmers have mostly had the options of going broke, or going six ways to Sunday for the sake of staying solvent. I know former tobacco growers who now raise certified organic gourmet mushrooms, bison steaks, or asparagus and fancy salad greens for restaurants. On the bluegrass that famously nourished Man o' War and Secretariat, more modest enterprises with names like "Hard Times Farm" and "Mother Hubbard's" are now raising pasture-fed beef, pork, lamb, and turkeys. Kentucky farms produce flowers, garlic, organic berries and vegetables, emu and ostrich products, catfish, and rainbow trout. Right off the Paris Pike, a country lane I drove a hundred times in my teenage years, a farmer named Sue now grows freshwater shrimp.

If we could have imagined this when I was in high school, that our county's fields might someday harbor prawn ponds and shiitakes, I suppose we would have laughed our heads off. The first time I went to a party where "Kentucky caviar" was served, I suspected a trick (as in "Rocky Mountain oyster"). It wasn't; it was Louisville-grown fish eggs. Innovative cottage industries are life and death for these farmlands. Small, pioneering agricultural ventures are the scene of more hard work, risk-taking, and creative management than most people imagine.

Among other obstacles, these farmers have to contend with a national press that is quick to pronounce them dead. Diversified food-producing farms on the outskirts of cities are actually the fastest-growing sector of U.S. agriculture. The small farm is at the moment very busy thinking its way out of a box, working like mad to protect the goodness and food security of a largely ungrateful nation.

These producers can't survive by catering only to the upscale market, either. The majority of farmers' market customers are people of ordinary means, and low-income households are not necessarily excluded. An urban area in eastern Tennessee has a vegetable equivalent of a bookmobile, allowing regional farmers to get produce into neighborhoods whose only other food-purchasing option might be a liquor store. Though many eligible mothers may not know it, the U.S. nutritional assistance program for women with infants and children (WIC) gives coupons redeemable at farmers' markets to more than 2.5 million participants in forty-four states.

Likewise, the Seniors Farmers Market Nutrition Program (SFMNP) awards grants to forty states and numerous Indian tribal governments to help low-income seniors buy locally grown fruits, vegetables, and herbs. Citizen-led programs from California to New York are linking small farmers with school lunch programs and food banks.

Even so, a perception of organic food as an elite privilege is a considerable obstacle to the farmer growing food for middle-income customers whose highest food-shopping priority is the lowest price. Raising food without polluting the field or the product will always cost more than the conventional mode that externalizes costs to taxpayers and the future. To farm sustainably and also stay in business, these market gardeners have to bridge the psychological gap between what consumers *could* pay, and what we will actually shell out.

Grocery money is an odd sticking point for U.S. citizens, who on average spend a lower proportion of our income on food than people in any other country, or any heretofore in history. In our daily fare, even in school lunches, we broadly justify consumption of tallow-fried animal pulp on the grounds that it's cheaper than whole grains, fresh vegetables, hormone-free dairy, and such. Whether on school boards or in families, budget keepers may be aware of the health tradeoff but still feel compelled to economize on food—in a manner that would be utterly unacceptable if the health risk involved an unsafe family vehicle or a plume of benzene running through a school basement.

It's interesting that penny-pinching is an accepted defense for toxic food habits, when frugality so rarely rules other consumer domains. The majority of Americans buy bottled drinking water, for example, even though water runs from the faucets at home for a fraction of the cost, and government quality standards are stricter for tap water than for bottled. At any income level, we can be relied upon for categorically unnecessary purchases: portable-earplug music instead of the radio; extra-fast Internet for leisure use; heavy vehicles to transport light loads; name-brand clothing instead of plainer gear. "Economizing," as applied to clothing, generally means looking for discount name brands instead of wearing last year's clothes again. The dread of rearing unfashionable children is understandable. But as a priority, "makes me look cool" has passed up "keeps arteries functional" and left the kids huffing and puffing (fashionably) in the dust.

Nobody should need science to prove the obvious, but plenty of studies do show that regularly eating cheaply produced fast food and processed snack foods slaps on extra pounds that increase the risks of diabetes, cardiovascular harm, joint problems, and many cancers. As a country we're officially over the top: the majority of our food dollars buy those cheap calories, and most of our citizens are medically compromised by weight and inactivity. The incidence of obesity-associated diabetes has more than doubled since 1990, with children the fastest-growing class of victims. (The name had to be changed from "adult-onset" to "Type II" diabetes.) One out of every three dollars we spend on health care, by some recent estimates, is paying for the damage of bad eating habits. One out of every seven specifically pays to assuage (but not cure) the multiple heartbreaks of diabetes—kidney failure, strokes, blindness, amputated limbs.

An embarrassing but arguable point is that we're applying deadly priorities to our food budgets because we believe the commercials. Industrial agriculture can promote its products on a supersized scale. Eighty percent of the beef-packing industry is controlled by four companies; the consolidation is exactly the same for soybean processing. With such vast corporate budgets weighing in on the side of beef and added fats, it's no surprise that billions of dollars a year go into advertising fast food. The surprise is how handsomely marketers recoup that investment: how successfully they convince us that cheap food will make us happy.

How delusional are we, exactly? Insisting to farmers that our food has to be *cheap* is like commanding a ten-year-old to choose a profession and move out of the house *now*. It violates the spirit of the enterprise. It guarantees bad results. The economy of the arrangement will come around to haunt you. Anyone with a working knowledge of children would see the flaw in that parenting strategy. Similarly, it takes a farmer to understand the analogous truth about food production—that time and care yield quality that matters—and explain that to the rest of us. Industry will not, but individual market growers *can* communicate concern that they're growing food in a way that's healthy and safe, for people and a place. They can educate consumers about a supply chain that's as healthy or unhealthy as we choose to make it.

That information doesn't fit in a five-syllable jingle. And those growers will never win a price war either. The best they can hope for is a marketing tactic known as friendship, or something like it. Their task is to com-

Paying the Price of Low Prices

A common complaint about organic and local foods is that they're more expensive than "conventional" (industrially grown) foods. Most consumers don't realize how much we're already paying for the conventional foods, before we even get to the supermarket. Our tax dollars subsidize the petroleum used in growing, processing, and shipping these products. We also pay direct subsidies to the large-scale, chemical-dependent brand of farming. And we're being forced to pay more each year for the environmental and health costs of that method of food production.

Here's an exercise: add up the portion of agricultural fuel use that is paid for with our taxes ($22 billion), direct Farm Bill subsidies for corn and wheat ($3 billion), treatment of food-related illnesses ($10 billion), agricultural chemical cleanup costs ($17 billion), collateral costs of pesticide use ($8 billion), and costs of nutrients lost to erosion ($20 billion). At minimum, that's a national subsidy of at least $80 billion, about $725 per household each year. That plus the sticker price buys our "inexpensive" conventional food.

Organic practices build rather than deplete the soil, using manure and cover crops. They eliminate pesticides and herbicides, instead using biological pest controls and some old-fashioned weeding with a hoe. They maintain and apply knowledge of many different crops. All this requires extra time and labor. Smaller farms also bear relatively higher costs for packaging, marketing, and distribution. But the main difference is that organic growers aren't forcing us to pay expenses they've shifted into other domains, such as environmental and health damage. As they're allowed to play a larger role in the U.S. agricultural economy, our subsidy costs to industrial agriculture will decrease. For a few dollars up front, it's a blue-chip investment.

STEVEN L. HOPP

municate the consumer value of their care, and how it benefits the neighborhood. This may seem like a losing battle. But the "Buy Cheap Eats" crusade is assisting the deaths of our compatriots at the rate of about 820 a day; somebody's bound to notice *that.* We are a social animal. The cost-benefit ratios of neighborliness are as old as our species, and probably inescapable in the end.

Ashfield, Massachusetts, is as cute as it gets, even by the standards of small-town New England. Downtown is anchored by a hardware store with rocking chairs on the front porch. The big local social event where folks catch up with their neighbors is the weekly farmers' market.

I didn't know this when we arrived there to stay at a friend's house. We brought our cooler in with the luggage, planning to give our hostess some of our little fist-sized tomatoes. These carefully June-ripened treasures would wow the New Englanders, I thought. Oops. As I started to pull them out of the cooler I spied half a dozen *huge* red tomatoes, languidly sunning their shapely shoulders in our friend's kitchen window. These bodacious babes made our Early Siberians look like Miss Congeniality. I pulled out some blackheart cherries instead, presenting them along with an offhand question: Um, so, where did those *tomatoes* come from?

"Oh, from Amy at the farmers' market," she said. "Aren't they nice?"

Nice, I thought. In the third week of June, in western Mass, if they taste as good as they look they're a doggone miracle. I was extremely curious. Our host promised that during our visit she would take us to see Amy, the tomato magician.

On the appointed morning we took a narrow road that led from Ashfield up through wooded hills to a farm where Amy grows vegetables and her partner Paul works as a consultant in the design and construction of innovative housing. Their own house is pretty much the definition of innovative: a little round, mushroom-shaped structure whose sod-and-moss roof was covered in a summer pelt of jewelweeds. It was the kind of setting that leads you to expect an elf, maybe, but Paul and Amy stepped out instead. They invited us up to the roof where we could sit on a little bench. Ulan the dog followed us up the ladder stairs and sat panting happily as we took in the view of the creek valley below. Part of Paul's work in dynamic housing design is to encourage people to think more broadly about both construction materials (walls of stacked straw bales are his specialty), and how to use space creatively (e.g., dog on the roof). I couldn't wait to see the gardens.

First, though, we had to eat the breakfast they'd made in their tiny, efficient kitchen. Everything locally produced: yogurt and strawberries, eggs, salsa made with Amy's enviable tomatoes. We lingered, talking farming and housing, but the day called us out to the fields, where rows of produce were already gulping morning sun. Amy, a self-described perfectionist, apologized for the state of what looked to me like the tidiest rows imaginable—more weedless than our garden on the best of days. Part-time interns sometimes help out, but the farm runs on Amy's full-time dedication.

A mid-June New England garden, two weeks past the last frost, is predominantly green: lacy bouquets of salad greens, Chinese cabbage, cilantro, broccoli, and peas. A tomato of any type seemed out of the question, until we crested a hill and came upon two long greenhouses. These are the sturdy workhorses of the farm, with heavy-duty plastic skins supported by wooden trusses. Amy no longer grows tomatoes anywhere except in a greenhouse. Cool spring soil, late frosts, and iffy New England weather make the season too short for noteworthy harvests of outdoor-planted tomatoes. But she doesn't grow them hydroponically, as is the norm for large-scale tomato houses. Her greenhouses are built over garden soil, her tomatoes grow in the ground. "They taste better," she said. "It's probably the micronutrients and microfauna in the soil that give them that garden taste. So many components of soil just aren't present in a more sterile environment."

Heating greenhouses through the Massachusetts winter didn't appeal to Paul and Amy either. After a few years of experiments, they've found it most cost-effective to heat with a combination of propane and woodstoves—or not at all. One of the houses is exclusively a cold frame, extending the season for salad greens, spinach, and other crops that can take temperatures down to the mid-twenties. Amy's greens will sell all winter for about $7 a pound. (In New York City, midwinter mesclun can bring $20.)

Her second greenhouse is heated, she said, but only in spring. As we approached that one, I peered in the door and actually gasped. Holy tomato. I've never seen healthier, more content-looking plants: ten feet tall, leafy, rising toward heaven on strings stretched from the ground to the rafters. If there were an Angel Choir of tomatoes, these would be singing. The breed she grows is one meant especially for greenhouses, a variety (perfectly enough) called Trust.

Amy was inspiring to watch, a knowledgeable farmer in her element as she narrowed her eyes for signs of pests, pausing to finger a leaf and study its color. We walked among the tall plants admiring the clusters of fruits hanging from bottom to top in a color gradient from mature red fruits below to the new, greenish white ones overhead. The support strings were rolled around spindles up above that could be cranked to lower the plant down gradually, as the top continues to climb. Tomato plants habitually lose their lower leaves as they grow; the point of this system is to coil the leafless stems on the ground and let the healthy growing part keep twining upward. But these plants were so healthy they refused to lose any lower leaves. Lily played hide-and-seek in the tomato wilderness while Amy showed me her growing system.

She fine-tunes it a little more each year, but already her operation is an obvious success. The last time she had soil tests, the technicians who came to evaluate her compost-built organic dirt had never seen such high nutrient values. The greatest limitation here is temperature; she could keep tomatoes growing all winter, but the cost of fuel would pass her profit margin. In early spring, when she's starting the plants, she economizes by heating the soil under the seedlings (woodstove-heated water runs through underground pipes) while letting air temperatures drop fairly low. Pushing the season early is more important than late, she says. People will pay more for a June tomato than one in October, when tomatoes are old hat, dropping off the vines in gardens everywhere. By starting in the spring, she can bring bushels of Trust out of her greenhouse to thrilled customers long before anything red is coming out of local gardens.

I observed that in the first week of June, she could charge anything she wanted for these. She laughed. "I could. But I don't. I belong to this community, people know me. I wouldn't want to take advantage."

She is more than just part of the community—she was a cofounder of the hugely successful Ashfield farmers' market. She sells to individuals and restaurants, and enjoys a local diet by relying on other producers for the things she doesn't grow. "We don't have chickens, for example, because so many other people do. We trade vegetables for eggs and meat." Their favorite local restaurant buys Amy's produce all summer, the last two months for credit, so that in winter she and Paul can eat there whenever they want. He deems the arrangement "a great substitute for canning."

Like many small farms, this perfectly organic operation is not certified organic. Amy estimates certification would cost her $700 a year, and she wouldn't gain that much value from it. Virtually everyone who buys her food knows Amy personally; many have visited the farm. They know she is committed to chemical-free farming because she values her health, her products, the safety of interns and customers, and the profoundly viable soil of her fields and greenhouses.

Farmers like Amy generally agree that organic standards are a good thing on principle. When consumers purchase food at a distance from where it's grown, certification lets them know it was grown in conditions that are clearly codified and enforced. Farmers who sell directly to their customers, on the other hand, generally don't need watchdogs—their livelihood tends to be a mission as well as a business. Amy's customers trust her methods. No federal bureaucracy can replace that relationship.

Furthermore, the paper trail of organic standards offers only limited guarantees to the consumer. Specifically, it certifies that vegetables were grown without genetic engineering or broadly toxic chemical herbicides or pesticides; animals were not given growth-promoting hormones or antibiotics. "Certified organic" does *not* necessarily mean sustainably grown, worker-friendly, fuel-efficient, cruelty-free, or any other virtue a consumer might wish for.

The rising consumer interest in organic food has inspired most of the country's giant food conglomerates to cash in, at some level. These big players have successfully moved the likes of bagged salads and hormone-free milk from boutique to mainstream markets and even big box stores.

But low price has its costs. In order to meet federal organic standards as cheaply as possible and maximize profits, some industrial-scale organic producers (though not all) cut every corner that's allowed, and are lobbying the government to loosen organic rules further. Some synthetic additives are now permitted, thanks to pressure from industrial organics. So is animal confinement. A chicken may be sold as "free range" if the house in which it's confined (with 20,000 others) has a doorway leading out to a tiny yard, even though that doorway remains shut for so much of the chickens' lives, they never learn to go outside. This is not a theoretical example. A national brand of organic dairy products also uses confined animals—in this case, cows whose mandated "free range" time may find them at home in crowded pens without water, shade, or anything resembling "the range." The larger the corporation, the more distant its motives are apt to be from the original spirit of organic farming—and the farther the products will likely be shipped to buyers who will smile at the happy farm picture on the package, and never be the wiser.

Because organic farming is labor-intensive, holding prices down has even led some large-scale organic growers into direct conflict with OSHA and the United Farm Workers. Just over half of U.S. farm workers are undocumented, and all are unlikely to earn more than minimum wage. Those employed by industrial-scale organic farms are spared direct contact with pesticides in their work, at least, but often live with their families in work-camp towns where pesticide drift is as common as poverty.

The original stated purpose of organic agriculture was not just to protect the quality of food products, but also to safeguard farm environments and communities through diversified, biologically natural practices that remain healthy over time. This was outlined by J. I. Rodale, Sir Albert Howard, Lady Eve Balfour, and all other significant contributors to the theory and practice of modern organic agriculture. Implicitly, these are values that many consumers still think they're supporting with their purchase of organic products. Increasingly, small-scale food farmers like Amy feel corporate organics may be betraying that confidence, extracting too much in the short term from their biotic and human communities, stealing the heart of a movement.

The best and only defense, for both growers and the consumers who care, is a commitment to more local food economies. It may not be possible to prevent the corruption of codified organic standards when they are so broadly applied. A process as complex as sustainable agriculture can't be fully mandated or controlled; the government might as well try to legislate happy marriage. Corporate growers, if their only motive is profit, will find ways to follow the letter of organic regulations while violating their spirit.

But "locally grown" is a denomination whose meaning is incorruptible. Sparing the transportation fuel, packaging, and unhealthy additives is a compelling part of the story, but the plot goes well beyond that. Local food is a handshake deal in a community gathering place. It involves farmers with first names, who show up week after week. It means an open-door policy on the fields, where neighborhood buyers are welcome to come have a look, and pick their food from the vine. *Local* is farmers growing trust.

9 · SIX IMPOSSIBLE THINGS BEFORE BREAKFAST

Late June

When I was in college, living two states away from my family, I studied the map one weekend and found a different route home from the one we usually traveled. I drove back to Kentucky the new way, which did turn out to be faster. During my visit I made sure all my relatives heard about the navigational brilliance that saved me thirty-seven minutes.

"Thirty-seven," my grandfather mused. "And here you just used up fifteen of them telling all about it. What's your plan for the other twenty-two?"

Good question. I'm still stumped for an answer, whenever the religion of time-saving pushes me to zip through a meal or a chore, rushing everybody out the door to the next point on a schedule. All that hurry can blur the truth that life is a zero-sum equation. Every minute I save will get used on something else, possibly no more sublime than staring at the newel post trying to remember what I just ran upstairs for. On the other hand, attending to the task in front of me—even a quotidian chore—might make it into part of a good day, rather than just a rock in the road to someplace else.

I have a farmer friend who would definitely side with my grandfather on the subject of time's economies. He uses draft animals instead of a tractor. Doesn't it take an eternity to turn a whole field with a horse-driven plow? The answer, he says, is yes. *Eternal* is the right frame of mind. "When I'm out there cultivating the corn with a good team in the quiet of the afternoon, watching the birds in the hedgerows, oh my goodness, I could just keep going all day. Kids from the city come out here and ask, 'What do you do for fun around here?' I tell them, 'I cultivate.' "

Now that I'm decades older and much less clever than I was in college, I'm getting better at facing life's routines the way my friend faces his cornfield. I haven't mastered the serene mindset on all household chores (What do you do for fun around here? *I scrub pots and pans, okay??*), but I might be getting there with cooking. Eternal is the right frame of mind for making food for a family: cooking down the tomatoes into a red-gold oregano-scented sauce for pasta. Before that, harvesting sun-ripened fruits, pinching oregano leaves from their stems, growing these things from seed—*yes.* A lifetime is what I'm after. Cooking is definitely one of the things we do for fun around here. When I'm in a blue mood I head for the kitchen. I turn the pages of my favorite cookbooks, summoning the prospective joyful noise of a shared meal. I stand over a bubbling soup, close my eyes, and inhale. From the ground up, everything about nourishment steadies my soul.

Yes, I have other things to do. For nineteen years I've been nothing but a working mother, one of the legions who could justify a lot of packaged, precooked foods if I wanted to feed those to my family. I have no argument with convenience, on principle. I'm inordinately fond of my dishwasher, and I like the shiny tools that lie in my kitchen drawers, ready to make me a menace to any vegetable living or dead. I know the art of the quickie supper for after-a-long-day nights, and sometimes if we're too weary we'll go out to a restaurant, mainly to keep the kitchen clean.

But if I were to define my style of feeding my family, on a permanent basis, by the dictum, "Get it over with, *quick,*" something cherished in our family life would collapse. And I'm not just talking waistlines, though we'd miss those. I'm discussing dinnertime, the cornerstone of our family's mental health. If I had to quantify it, I'd say 75 percent of my crucial parenting effort has taken place during or surrounding the time our family convenes for our evening meal. I'm sure I'm not the only parent to think so. A survey of National Merit scholars—exceptionally successful eighteen-year-olds crossing all lines of ethnicity, gender, geography, and class—turned up a common thread in their lives: the habit of sitting down to a family dinner table. It's not just the food making them brilliant. It's probably the parents—their care, priorities, and culture of support. The words: "I'll expect you home for dinner."

I understand that most U.S. citizens don't have room in their lives to grow food or even see it growing. But I have trouble accepting the next step in our journey toward obligate symbiosis with the packaged meal and takeout. Cooking is a dying art in our culture. *Why* is a good question, and an uneasy one, because I find myself politically and socioeconomically entangled in the answer. I belong to the generation of women who took as our youthful rallying cry: Allow us a good education so we won't have to slave in the kitchen. We recoiled from the proposition that keeping a husband presentable and fed should be our highest intellectual aspiration. We fought for entry as equal partners into every quarter of the labor force. We went to school, sweated those exams, earned our professional stripes, and we beg therefore to be excused from manual labor. Or else our full-time job is manual labor, we are carpenters or steelworkers, or we stand at a cash register all day. At the end of a shift we deserve to go home and put our feet up. Somehow, though, history came around and bit us in the backside: now most women have jobs *and* still find themselves largely in charge of the housework. Cooking at the end of a long day is a burden we could live without.

It's a reasonable position. But it got twisted into a pathological food culture. When my generation of women walked away from the kitchen we were escorted down that path by a profiteering industry that knew a tired, vulnerable marketing target when they saw it. "Hey, ladies," it said to us, "go ahead, get liberated. *We'll* take care of dinner." They threw open the door and we walked into a nutritional crisis and genuinely toxic food supply. If you think *toxic* is an exaggeration, read the package directions for handling raw chicken from a CAFO. We came a long way, baby, into bad eating habits and collaterally impaired family dynamics. No matter what else we do or believe, food remains at the center of every culture. Ours now runs on empty calories.

When we traded homemaking for careers, we were implicitly promised economic independence and worldly influence. But a devil of a bargain it has turned out to be in terms of daily life. We gave up the aroma of warm bread rising, the measured pace of nurturing routines, the creative task of molding our families' tastes and zest for life; we received in exchange the minivan and the Lunchable. (Or worse, convenience-mart hot dogs and latchkey kids.) I consider it the great hoodwink of my generation.

Now what? Most of us, male or female, work at full-time jobs that seem organized around a presumption that some wifely person is at home picking up the slack—filling the gap between school and workday's end, doing errands only possible during business hours, meeting the expectation that we are *hungry* when we get home—but in fact June Cleaver has left the premises. Her income was needed to cover the mortgage and health insurance. Didn't the workplace organizers notice? In fact that gal Friday is *us,* both moms and dads running on overdrive, smashing the caretaking duties into small spaces between job and carpool and bedtime. Eating preprocessed or fast food can look like salvation in the short run, until we start losing what real mealtimes give to a family: civility, economy, and health. A lot of us are wishing for a way back home, to the place where care-and-feeding isn't zookeeper's duty but something happier and more creative.

"Cooking without remuneration" and "slaving over a hot stove" are activities separated mostly by a frame of mind. The distinction is crucial. Career women in many countries still routinely apply passion to their cooking, heading straight from work to the market to search out the freshest ingredients, feeding their loved ones with aplomb. In France and Spain I've sat in business meetings with female journalists and editors in which the conversation veered sharply from postcolonial literature to fish markets and the quality of this year's mushrooms or leeks. These women had no apparent concern about sounding unliberated; in the context of a healthy food culture, fish and leeks are as respectable as postcolonial literature. (And arguably more fun.)

Full-time homemaking may not be an option for those of us delivered without trust funds into the modern era. But approaching mealtimes as a creative opportunity, rather than a chore, *is* an option. Required participation from spouse and kids is an element of the equation. An obsession with spotless collars, ironing, and kitchen floors you can eat off of—not so much. We've earned the right to forget about stupefying household busywork. But kitchens where food is cooked and eaten, those were really a good idea. We threw that baby out with the bathwater. It may be advisable to grab her by her slippery foot and haul her back in here before it's too late.

It's easy for any of us to claim no time for cooking; harder to look at what we're doing instead, and why every bit of it is presumed more worthy. Some people really do work double shifts with overtime and pursue no recreational activities, ever, or they are homeless or otherwise without access to a stove and refrigerator. But most are lucky enough to do *some* things for fun, or for self-improvement or family entertainment. Cooking can be one of those things.

Working people's cooking, of course, will develop an efficiency ethic. I'm shameless about throwing out the extraneous plot twists of a hoity-toity recipe and getting to its main theme. Or ignoring cookbooks altogether during the week, relying mostly on simple meals I've made a thousand times before, in endless variation: frittata, stir-fry, pasta with one protein and two vegetables thrown in. Or soups that can simmer unattended all day in our Crock-Pot, which is named Mrs. Cleaver. More labor-intensive recipes we save for weekends: lasagna, quiches, roasted chicken, desserts of any kind. I have another rule about complicated dishes: *always* double the recipe, so we can recoup the investment and eat this lovely thing again later in the week.

Routines save time, and tempers. Like a mother managing a toddler's mood swings, our family has built some reliable backstops for the times in our week when work-weary, low-blood-sugar blowouts are most likely. Friday nights are always pizza-movie nights. Friends or dates are welcome; we rent one PG feature and one for after small children go to bed. We always keep the basic ingredients for pizza on hand—flour and yeast for the dough, mozzarella, and tomatoes (fresh, dried, or canned sauce, depending on the season). All other toppings vary with the garden and personal tastes. Picky children get to control the toppings on their own austere quadrant, while the adventurous may stake out another, piling on anything from smoked eggplant to caramelized onions, fresh herbs, and spinach. Because it's a routine, our pizzas come together without any fuss as we gather in the kitchen to decompress, have a glass of wine if we are of age, and talk about everybody's week. I never have to think about what's for dinner on Fridays.

I like cooking as a social event. Friends always seem happy to share the work of putting together a do-it-yourself pizza, tacos, or vegetarian wraps. Potluck dinner parties are salvation. Takeout is not the only easy way out. With a basic repertoire of unfussy recipes in your head, the better part of valor is just turning on the burner and giving it a shot. With all due respect to Julia, I'm just thinking *Child* when I hazard a new throw-in-everything stew. I also have a crafty trick of inviting friends over for dinner whose cooking I admire, offering whatever ingredients they need, and myself as sous-chef. This is how I finally learned to make paella, pad Thai, and sushi, but the same scheme would work for acquiring basic skills and recipes. For a dedicated non-cook, the first step is likely the hardest: convincing oneself it's worth the trouble in terms of health and household economy, let alone saving the junked-up world.

It really is. Cooking is the great divide between good eating and bad. The gains are quantifiable: cooking and eating at home, even with quality ingredients, costs pennies on the dollar compared with meals prepared by a restaurant or factory. Shoppers who are most daunted by the high price of organics may be looking at bar codes on boutique-organic prepared foods, not actual vegetables. A quality diet is not an elitist option for the do-it-yourselfer. Globally speaking, people consume more soft drinks and packaged foods as they grow more affluent; home-cooked meals of fresh ingredients are the mainstay of rural, less affluent people. This link between economic success and nutritional failure has become so widespread, it has a name: the nutrition transition.

In this country, some of our tired and poorest live in neighborhoods where groceries are sold only in gas station mini-marts. Food stamp allowances are in some cases as low as one dollar a person per meal, which will buy beans and rice with nothing thrown in. But many more of us have substantially broader food options than we're currently using to best advantage. Home-cooked, whole-ingredient cuisine *will* save money. It will also help trim off and keep off extra pounds, when that's an issue—which it is, for some two-thirds of adults in the U.S. Obesity is our most serious health problem, and our sneakiest, because so many calories slip in uncounted. Corn syrup and added fats have been outed as major ingredients in fast food, but they hide out in packaged foods too, even presumed-innocent ones like crackers. Cooking lets you guard the door, controlling not only what goes into your food, but what stays out.

Finally, cooking is good citizenship. It's the only way to get serious about putting locally raised foods into your diet, which keeps farmlands healthy and grocery money in the neighborhood. Cooking and eating with children teaches them civility and practical skills they can use later on to save money and stay healthy, whatever may happen in their lifetimes to the gas-fueled food industry. Family time is at a premium for most of us, and legitimate competing interests can easily crowd out cooking. But if grabbing fast food is the only way to get the kids to their healthy fresh-air soccer practice on time, that's an interesting call. Arterial-plaque specials that save minutes now can cost years, later on.

Households that have lost the soul of cooking from their routines may not know what they're missing: the song of a stir-fry sizzle, the small talk of clinking measuring spoons, the yeasty scent of rising dough, the painting of flavors onto a pizza before it slides into the oven. The choreography of many people working in one kitchen is, by itself, a certain definition of family, after people have made their separate ways home to be together. The nurturing arts are more than just icing on the cake, insofar as they influence survival. We have dealt to today's kids the statistical hand of a shorter life expectancy than their parents, which would be *us,* the ones taking care of them. Our thrown-away food culture is the sole reason. By taking the faster drive, what did we save?

⚘

Once you start cooking, one thing leads to another. A new recipe is as exciting as a blind date. A new *ingredient,* heaven help me, is an intoxicating affair. I've grown new vegetables just to see what they taste like: Jerusalem artichokes, edamame, *potimarrons.* A quick recipe can turn slow in our kitchen because of the experiments we hazard. We make things from scratch just to see if we can. We've rolled out and cut our pasta, raised turkeys to roast or stuff into link sausage, made chutney from our garden. On high occasions we'll make cherry pies with crisscrossed lattice tops and ravioli with crimped edges, for the satisfaction of seeing these storybook comforts become real.

A lot of human hobbies, from knitting sweaters to building model airplanes, are probably rooted in the same human desire to control an entire process of manufacture. Karl Marx called it the antidote to alienation. Modern business psychologists generally agree, noting that workers will build a better a car when they participate in the whole assembly rather than just slapping on one bolt, over and over, all the tedious livelong day. In the case of modern food, our single-bolt job has become the boring act of poking the thing in our mouths, with no feeling for any other stage in the process. It's a pretty obvious consequence that one should care little about the product. When I ponder the question of why Americans eat so much bad food on purpose, this is my best guess: alimentary alienation. We can't feel how or why it hurts. We're dying for an antidote.

If you ask me, that's reason enough to keep a kitchen at the center of a family's life, as a place to understand favorite foods as processes, not just products. It's the prime motivation behind our vegetable garden, our regular baking of bread, and other experiments that ultimately become household routines. Our cheesemaking, for example.

Okay, I know. You were with me right up to that last one. I'm not sure why, since it takes less time to make a pound of mozzarella than to bake a cobbler, but most people find the idea of making cheese at home to be preposterous. If the delivery guy happens to come to the door when I'm cutting and draining curd, I feel like a Wiccan.

What kind of weirdo makes *cheese*? It's too hard to imagine, too homespun, too something. We're so alienated from the creation of even ordinary things we eat or use, each one seems to need its own public relations team to calm the American subservience to hurry and bring us back around to doing a thing ourselves, at home. Knitting clothes found new popularity among college girls, thanks largely to a little book called *Stitch and Bitch*. Homemaking in general has its Martha. French cuisine had its immortal Julia. Grilling, Cajun cooking, and cast-iron stewing all have their celebrity gurus. What would it take to convince us that an hour spent rendering up cheese in our kitchens could be worth the trouble? A motivational speaker, a pal, an artisan—a Cheese Queen, maybe?

Yes, all of the above, and she exists. Her name is Ricki Carroll. Since 1978, when she founded New England Cheesemaking Supply and began holding workshops in her kitchen, she has directly taught more than 7,000 people how to make cheese. That's face to face, not counting those of us who ordered supplies online and worked our way through her book, *Cheesemaking Made Easy,* which has sold over 100,000 copies. An Internet search for Cheese Queen will pop her right up.

When I went to see Ricki, it was equal parts admiration and curiosity. If my family is into reconnecting with the processes that bring us our foods, if we've taken it upon ourselves to be a teeny bit evangelical about this, we have a lot to learn from Ricki Carroll. We're just small-time country preachers. This woman has inspired artisans from the Loire to Las Vegas. She's the Billy Graham of Cheese.

Okay, not really. She's just Ricki. She starts to win you over when you step onto the porch of her Massachusetts farmhouse, a colorful Queen Anne with lupines and lilies blooming around the stoop. Then you walk through the door and fall through the looking glass into a space where cheesemaking antiques blend with the whimsy of handmade dolls winking at African masks, unusual musical instruments and crazy quilts conversing quietly in several languages. The setting prepares you to meet the Queen, greeting her workshop guests with a smile, waving everyone into the big kitchen as she pins up her wildly curly hair with a parrot-shaped barrette.

Ricki had invited our family to come for a visit, after hearing of our interest in local and artisanal foods. Generously, she let us and half a dozen of our friends sit in as her guests at an all-day workshop for beginning cheesemakers. Now we sat down at long tables and introduced ourselves to the twenty other workshoppers. I was already taking notes, not on cheesemaking, but on *who in the heck comes here and does this thing?*

Anybody. For several men it was an extremely original Father's Day gift. A chef hoped to broaden her culinary range; mothers were after healthy, more local diets for their families. Martha, from Texas, owns water buffalo and dreams of a great mozzarella. (Their names are Betsy and Beau; she passed around photos.) Maybe we were all a little nuts, but being there made us feel like pilgrims of a secret order. We had turned our backs on our nation's golden calf of cellophane-wrapped Cheese Product Singles. Our common wish was to understand a food we cared about, and take back one more measure of control over our own care and feeding.

We examined the stainless steel bowls, thermometers, and culture packets assembled before us while Ricki began to talk us into her world. Cheese is a simple idea: a way to store milk, which goes bad quickly without refrigeration but keeps indefinitely—improves, even—in the form of cheese. From humble beginnings it has become a global fascination. "Artisanal cheesemakers combine science and art. All over the world, without scientific instruments, people make cheeses the way their grandparents did." In the Republic of Georgia, she told us, she watched cheesemakers stir their curd with a twig and then swaddle the warm pot (in lieu of monitoring it with a thermometer) in a kitty-print sweater, a baby blanket, and a cape.

Forging ahead, Ricki announced we were making *queso blanco,* whole milk ricotta, mascarpone, mozzarella, and farmhouse cheddar. Yes, *us,* right here, today. We looked on in utter doubt as she led us into our first cheese, explaining that we'd make all this with ordinary milk from the grocery. Raw milk from a farm is wonderful to work with, unhomogenized is great, but any milk will do, so long as it's not labeled "ultra-pasteurized." Ultra-high-temperature pasteurization, Ricki explained, denatures proteins and destroys the curd. The sole purpose of UHP is to ship milk over long distances; after this process it can sit for many weeks without any change in its chemistry.

Because its chemistry is already so altered, though, UHP milk will not make cheese, period. This discussion confirmed what I'd learned the hard way at home in my earliest efforts. Before I knew to look for the term "UHP" and avoid it, I'd used some to create a messy mozzarella failure. The curd won't firm up, it just turns into glop.

"Ask your grocer where your milk comes from," Ricki instructed us; the closer to home your source, the better. Reading labels in your dairy case may lead you to discover a dairy that isn't too far away, and hasn't ultra-pasteurized the product for long-distance travel. Better yet, she suggested, ask around to find a farmer who has fresh milk. It may not be for sale, since restrictions in most states make it impossible for small dairies to sell directly to the consumer. But some allow it, or have loopholes the farmer can advise you about. You may be able to buy raw milk for your pets, for example. (Those kitties will love your mozzarella.) You can pasteurize raw milk yourself if you like, but most outbreaks of listeria and other milk-borne diseases occur in factory-scale dairies, Ricki said, not among small dairies and artisans where the center of attention is product quality.

The subject of regulations touched a nerve for several small milk producers in our workshop. Anne and Micki, two mothers raising families on neighboring New England farms, got interested in home dairying after their pediatrician suggested switching to organic milk. If a family can put one organic choice on their shopping list, he'd said, it should be dairy. The industry says growth hormones in milk are safe; the pediatrician (and for the record, he's not alone) said he had seen too many girls going through early puberty.

So Micki and Anne acquired their own Jersey cows, happily guaranteeing their families a lifetime supply of hormone-free milk. Anne also makes kefir, which she would like to sell at her farmers' market, but can't. Micki's daughter makes ice-cream-and-cookie sandwiches using their own milk and eggs—a wildly popular item she could sell to build her college fund, except it's illegal. "We're not licensed," Micki said, "and we never will be. The standards are impossible for a small dairy."

She wasn't exaggerating. Most states' dairy codes read like an obsessive compulsive's to-do list: the milking house must have incandescent fixtures of 100 watts or more capacity located near but not directly above any bulk milk tank; it must have employee dressing rooms and a separate, permanently installed hand-washing facility (even if a house with a bathroom is ten steps away) with hot and cold water supplied through a mix valve; all milk must be pasteurized in a separate facility (not a household kitchen) with its own entrance and separate, paved driveway; processing must take place daily; every batch must be tested for hormones (even if it's your cow, and you gave it no hormones) by an approved laboratory.

Pasteurization requires three pieces of equipment: a steel pot, a heat source, and a thermometer that goes up to 145°F. Add to this list, I suppose, the brainpower to read a thermometer. I've done it many times without benefit of extra driveways and employee lockers, little knowing I was a danger to the public. In fact, later on when I went poking into these codes, I learned I might stand in violation of Virginia State Law 2VAC5-531-70 just by making cheese for my own consumption. It takes imagination to see how some of these rules affect consumer safety. Many other raw food products—notably poultry from CAFOs—typically carry a much higher threat to human health in terms of pathogen load, and yet the government trusts us to render it safe in our own humble kitchens. But it's easy to see how impossibly strict milk rules might gratify industry lobbyists, by eliminating competition from family producers.

Ricki was sympathetic to that position, having traveled the world and seen a lot of people working without major milking-room specs. In Greece, for example, she watched shepherds make cheese in a cinderblock shed right after they milked, making feta over a fire, pouring out the whey over the stone floor to wash it. The specific bacteria that thrived there created a good environment for making the cheese, while crowding out other, potentially harmful microorganisms. French winemakers apply the same principle when using their grapes' leftover yeasty pulp as compost in their vineyards. Over the centuries, whole valleys become infused with the right microbes to make the wine ferment properly and create its flavorful terroir.

Many of our most useful foods—yogurt, wine, bread, and cheese—are products of controlled microbe growth. We may not like thinking about it, but germs crawl eternally over every speck of our planet. Our own bodies are bacterial condos, with established relationships between the upstairs and downstairs neighbors. Without these regular residents, our guts are easily taken over by less congenial newcomers looking for low-rent space. What keeps us healthy is an informed coexistence with microbes, rather than the micro-genocide that seems to be the rage lately. Germophobic parents can now buy kids' dinnerware, placemats, even clothing imbedded with antimicrobial chemicals. Anything that will stand still, if we mean to eat it, we shoot full of antibiotics. And yet, more than 5,000 people in the United States die each year from pathogens in our food. Sterility is obviously the wrong goal, especially as a substitute for careful work.

That was our agenda here: careful work. Ricki moved in a flash from terroir to bacterial cheese cultures to warming our own pots of milk to the right temperature. While waxing poetic in praise of slowness, she moved fast. By the time we'd added the culture to set our cheddar, she was on to the next cheese. With a mirror propped over the stove so we could see down into the pot, she stirred in vinegar to curdle the *queso blanco,* laughing as she guessed on the quantity. There's no perfect formula, she insisted, just some basic principles and the confidence to give it a try.

Confidence was not yet ours, but we got busy anyway, we maverick dairywomen, fathers, buffalo ranchers, and dreamers. It does feel subversive to flout the professionals and make a thing yourself. Our nostrils inhaled the lemony-sweet scent of boiling whey. The steamy heat of the kitchen curled our hair, as new textures and flavors began to rise before us as possibilities: mascarpone, fromagina, mozzarella. *Remote* possibilities, maybe. That many successes in one day still seemed unlikely.

At lunch break I checked out the wildly colorful powder room, where a quote from *Alice in Wonderland* was painted on the wall:

" 'There's no use trying,' Alice said. 'One *can't* believe impossible things.'

" 'I daresay you haven't had much practice,' said the Queen. 'When I was your age, I always did it for half-an-hour a day. Why, sometimes I've believed as many as six impossible things before breakfast.' "

✤

It's fair to admit, I wasn't a complete novice. I had already been making cheese for a few years, ordering supplies and cultures from Ricki and following the recipes in her book. It wasn't only a spirit of adventure that led my family into this line of cooking, but also bellyaches. Lactose intolerance is a common inherited condition in which a person's gut loses, after childhood, its ability to digest the milk sugar called lactose. The sugary molecules float around undigested in the intestine, ferment, and create a gassy havoc. The effect is somewhat like eating any other indigestible carbohydrate, such as cardboard or grass.

This is not an allergy or even, technically, a disorder. Physical anthropologists tell us that age four, when lactose intolerance typically starts, is about when nature intended for our kind to be wholly weaned onto solid food; in other words, a gradual cessation of milk digestion is normal. In all other mammals the milk-digesting enzyme shuts down soon after weaning. So when people refer to this as an illness, I'm inclined to point out we L.I.'s can very well digest the sugars in grown-up human foods like fruits and vegetables, thank you, we just can't *nurse. From a cow.* Okay?

But there is no animal weirder than *Homo sapiens.* Over thousands of years of history, a few isolated populations developed intimate relationships with their domestic animals and a genetic mutation gave them a peculiar new adaptation: they kept their lactose-digesting enzymes past childhood. Geneticists have confirmed that milk-drinking adults are the exception to the norm, identifying a deviant gene on the second chromosome that causes lactase persistence. (The gene is SNP C/T13910, if you care.) This relatively recent mutation occurred about ten thousand years ago, soon after humans began to domesticate milk-producing animals. The gene rapidly increased in these herding populations because of the unique advantage it conferred, allowing them to breast-feed for life from another species.

The gene for lifelong lactose digestion has an 86 percent frequency among northern Europeans. By contrast, it shows up in only about one-third of southern Europeans, who historically were not big herders. In the Far East, where dairy cattle were unknown, the gene is absent. Even now, Southeast Asians have virtually zero tolerance for lactose. Only about 10 percent of Asian Americans can digest milk as adults, along with fewer than half of American Jews and about a quarter of rural Mexicans. Among Native Americans it's sketchily documented—estimates range from 20 to 40 percent. Among African Americans, adult milk-drinking tolerance is high, nearly 50 percent, owing to another interesting piece of human history. The mutation for lactase persistence emerged several times independently, alongside the behavior of adult milk-drinking. It shows up in populations that have little else in common other than cows: the tall, lean Fulani of West Africa; the Khoi pastoralists of southern Africa; and the fair-skinned Northern Europeans.

And then, to make a long story short, one of those populations proceeded to take over the world. If that's a debatable contention, let's just say they've gotten their hands on most of the planet's billboards and commercials. And so, whether or not we were born with the La Leche for Life gene, we're all hailed with a steady song and dance about how we ought to be drinking tall glasses of it every day. And we believe it, we want those strong bones and teeth. Oh, how we try to behave like baby cows. Physicians will tell you, the great majority of lactose-intolerant Americans don't even know it. They just keep drinking milk, and having stomachaches.

White though we are, my redheaded elder daughter and me, some sturdy, swarthy gene has come down through the generations to remind us that "white" is relative. We're lactose intolerant. But still, like most everyone else, we include some dairy products in our diet. I can't blame dairy-industry propaganda, purely, for our behavior. The milk of mammals is a miraculously whole food for the babes it was meant to nourish; it's the secret of success for the sheep, oxen, bison, kangaroos, seals, elephants, whales, and other mammals that have populated every corner of the blue-green world with their kind and their suckling young. For the rest of us it's a tempting source of protein, calcium, minerals, and wholesome fats.

It's no surprise that cultures the world over have found, through centuries of experimentation, countless ways to make it more digestible. Yogurt, kefir, paneer, queso fresco, butter, mascarpone, montasio, parmesan, haloumi, manchego, bondon, emmental, chenna, ricotta, and quark: the forms of altered milk are without number. Taste is probably not the main point. They all keep longer than fresh milk, and their production involves reducing the lactose sugars.

The chemistry is pretty simple. Milk is about 85 percent water; the rest is protein, minerals, butterfat, vitamins and trace elements, and sugars (lactose)—which are dissolved in the water. When the whole caboodle is made more acidic, the protein solids coagulate into a jellylike curd. When gently heated, this gel releases the liquid whey (lactose and water). Traditionally the milk is curdled by means of specific bacteria that eat—

guess what?—lactose. These selective bugs munch through the milk, turning the lactose into harmless lactic acid, which causes the curdling.

The sugars that still remain are dissolved in the whey. As this liquid separates and is drained off from the curd, lactose goes with it. Heating, pressing, and aging the curd will get rid of still more whey, making it harder and generally sharper-flavored. As a rule, the harder the cheese, the lower the lactose content. (Anything less than 2 percent lactose is tolerable for just about everybody.) Also, higher fat content means less lactose—butter has none. Conversely, sweet condensed milk is 12 percent lactose. For other products, the amount of lactose removed depends on the bacterial cultures used for fermentation. A good live-culture yogurt contains as many as five different sugar-eating bacteria. A little biochemistry goes a long way, in safely navigating the dairy path.

At our house soft cheeses were the tricky terrain. Factory-made cheeses can vary enormously in lactose content. Fermentation and whey removal take time that mass production doesn't always allow. Some soft cheeses are not cultured at all, but curdled simply by adding an acid. For whatever reason, store-bought cream cheese proved consistently inedible for us. But I don't like to give up. If I could monitor the process myself, seeing personally to lactose removal, I wondered if I might get something edible.

Soft cheeses are ridiculously easy to make, it turns out. The hardest part is ordering the cultures (by catalog or online). With these packets of cheesemaking bugs in your freezer and a gallon of good milk, plus a thermometer, colander, and some cheesecloth, soft cheeses are at your command: in a stainless steel pot, warm the milk to 85 degrees, open the culture packet, and stir the contents into the milk. Take the pot off the stove, cover, let it stand overnight. By the next morning it will have gelled into a soft white curd. Spoon this into a cheesecloth-lined colander and let the whey run off. Salt it, spread it on bread, smile. Different bacterial cultures make different cheeses. The bugs stay up all night doing the work, not you. You just sleep. Is that not cool?

Our chevre and fromagina were so tasty, and digestible, we were inspired to try hard cheeses. These are more work, but it's basically the same process. Most recipes call for both a bacterial culture and rennet (a natural enzyme), which together cause the milk to set up into a very firm curd in just minutes, rather than overnight. For mozzarella, this curd is kneaded like dough, heated until almost untouchably hot, then stretched like taffy, which is a lot of fun. The whole process—from cold milk to a beautiful braided, impress-your-guests mozzarella on the plate—takes less than an hour. For hard cheeses like cheddar, the firm curd is sliced into little cubes, stirred and heated gently, then pressed into a round wheel and, ideally, aged for weeks or months. We have to hide our cheeses from ourselves to keep them around this long. Over time, we've converted a number of our friends to the coven of cheesemaking.

At Ricki's workshop we really did make six impossible things, but only half of them by noon. Lunch included our *queso blanco* stir-fried with vegetables, sliced tomatoes with our mozzarella, and mascarpone-filled dates. We tasted, congratulated ourselves, and headed back for the next round. We put our cheddar into a mechanical press to squeeze extra moisture out of the curd, while Ricki talked about aging and waxing as if these really lay ahead of us—as if we were all going home to make cheeses. I'd be willing to bet we all did. At the workshop's end, everyone gathered in Ricki's office to order the cultures and supplies we'd need for our next efforts. A few dollars' worth of packaged bacteria will curdle many gallons of milk. A cheese thermometer costs ten dollars, and the rest of the basics—stainless steel bowls and pots—already reside in the kitchen of any earnest cook. We left with the confidence to strike out on our own. Our friends who'd shared the workshop went back to their homes in Virginia, New York, and Boston. They all called me within the week with exciting cheese updates.

Why do we do this? It's hard to say. Some are refining exquisite products, while others of us are just shooting for edible, but we're all dazzled by the moment of alchemy when the milk divides into clear whey and white curd, or the mozzarella stretches in our hands to a glossy golden skein. We're connecting across geography and time with the artisans of Camembert, the Greek shepherds, the Mongols on the steppes who live by milking their horses—everybody who ever looked at a full-moon pot of white milk and imagined cheese. We're recalling our best memories infused with scents, parental love, and some kind of food magically coming together in the routines of childhood. We're hoping *our* kids will remember us somewhere other than in the driver's seat of the car.

Later in the summer when this workshop and trip were behind us, Steven's mother came for a long visit. She served grandma duty on many fronts, but seemed happiest in the kitchen. She told us stories I hadn't heard before, mostly about her mother, who at age fifteen was sent out from her hardscrabble village in the mountains of Italy to seek her fortune in America. In the dusty town of Denver she married a handsome Sicilian vegetable farmer and raised five daughters with a good working knowledge of gardening, pasta, and other fundamentals. She made ricotta routinely, to the end of her life.

Laura was her name, ultimately known as Nonnie, and I suppose she'd have loved to see us on a summer Saturday making mozzarella together: daughter, grandson, great-granddaughters, and me, all of us laughing, stretching the golden rope as far as we could pull it. Three more generations answering hunger with the oldest art we know, and carrying on.

Growing Up in the Kitchen

BY CAMILLE

In our house, the kitchen is the place to be. The time we spend making dinner is hugely important because it gets us together after all our separate agendas, and when we sit down to eat we have a sense that the food in front of us is special. Growing vegetables from seed and raising poultry from hatchlings obviously makes us especially grateful for our food. But just making dinner from scratch gives us a little time to anticipate its flavor, so we'll notice every bite.

Cooking in our family helped me cultivate certain food habits that I later found out are a little unusual for my generation—for example, I can't stand to eat anything while I'm standing up. I sit down, even if it's just a quick snack, to make sure this will be a thoughtful munching instead of a passive grab. I'll probably carry that habit through my whole life, and nag my kids about it.

I know plenty of families that have dinner together, and some that cook, but very few that take "cooking from scratch" to the level mine does. I've never had any illusions about how unique it is to have one parent who makes cheese and another who bakes bread almost daily. The friends I've brought home over the years have usually been impressed and intrigued by the wacky productions taking place in our kitchen. They definitely enjoyed eating fresh, warm bread at dinner and homemade cream cheese at lunch. It was a little awkward, though, when one of my vegetarian friends and I arrived at my house one Saturday when my parents were in the middle of making turkey sausage.

"What are they doing?" she whispered, as she stared at the tube of encased raw meat that was steadily growing longer on our countertop. "Oh, that's just sausage. Don't worry about it." I nudged her past the kitchen toward my room. The scenario was a little embarrassing, but it probably would have been more uncomfortable to come home to parents who used the kitchen for screaming and throwing dishes at each other. Anyway, whose parents aren't embarrassing sometimes?

The hardest thing about being raised in a household where most everything is made from scratch is that someday you move out and have to deal with store-bought bread and yogurt. My mom was quick to catch on to the leverage she got out of that. "I guess you'll have to come home more often if you want good food," she would tell me. Away from home, I realized I missed more about mealtimes than just the food. I missed picking fresh greens from the garden, or taking a jar of dried tomatoes from the pantry, as the starting point of a meal. It's obviously convenient to grab a salad or package of sushi from the dining hall between classes, but eating on the fly seems like cheating to me.

Maybe I feel this way because my make-it-yourself upbringing drummed into me the ethic of working for the things I want. I've been involved in growing and cooking the food that feeds me since I was a little kid, and it has definitely given me a certain confidence about relying on myself. Just as meals don't materialize in the grocery store, I realize a new car and a good education won't just spring into my life on their own, but hopefully I will get there. If everything my heart desired was handed to me on a plate, I'd probably just want something else.

Cooking meals doesn't have to be that complicated. Most of the recipes in this book take less than an hour to prepare. The average American spends three and a half hours watching TV every day. Even if a family can only manage to eat a meal together a couple of times a week, whether it's breakfast, lunch, or dinner, my vote would be that it's worth the effort. For most of my high school years, dinner was often the only chance I'd have to see my sister or parents all day. We could check up on each other and recount the traumas and victories of our days. We might end up laughing through the whole meal. A choking hazard, maybe, but also a pretty good way to relieve stress.

Cheese is one of our favorite special foods to make from scratch. This recipe for homemade mozzarella is from *Home Cheese Making* by Ricki

Carroll and really does take only thirty minutes. For the rennet, plus the cultures for making other cheeses, contact New England Cheesemaking Supply Company.

❋ 30-MINUTE MOZZARELLA

Measure out all additives before you start, in clean glass or ceramic cups. Use unchlorinated water.

1 gallon pasteurized milk (NOT ultra-pasteurized)

1½ level teaspoons citric acid dissolved in ¼ cup cool water

Stir the milk on the stove in a stainless steel kettle, heating very gently. At 55° add the citric acid solution and mix thoroughly. At 88° it should begin to curdle.

¼ teaspoon liquid rennet, diluted in ¼ cup cool water

Gently stir in diluted rennet with up-and-down motion, and continue heating the milk to just over 100°, then turn off heat. Curds should be pulling away from sides of pot, ready to scoop out. The whey should be clear. (If it's still milky, wait a few minutes.) Use a large slotted spoon or ladle to move curds from pot to a 2-quart microwaveable bowl. Press curds gently with hands to remove as much whey as possible, and pour it off. Microwave the curds on high for one minute, then knead the cheese again with hands or a spoon to remove more whey. (Rubber gloves help—this gets hot!) Microwave two more times (about 35 seconds each), kneading between each heating. At this point, salt the cheese to taste, then knead and pull until it's smooth and elastic. When you can stretch it into ropes like taffy, you are done. If the curds break instead, they need to be reheated a bit. Once cheese is smooth and shiny, roll it into small balls to eat warm or store for later in the refrigerator.

Lacking a microwave, you can use the pot of hot whey on the stove for the heating-and-kneading steps. Put the ball of curd back in with a big slotted spoon, and heat it until it's almost too hot to touch. Good stretching temperature is 175 degrees.

Here are three great ways to eat your mozzarella:

❋ SUMMERTIME SALAD

2 large tomatoes
1 ball of mozzarella
Basil leaves
Olive oil
Salt to taste

Slice tomatoes and spread them out on a large platter. Place a thin slice of cheese and a basil leaf on each slice of tomato. Drizzle olive oil over top, sprinkle with salt, and serve.

❋ EGGPLANT PAPOUTZAKIA

2 pounds eggplant
Olive oil

Slice eggplant lengthwise and sauté lightly in olive oil. Remove from skillet and arrange in a baking dish.

2 medium onions, garlic to taste
2 large tomatoes, diced
2 teaspoons nutmeg
Salt and pepper to taste
6 ounces grated or sliced mozzarella

Chop onions and garlic and sauté in olive oil. Add diced tomato and spices and mix thoroughly. Spread mixture over the eggplant and sprinkle an even layer of cheese over top. Bake at 350° for 20 minutes, until golden on top.

❋ FRIDAY NIGHT PIZZA

(Makes two 12-inch pizzas: enough for family, friends, and maybe tomorrow's lunch.)

3 teaspoons yeast
1½ cups WARM water
3 tablespoons olive oil
1 teaspoon salt
2½ cups white flour
2 cups whole wheat flour

To make crust, dissolve the yeast into the warm water and add oil and salt to that mixture. Mix the flours and knead them into the liquid mixture. Let dough rise for 30 to 40 minutes.

1 cup sliced onions
2 peppers, cut up

While the dough is rising, prepare the sliced onions: a slow sauté to caramelize their sugars makes fresh onions into an amazing vegetable. First sizzle them on medium heat in a little olive oil, until transparent but not browned. Then turn down the burner, add a bit of water if necessary to keep them from browning, and let them cook ten to fifteen minutes more, until they are glossy and sweet. Peppers can benefit from a similar treatment.

Once the dough has risen, divide it in half and roll out two round 12-inch pizza crusts on a clean, floured countertop, using your fingers to roll the perimeter into an outer crust as thick as you like. Using spatulas, slide the crusts onto well-floured pans or baking stones and spread toppings.

16 ounces mozzarella, thinly sliced
2 cups fresh tomatoes in season (or sauce in winter)
Other toppings
1 tablespoon oregano
1 teaspoon rosemary
Olive oil

Layer the cheese evenly over the crust, then scatter the toppings of the week on your pizza, finishing with the spices. If you use tomato sauce (rather than fresh tomatoes), spread that over crust first, then the cheese, then other toppings. Bake pizzas at 425° for about 20 minutes, until crust is browned on the edge and crisp in the center.

Download these and all other *Animal, Vegetable, Miracle* recipes at www.AnimalVegetableMiracle.com

SOME OF OUR FAVORITE COMBINATIONS FOR SUMMER ARE:

Mozzarella, fresh tomato slices, and fresh basil, drizzled with olive oil

Mozzarella, chopped tomatoes, caramelized onions and peppers, mushrooms

Chopped tomatoes, crumbled feta, finely chopped spinach or chard, black olives

GOOD WINTER COMBINATIONS INCLUDE:

Farmer cheese, chicken, olives, and mushrooms

Tomato sauce, mozzarella, dried peppers, mushrooms, and anchovies

10 · EATING NEIGHBORLY

Late June

Just a few hours north of Massachusetts lie the working-class towns of central Vermont, where a granite statue on Main Street is more likely to celebrate an anonymous stonecutter than some dignitary in a suit. Just such a local hero stood over us now, and we admired him as we drove past: stalwart as the mallet in his hand, this great stone man with his rolled-up sleeves reminded us of Steven's Italian grandfather.

We were still on vacation, headed north, now hungry. We pulled in for lunch at a diner with a row of shiny chrome stools at the long counter, and booths lining one wall. Heavy white mugs waited to be filled with coffee. Patsy Cline and Tammy Wynette sang their hearts out for quarters. A handmade sign let us know the jukebox take is collected at the end of every month and sent to Farm Aid. The lunch crowd had cleared out, so we had our pick of booths and our order was up in a minute. The hamburgers were thick, the fries crispy, the coleslaw cool. The turkey wrap came with mashed sweet potatoes. Lily seemed so lost in her milkshake, we might never get her back.

The owners, Tod Murphy and Pam Van Deursen, checked by our booth to see how we liked everything—and to tell us which of their neighbors produced what. Everything on our plates was grown a stone's throw from right here. The beef never comes from Iowa feedlots, nor do the fries come in giant frozen packages shipped from a factory fed by the world's cheap grower of the moment. In a refreshing change of pace, the fries here are made from potatoes. This is the Farmers Diner, where it's not just quarters in the jukebox that support farming, but the whole transaction.

It is the simplest idea in the world, really: a restaurant selling food produced by farms within an hour's drive. So why don't we have more of them? For the same reason that statue down the street clings to his hammer while all the real stonecutters in this granite town have had to find other jobs, in a nation that now imports its granite from China. The giant building directly behind this diner, formerly a stonecutting works, is now a warehouse for stone that is cut, worked, and shipped here from the other side of the planet. If ever a town knew the real economics of the local product versus the low-cost import, this ought to be it.

Buying your goods from local businesses rather than national chains generates about three times as much money for your local economy. Studies from all over the country agree on that, even while consumers keep buying at chain stores, and fretting that the downtown blocks of cute mom-and-pop venues are turning into a ghost town. Today's bargain always seems to matter more.

The Farmers Diner is therefore a restaurant for folks who want to fill up for under ten bucks, and that is what they get: basic diner food, affordable and not fancy. The Farmers Breakfast—two eggs, two pancakes, your choice of sausage or bacon—is $6.75. The Vermont-raised hamburger with a side of slaw, home fries, or a salad is $6.50. At any price, it's an unusual experience to order a diner burger that does not come with a side of feedlot remorse. For our family this was a quiet little red-letter occasion, since we'd stopped eating CAFO-produced beef about ten years earlier. Virtually all beef in diners and other standard food services comes from CAFOs. Avoiding it is one pain in the neck, I'll tell you, especially on hectic school mornings when I glance at the school lunchroom calendar and see that, once again, it's hamburgers or tacos or "manager's choice." (The manager always chooses cow meat.) But I slap together the peanut butter sandwich; our reasons are our reasons. In Lily's life, this was the first time we'd ever walked into a diner and ordered burgers. Understandably, she kept throwing me glances—*this is really okay?* It was. The cattle were raised on pasture by an acquaintance of the owner. When Tod asked, "How's your burger?" it was not a restaurant ritual but a valid question. We told him it was great.

Tod Murphy's background was farming. The greatest economic challenge he and his farming neighbors faced was finding a market for their good products. Opening this diner seemed to him like a red-blooded American kind of project. Thomas Jefferson, Tod points out, presumed on the basis of colonial experience that farming and democracy are intimately connected. Cultivation of land meets the needs of the farmer, the neighbors, and the community, and keeps people independent from domineering centralized powers. "In Jefferson's time," he says, "that was the king. In ours, it's multinational corporations." Tod didn't think he needed to rewrite the Declaration of Independence, just a good business plan. He found investors and opened the Farmers Diner, whose slogan is "Think Locally, Act Neighborly."

For a dreamer, he's a practical guy. "Thinking globally is an abstraction. What the world needs now isn't love sweet love—that's a slogan." What the world needs now, he maintains, is more compassionate local actions: "Shopping at the hardware store owned by a family living in town. Buying locally raised tomatoes in the summer, and locally baked bread. Cooking meals at home. Those are all acts of love for a place."

The product of his vision is a place that's easy to love, where a person can sit down and eat two eggs sunny side up from a chicken that is having a good life, and a farmer that will too, while Tammy Wynette exhorts us all to stand by our man. It's also an unbelievable amount of work, I suspect, for Tod, Pam, and their kids Grace and Seamus, who start the day early on their farm and keep things running here until closing time. The diner has had to create a network of reliable year-round producers, facilitating local partnerships and dealing with human problems, for better and for worse. Supplies have to keep running even if a potato grower falls ill or the onion farm gets a divorce.

Trying to make a small entrepreneurial economy competitive with the multinationals is an obvious challenge. Tod has met it, in part, by creating an allied business that processes all their breakfast sausage, bacon, smoked ham, and turkey, and also sells these products in regional stores.

With the Farmers Diner Smokehouse and the diner itself both doing half a million dollars in business annually, they can create a market for 1,500 hogs per year. That's just about how many it takes to keep a processing plant running. A nearby bakery stays busy making their burger buns and bread. The stonecutting jobs have all gone to China, but Tod taps every channel he can think of to make sure it's Vermont farmers' hogs, grain, potatoes, and eggs that end up on the white porcelain plates of his diner.

His unusual take on the ordinary has recently made the place world-famous, at least among those who pay attention to food economies. Here in town, though, it's just the diner. The average customer comes in for the atmosphere and the food: the NASCAR crowd, or elderly Italians and Ukranians from a nearby retirement home. The old folks love the Chioggia beets and greens, farmstead fare that reminds them of home. Some of his customers also enthusiastically support the idea of keeping local businesses in business. But whether they care or not, they'll keep coming back for the food.

How is *local* defined, in this case? "An hour's drive," Tod said. Their longest delivery run is seventy miles. Maintaining a year-round supply of beef, pork, chicken, and turkey from nearby farms is relatively simple, because it's frozen. Local eggs, milk, ice cream, and cheese are also available all year, as are vegetal foods that store well, such as potatoes, beets, carrots, onions, sauerkraut, and maple syrup. The granola is made in Montpelier, the spaghetti and ravioli right here in town. Fresh vegetables are a challenge. The menu doesn't change much seasonally, but ingredients do; there's less green stuff on the plate when the ground outside is white. The beer is locally brewed except for Bud and Bud Light, which, according to Tod, "you've gotta have. We're not selling to purists." Obviously, at a diner you've also gotta have coffee, and it's fair-trade organic.

The Farmers Diner does not present itself as a classroom, a church service, or a political rally. For many regional farmers it's a living, and for everybody else it's a place to eat. Tod feels that the agenda here transcends politics, in the sense of Republican or Democrat. "It's oligarchy vs. non-oligarchy," Tod says—David vs. Goliath, in other words. Tom Jefferson against King George. It's people trying to keep work and homes together, versus conglomerates that scoop up a customer's money and move it out of town to a corporate bank account far away. Where I grew up, we used to call that "carpetbagging." Now it seems to be called the American way.

Marketing jingles from every angle lure patrons to turn our backs on our locally owned stores, restaurants, and farms. And nobody considers that unpatriotic. This appears to aggravate Tod Murphy. "We have the illusion of consumer freedom, but we've sacrificed our community life for the pleasure of purchasing lots of cheap stuff. Making and moving all that stuff can be so destructive: child labor in foreign lands, acid rain in the

Speaking Up

The increased availability of local food in any area is a direct function of the demand from local consumers. Most of us are not accustomed to asking about food origins, but it's easy enough to do.

First: in grocery stores, when the cashier asks if you found everything you were looking for, you could say, "Not really, I was looking for local produce." The smaller the store, the more open a grocer may be to your request. Food co-ops should be especially receptive. Restaurants may also be flexible about food purchasing, and your exchanges with the waitstaff or owner can easily include questions about which entrees or wines are from local sources. Restaurateurs do understand that local food is the freshest available, and they're powerful participants in the growing demand for local foods. You can do a little homework in advance about what's likely to be available in your region.

Local and regional policymakers need to hear our wishes. Many forums are appropriate for promoting local food: town and city hall meetings, school board meetings, even state commissioner meetings. It makes sense to speak up about any venue where food is served, or where leaders have some control over food acquisition, including churches, social clubs, and day-care centers. Federal legislators also need to hear about local food issues. Most state governments consider farming-related legislation almost weekly. You can learn online about what issues are being considered, to register your support for laws that help local farms. In different parts of the country the specifics change, but the motives don't. As more people ask, our options will grow.

STEVEN L. HOPP

Northeast, depleted farmland, communities where the big economic engine is crystal meth. We often have the form of liberty, but not the substance."

It does not seem exactly radical to want to turn this tide, starting with lunch from the neighborhood. Nor is it an all-or-nothing proposition. "If every restaurant got just ten percent of its food from local farmers," Tod boldly proposed, "the infrastructure of corporate food would collapse."

Ten percent seemed like a small pebble to aim at Goliath's pate. Lily picked up her spoon and dipped into Rock Bottom Farm's maple ice cream. We could hear the crash of corporate collapse with every bite. Tough work, but somebody's got to do it.

11 · SLOW FOOD NATIONS

Late June

North of the border, in Petite Italie, where everyone speaks French, it can be hard to remember where you are exactly. This was Montreal, outermost point on our elliptical vacation. Our Canadian relatives gamely asked what we wanted to see in their city, and we answered: *Food!* We wondered what was available locally here at the threshold (to our southerly way of thinking) of the frozen tundra. We lit out for Chinatown and Little Italy. Here, as in the United States, the best shot at finding locally based cuisine seems to involve seeking out the people who recently moved here from someplace else.

We passed a few restaurants that advertised "Canadian food" along with the principal ethnic fare. Our hosts explained this meant something like "American" food, more an absence than a presence of specific character: *not* Chinese, *not* Italian. Is it true that "American food" means "nothing?" I pondered this as we walked down a street of Chinese shops where butchers pinned up limp, plucked ducks like socks on clotheslines (if your mind's eye can handle socks with feet and bills). It's easy enough to say what's *not* American cuisine: anything with its feet still on, for starters. A sight like this on Main Street USA would send customers running the other way, possibly provoking lawsuits over psychological damage to children. As a concept, our national cuisine seems to be food without obvious biological origins, chosen for the color and shape of the sign out front: arches, bucket, or cowboy hat. That's the answer to the question, "Where did it come from?"

Of course that's not the whole story. We have our New England clam chowder, Louisiana gumbo, southern collards and black-eyed peas, all regionally specific. But together they don't add up to any amalgamated themes or national guidelines for enjoying what grows near us. The food cultures of other geographically diverse nations are not really one thing either; Italy is particularly famous for its many distinct regional specialties. But still that whole country manages to export a cuisine that is recognizably "Italian," unified by some basic ingredients (i.e. pasta), and an intrinsic attitude. We recognize the origins of other countries' meals when we see them, somehow sensing their spirit: *Mama mia! Bon appétit.* Pass the salsa.

If you ask a person from Italy, India, Mexico, Japan, or Sweden what food the United States has exported to them, they will all give the same answer, and it starts with a Mc. And it must be said, they're swallowing it. Processed food consumption is on the rise worldwide, proportional to growing affluence. French metro stations are plastered with ads for convenience foods. On a recent trip there I queried audiences about the danger of France losing its traditional foodways, and found them evenly divided between "Never!" and "Definitely!" Working women my age and younger confessed to giving in to convenience, even though (as they put it) they knew better. They informed me that even the national culinary institute was going soft, having just announced that its chicken courses would no longer begin with "Feathers, Feet, and Viscera 101." A flutter of conversation ran through the crowd over this point, a major recent controversy that had created radio call-in riots. I got an inkling that "giving in to convenience" means something different on that side of the pond. But still, these are real signs of change. Plenty of Parisians visit "MacDo" every day, even though it's probably not the same customers going back every day. They're in for the novelty, not the food value.

We are all, I suppose, dazzled by the idea of things other people will eat. There on the Montreal Chinatown sidewalk we stopped to admire what must have been twenty-five-pound fish chasing each other's tails in slow motion in a half barrel of water. Lily and my young nieces inspected them closely, then looked up at me with eyebrows raised in the age-old question: dinner, or pets? I had no idea. We poked into shops that sold tea, dried mushrooms, and fabulous dresses that zip up the side so tightly they look painted on. We ate lunch in a bustling cafeteria where the goods ranged from fried squid to Jell-O.

Later we stopped in at a Lebanese market, which the kids also considered fine entertainment. They kept running up to show me intriguing edibles: powdered flowers in bottles; some kind of cola apparently made from beans; "Greek Mountain tea," which looked to me like a bunch of weeds in a cellophane bag. An enormous glass case ran the full width of the store across the back, displaying cheeses. No modest yellow blocks or wheels were these, but gigantic white tablets of cheese, with the shape and heft of something Moses might have carried down from the mountain. Serious cheesemaking happened here, evidently. A young woman in a white apron stood ready to saw off a bit of goat, cow, or sheep cheese for me. We chatted, and she confirmed that these products were made in a kitchen nearby. I was curious about what kind of rennet and cultures were used for these Middle Eastern cheeses. She answered but seemed puzzled; most customers weren't interested in the technicalities. I confessed I'd tried this at home.

"You make cheese *yourself,*" she repeated reverently. "You are a *real* housewife."

It has taken me decades to get here, but I took that as a compliment.

❦

Our search kept us moving through Montreal's global neighborhoods until we arrived at the grand farmers' market of Petite Italie. An arrangement of flowering plants near the entrance spelled out "Benvenuto." Under an awning that covered several blocks, matrons with bulging bags crowded the aisles between open stalls spilling over with fresh goods. This was the place to shop, in any language.

I tried French, since I don't speak Italian. *Elles sont d'où, les tomates?*

D'ici, madame! From right here, Quebec, the vendors replied proudly, again and again. (Except for one sardonic farmer who answered, when

I asked about his eggs, "From chickens, madame.") We were flat-out amazed to see what enterprising Quebecois growers had managed to bring out already, on the first official day of summer here in the recently frozen north: asparagus, carrots, lettuce, rhubarb, hothouse tomatoes, and small, sweet strawberries. Maple syrup and countless other maple products were also abundant, of course, here where a maple leaf is literally the flag. More surprising were the local apples, plenty of them, that had been stored since their harvest late last fall but still burst sweet and crisp under our teeth when we sampled them. English orchardists once prized certain apples for their late-bearing and good storage qualities—varieties now mostly lost from the British Isles, crowded out by off-season imports from New Zealand. Evidently the good storage heirlooms have not been lost from Quebec.

I picked up a gargantuan head of broccoli. It looked too good to be true, but the cabbage family are cool-season crops. I asked the vendor where it came from.

"L'Amérique du Sud, madame," he replied. South America.

Too bad, I thought. But really, South America, where it's either tropical or now wintertime? "Quel pays?" I asked him—which country?

"La Californie, madame."

I laughed. It was a natural mistake. In the world map of produce, California might as well be its own country. A superpower in fact, one state that exports more fresh produce than most countries of the world. If not for the fossil fuels involved, this culinary export could have filled me with patriotic pride. Our country is not only arches and cowboy hats, after all. We just don't get credit for this as "American food" because vegetables are ingredients. The California broccoli would be diced into Asian stir-fries, tossed with Italian pasta primavera, or served with a bowl of mac-and-cheese, according to the food traditions of us housewives.

Still, whether we get cultural points for them or not, those truckloads of California broccoli and artichokes bring winter cheer and vitamins to people in drearier climes all over the world. From now until September the Quebecois would have local options, but in February, when the snow is piled up to the windowsills and it takes a heating pad on the engine block to get the car started, fresh spinach and broccoli would be a welcome sight. I'd buy it if I lived here, and fly the flag of La Californie in my kitchen. Even down in Dixie I'd bought winter cucumbers before, and would probably do it again. I wondered: once I was out of our industrial-food dry-out, would I be able to sample the world's vegetable delights responsibly—as a *social* broccoli buyer—without falling into dependency? California vegetables are not the serpent, it's all of us who open our veins to the flow of gas-fueled foods, becoming yawning addicts, while our neighborhood farms dry up and blow away. We seem to be built with a faulty gauge for moderation.

In the market we bought apples, maple syrup, bedding plants for our hosts' garden, and asparagus, because the season was over at home. Like those jet-setters who fly across the country on New Year's Eve, we were going to cheat time and celebrate the moment more than once. Asparagus season, twice in one year: the dream vacation.

❧

We left Canada by way of Niagara Falls, pausing to contemplate this churning cataract that has presented itself to humans down through the ages as inspiration, honeymoon destination, and every so often a rip-snorting carnival ride. We got ourselves soaked on the Maid of the Mist, and pondered the derring-do of the fourteen men, two women, and one turtle who have plunged over this crashing waterfall in conveyances including wooden barrels, a giant rubber ball, a polyethylene kayak, a diving bell, a jet-ski, and in one case only jeans and a lightweight jacket. Both women and the turtle (reputed to be 105 years old) survived, as did nine of the men, though the secret of success here is hard to divine. The jacket-and-jeans guy made it; the jet-skier and the kayaker did not, nor did William "Red" Hill in his fancy rubber bubble. And a half-ton iron-bumpered barrel that safely delivered the turtle failed to save his inventive human companion.

If there seems to be, running through this book, a suggestion that humans are a funny animal when it comes to respecting our own best interests, I rest my case.

⁂

From the border we traveled southwest across the wine country of New York and northern Pennsylvania, where endless vineyards flank the pebbled shore of Lake Erie. Another day's drive brought us into the rolling belly of Ohio, where we would be visiting friends on their dairy farm. Their rural county looked like a postcard of America's heartland, sent from a time when the heart was still healthy. Old farmhouses and barns stood as quiet islands in the undulating seas of corn, silvery oats, and auburn spelt.

We pulled into our friends' drive under a mammoth silver maple. Lily sized up the wooden swing that twisted on twenty-foot ropes from one of its boughs. A platoon of buff-colored hens ignored us, picking their way over the yard, while three old dogs trotted out to warn their mistress of our arrival. Elsie came around the corner, beaming her pure-sunshine smile. "Rest on the porch," she said, drawing us glasses of water from the pump in the yard. "David is cultivating the corn, so there's no knowing when that will finish."

We offered to help with whatever she'd been doing, so Elsie rolled the wheelbarrow to her garden and returned with a tall load of pea plants she had just pulled. We pulled lawn chairs into a circle under the cherry trees, lifted piles of vines into our laps, and tackled the shelling. Peas are a creature of spring, content to germinate in cold soil and flourish in cool, damp days, but heat causes them to stop flowering, set the last of their pods, and check out. Though nutritionally similar, peas and beans inhabit different seasons; in most gardens the peas are all finished before the first bean pod is ready to be picked. That's a good thing for the gardener, since each of these plants in its high season will bring you to your knees on a daily basis. Tall, withered pea vines are a sigh from the end of spring, a pause before the beans, squash, and tomatoes start rolling.

We caught up on news while steadily popping peas from their shells. Over our heads hung Stark's Gold cherries the size of silver dollars. The central Ohio season was a week or so behind ours, and it was dryer here too. Elsie reported they'd had no rain for nearly a month—a fairly disastrous course for June, a peak growing time for crops and pastures. A few storms had gathered lately but then dissipated. The afternoon was still: no car passed on the road, no tractor churned a field within earshot. It's surprising how selectively the human ear attends to human-made sounds: speech, music, engines. An absence of those is what we call *silence*. Maybe in the middle of a city, or a chemically sterilized cornfield, it really is quiet when all the people and engines cease. But in that particular dot on the map I was struck with how full a silence could be: a Carolina wren sang from the eave of the shed; cedar waxwings carried on whispery bickerings up in the cherry; a mockingbird did an odd jerky dance, as if seized by the bird spirit, out on the driveway. The pea bowl rang like an insistent bell as we tossed in our peas.

We heard mooing as thirty caramel brown Jersey cows came up the lane. Elsie introduced her daughter Emily and son-in-law Hersh, who waved but kept the cows on course toward the milking barn. Lily and I shook pea leaves off our laps and followed. Emily and Hersh, who live next door, do the milking every day at 5:00 a.m. and p.m. Emily coaxed the cows like children into the milking parlor ("Come on, Lisette, careful with your feet") and warned me to step back from Esau, the bull. "He's very bossy and he doesn't like women," she said. "I don't think the cows care much for him either, for the same reason. But he sires good milkers."

While Emily moved the cows through, Hersh attached and moved the pipelines of the milking machine. During lulls the couple sat down together on a bench while their toddler Noah bumped through the milking parlor and adjacent rooms, bouncing off doorjambs and stall sides in his happy orbit. Lily helped him into a toddler swing that hung in the doorway. The milking machine made a small hum but otherwise the barn was quiet, save for the jostling cows munching hay. The wood of the barn looked a hundred years old, dusty and hospitable. I couldn't imagine, myself, having an unbreakable milking date with every five o'clock of this world, but Emily seemed not to mind it. "We're so busy the rest of the day, going different directions," she said. "The milking gives us a time for Hersh and me to sit a minute."

A busy little pride of barn cats gathered near the bench, tails waving, to lap up milk-pipe overflows collected in a pie plate. I watched a few hundred gallons of Jersey milk throbbing and flowing upward through the maze of clear, flexible pipes like a creamy circulatory system. A generator-powered pump drew the milk from the cows' udders into a refrigerated stainless steel tank. The truck from an organic cooperative comes to fetch it to the plant where it is pasteurized and packed into green-and-white cartons. Where it may go from there is anyone's guess. Our own supermarket back home stocks the brand, so over the years our family may have purchased milk that came from this barn, or at least some molecules of it mixed in with milk from countless other farms. As long as it meets the company's standards, with a consistent cream percentage and nominal bacterial counts, milk from this farm becomes just another part of the blend, an anonymous commodity. This loss of identity seemed a shame, given its origin. The soil minerals and sweetness of this county's grass must impart their own flavor to the milk, just as the regions of France flavor their named denominations of wine.

David came in from the cornfield shortly after milking time. He laughed at himself for having lost track of time—as Elsie predicted—while communing with his corn. We stood for a minute, retracting the distances between our lives. Both David and Elsie are possessed of an ageless, handsome grace. Elsie is the soul of unconditional kindness, while David sustains a deadpan irony about the world and its inhabitants, including his colleagues who wear the free caps with Cargill or Monsanto logos: "At least they let you know who's controlling their frontal lobes." David and Elsie live and work in exactly the place they were born, in his case the same house and farm. It's a condition lamented in a thousand country music ballads, but seems to have worked out well for this couple.

We carried dishes of food from the kitchen to a picnic table under the cherry trees. Hersh joined us, settling Noah into a high chair while Emily brought a pitcher of milk from the tank in the barn. Obviously this family had the genes to drink it. For the first time in awhile, I had C/T13910-gene envy. Dinner conversation roamed from what we'd seen growing in Canada to what's new for U.S. farmers. David was concerned about the National Animal Identification System, through which the USDA now plans to attach an ID number and global positioning coordinates to every domestic animal in the country. Anyone who owns even one horse, chicken, cow, or canary will be required by law to get onto the map and this federal database. Farmers aren't cottoning to the plan, to put it mildly. "Mark Twain's wisdom comes to mind," David observed. " 'Sometimes I wonder whether the world is being run by smart people who are putting us on, or by imbeciles who really mean it.' " But in truth he's not too worried, as he doubts the government will be up to the job. Forcing half a million farmers to register every chicken and cow, he predicts, will be tougher than getting Afghan farmers to quit growing poppies.

The steer that had contributed itself to the meatballs on our plates had missed the sign-up. Everything else on the table was also a local product: the peas we'd just shelled, the salad picked ten minutes earlier, the strawberries from their daughter. I asked Elsie how much food they needed from outside the community. "Flour and sugar," she said, and then thought a bit. "Sometimes we'll buy pretzels, for a splurge."

It crossed my mind that the world's most efficient psychological evaluation would have just the one question: Define *splurge.* I wondered how many more years I'd have to stay off Belgian chocolate before I could attain Elsie's self-possession. I still wanted the moon, really—and I wanted it growing in my backyard.

After dinner, the long evening of midsummer still stretched ahead of us. David was eager to show us the farm. We debated the relative merits of hitching up David's team and driving the wagon, versus our hybrid gas-electric vehicle, new to us, now on its first road trip. The horses had obvious appeal, but David and Hersh had heard about the new hybrids and were eager to check out this technology. David confessed to having long ago dreamed up (while cultivating his corn) the general scheme of harnessing the friction from a vehicle's braking, capturing that energy to assist with forward momentum. Turns out, Toyota was right behind him on that. We piled into the vehicle that does not eat oats, and rode up the dusty lane past the milking barn, up a small rise into the fields.

As Elsie had said, the drought here was manifest. The animal pastures looked parched, though David's corn still looked good—or fairly good, depending. The lane divided two fields of corn that betrayed different histories: the plot on our left had been conventionally farmed for thirty years before David took the helm; on our right lay soil that had never known anything but manure and rotation. The disparity between the two fields was almost comically dramatic, like a 1950s magazine ad, except that "new and improved" was not the winner here. Now David treated both sides identically, but even after a decade, the corn on the forever-organic side stood taller and greener.

The difference is an objective phenomenon of soil science; what we call "soil" is a community of living, mostly microscopic organisms in a nutrient matrix. Organic farming, by definition, enhances the soil's living and nonliving components. Modern conventional farming is an efficient reduction of that process that adds back just a few crucial nutrients of the many that are removed each year when biomass is harvested. At first, it works well. Over time, it's like trying to raise all children on bread, peanut butter, and the same bedtime story every night for ten years. (If they cry, give them *more* bread, *more* peanut butter, and the same story twice.) An observer from another planet might think all the bases were covered, but a parent would know skipping the subtleties adds up to slow starvation. In the same way, countless micronutrients are essential to plants. Chemicals that sterilize the soil destroy organisms that fight plant diseases, aerate, and manufacture fertility. Recent research has discovered that just adding phosphorus (the P in all "NPK" fertilizers) kills the tiny filaments of fungi that help plants absorb nutrients. The losses become most apparent in times of stress and drought.

"So many people were taken in by the pesticide-herbicide propaganda," David said. "Why would we fall for that?" He seemed to carry it like an old war wound, the enduring damage done to this field. By "we" he means farmers like himself, though he didn't apply the chemicals. He came of age early in the era of ammonia-based fertilizers and DDT, but still never saw the intrinsic logic in poisoning things to make a farm.

As we crested the hill he suddenly motioned for us to stop, get out, and look: we'd caught a horned lark in the middle of his courtship display. He shot straight up from the top of a little knoll in the corn and hovered high in the sky, singing an intense, quivery, question mark of a song, *Will she, please, will she?* He hung there above us against the white sky in a breathless suspense until—zoom—his flight dance climaxed suddenly in a nosedive. We stood in the bronze light, impressed into silence. We could hear other birds, which Elsie and David distinguished by their evening songs: vesper and grasshopper sparrows, indigo buntings, a wood thrush. Cliff swallows wheeled home toward their bottle-shaped mud nests under the eaves of the barn. Tree swallows, wrens, bluebirds, mockingbirds, great horned owls, and barn owls also nested nearby. All seemed as important to David and Elsie as the dairy cows that earn them their living.

Elsie and David aren't Audubon Club members with binoculars and a life list, nor are they hippie idealists trying to save the whales. They are practical farmers saving a livelihood that has been lost to many others who walked the same road. They spare the swallows and sparrows from death by pesticide for lots of reasons, not the least of which is that these creatures *are* their pesticides. Organic farming involves a level of biotic observation more commonly associated with scientists than with farmers.

Losing the Bug Arms Race

What could be simpler: spray chemicals to kill insects or weeds, increase yields, reap more produce and profits. Grow the bottom line by spraying the current crop. From a single-year perspective, it may work. But in the long term we have a problem. The pests are launching a counterattack of their own.

Within one field, an application of pesticides will immediately reduce insect populations but not eliminate them. Depending on the spray density and angle, wind, proximity to the edge of the field, and so forth, bugs get different doses of the poison. Those receiving a lethal dose are instant casualties.

Which bugs stay around? Obviously, those lucky enough to duck and cover. Also a few of those who did get a full, normally lethal dosage, but who have a natural resistance to the chemicals. If their resistance is genetic, that resistance will come back stronger in the next generation. Over time, with continued spraying, the portion of the population with genetic resistance will increase. Eventually the whole population will resist the chemicals.

This is a real-world example of evolution, and whether or not it's showing up in textbooks, it is going strong in our conventional agriculture. More than 500 species of insects and mites now resist our chemical controls, along with over 150 viruses and other plant pathogens. More than 270 of our recently devel-

David's communion with his cornfield is part meditation and part biology. The plants, insects, birds, mammals, and microbes interact in such complicated ways, he is still surprised by new discoveries even after a lifetime spent mostly outdoors watching. He has seen swallow populations fluctuate year by year, and knows what that will mean. He watches cliff swallows following the mower and binder in the fields, downwind, snapping up leafhoppers and grasshoppers, while the purple martins devour crane flies. The prospect of blanketing them all with toxic dust even once, let alone routinely, strikes him as self-destructive, like purposely setting fire to his crops or barn.

David and Elsie were raised by farmers and are doing the same for another generation. Throughout the afternoon their children, in-laws, and grandchildren flowed in and out of the yard and house, the adults consulting about shared work, the barefoot kids sharing a long game of cousins and summer. Now we stopped in to meet the son and daughter-in-law

---

oped herbicides have now become ineffective for controlling some weeds. Some 300 weed species resist all herbicides. Uh-oh, now what?

The standard approach has been to pump up the dosage of chemicals. In 1965, U.S. farmers used 335 million pounds of pesticides. In 1989 they used 806 million pounds. Less than ten years after that, it was 985 million. That's three and a half pounds of chemicals for every person in the country, at a cost of $8 billion. Twenty percent of these approved-for-use pesticides are listed by the EPA as carcinogenic in humans.

So, how are the bugs holding out? Just fine. In 1948, when pesticides were first introduced, farmers used roughly 50 million pounds of them and suffered about a 7 percent loss of all their field crops. By comparison, in 2000 they used nearly *a billion* pounds of pesticides. Crop losses? Thirteen percent.

Biologists point out that conventional agriculture is engaged in an evolutionary arms race, and losing it. How can we salvage this conflict? Organic agriculture, which allows insect predator populations to retain a healthy presence in our fields, breaks the cycle.

STEVEN L. HOPP

whose house stood just beyond David's cornfield. As we stayed into evening, conversation on the porch debated the most productive pasture-grass rotations for cattle; out in the yard it was mostly about whose turn it was for the tire swing.

Many months later when I described this visit to a friend, he asked acerbically, "What, not even a mosquito to bother heaven?" (He also mentioned the name Andy Griffith.) Probably there were mosquitoes. I don't remember. I do know this family has borne losses and grief, just like the rest of us. But if they are generally content, must such a life inevitably be dismissed as mythical, or else merely quaint? Urban people may be allowed success, satisfaction, and consequence, all at once. Members of David and Elsie's extended family share work they love, and impressive productivity. They are by no means affluent, but seem comfortable with their material lot, and more importantly—for farmers—are not torn by debt or drained by long commutes to off-the-farm jobs. They work long hours, but value a life that allows them to sit down to a midday lunch with family, or stand outside the barn after the evening milking and watch swallows come in to roost.

They belong to a surprisingly healthy community of similar small farms. Undoubtedly, some people in the neighborhood have their ludicrous grudges, their problem children, their disputed fencelines; they are human. But they are prospering modestly by growing food. At a vegetable auction in town, farmers sell their produce wholesale to restaurants and regional grocery chains. Nothing travels very far. The farmers fare well on the prices, and buyers are pleased with the variety and quality, starting with bedding plants in May, proceeding through all the vegetables, ending with October pumpkins and apples. The nearby food co-op sells locally made cheeses, affordably priced. The hardware store sells pressure canners and well-made tools, not mechanical singing fish.

It sounds like a community type that went extinct a generation ago. But it didn't, not completely. If a self-sufficient farming community has survived here, it remains a possibility elsewhere. The success of this one seemed to hinge on many things, including steady work, material thrift, flexibility, modest expectations, and careful avoidance of debt—but not including miracles, as far as I could see. Unless, of course, we live in a country where those qualities have slipped from our paddock of everyday virtues, over to the side of "miracle." I couldn't say.

It was dark by the time we headed back through the cornfield to Elsie and David's house. At low speeds our car runs solely on battery, so it's spookily quiet, as if the engine had died but you're still rolling along. We could hear night birds and the tires softly grinding dust as we turned into the field.

"Stop here," David said suddenly. "Pull ahead just a little, so the headlights are pointing up into the field. Now turn off the headlights."

The field sparkled with what must have been millions of fireflies—the most I've ever seen in one place. They'd probably brought their families from adjacent states into this atrazine-free zone. They blinked densely, randomly, an eyeful of frenzied stars.

"Just try something," David said. "Flash the headlights one time, on and off."

What happened next was surreal. After our bright flash the field went black, and then, like a wave, a million lights flashed back at us in unison.

Whoa. To convince ourselves this was not a social hallucination, we did it again. And again. Hooting every time, so pleased were we with our antics. It's a grand state of affairs, to fool a million brainless creatures all at the same time. After five or six rounds the fireflies seemed to figure out that we were not their god, or they lost their faith, or at any rate went back to their own blinky business.

David chuckled. "Country-kid fireworks."

We sat in the dark until after midnight, out in the yard under the cherries, talking about the Farm Bill, our kids, religion, the future, books, writing. In his spare time (a concept I can hardly imagine for this man) David is a writer and editor of *Farming Magazine,* a small periodical on sustainable agriculture. We could have talked longer, but thought better of it. People might sometimes wish to sleep in, but cows never do.

In the morning I woke up in the upstairs bedroom aware of a breeze coming through the tall windows, sunlight washing white walls, a horse clop-clopping by on the road outside. I had the sensation of waking in another country, far from loud things.

Lily went out to the chicken coop to gather eggs, making herself right at home. We ate some for breakfast, along with the farm's astonishingly good oatmeal sweetened with strawberries and cream. This would be a twenty-dollar all-natural breakfast on the room service menu of some hotels. I pointed this out to David and Elsie—that many people think of such food as an upper-class privilege. David laughed. "We eat fancy food all right. Organic oatmeal, out of the same bin we feed our horses from!"

We packed up to leave, reminding one another of articles we meant to exchange. We vowed to come again, and hoped they'd come our way too, though that is less likely, because they don't travel as much as we do. Nearly all their trips are limited by the stamina of the standardbred horses that draw their buggy. David and Elsie are Amish.

Before I had Amish friends, I imagined unbending constraints or categorical aversions to such things as cars (hybrid or otherwise). Like many people, I needed firsthand acquaintance to educate me out of religious bigotry. The Amish don't oppose technology on principle, only particular technologies they feel would change their lives for the worse. I have sympathy for this position; a good many of us, in fact, might wish we'd come around to it before so much noise got into our homes. As it was explained to me, the relationship of the Amish with their technology is to strive for what is "appropriate," making that designation case by case. When milking machines came up for discussion in David and Elsie's community, the dairy farmers pointed out that milking by hand involves repeatedly lifting eighty-pound milk cans, limiting the participation of smaller-framed women and children. Milking machines were voted in because they allow families to do this work together. For related reasons, most farmers in the community use tractors for occasional needs like pulling a large wagon or thresher (one tractor can thus handle the work on many farms). But for daily plowing and cultivating, most prefer the quiet and pace of a team of Percherons or Belgians.

David summarizes his position on technology in one word: boundaries. "The workhorse places a limit on the size of our farms, and the standardbred horse-drawn buggy limits the distances we travel. This is basically what we need. This is what keeps our communities healthy." It makes perfect sense, of course, that limiting territory size can yield dividends in appreciation for what one already has, and the ability to manage it without debt. The surprise is to find whole communities gracefully accepting such boundaries, inside a nation that seems allergic to limitations, priding itself instead on the freedom to go as far as we want, as fast as we can, and buy until we run out of money—or longer, if we have credit cards.

Farmers like Elsie and David are a link between the past and future. They've declined to participate in the modern century's paradigm of agriculture—and of family life, for that matter, as they place high value on nonmaterial things like intergenerational family bonds, natural aesthetics, and the pleasures of shared work. By restraining their consumption and retaining skills from earlier generations of farmers, they are succeeding. When the present paradigm of extractive farming has run its course, I don't foresee crowds of people signing up for the plain wardrobe. But I do foresee them needing guidance on sustainable agriculture.

I realized this several years ago when David and Elsie came to our county to give an organic dairy workshop, at the request of dairy farmers here who were looking for new answers. It was a discouraged lot who attended the meeting, most of them nearly bankrupt, who'd spent their careers following modern dairy methods to the letter: growth hormones, antibiotics, mechanization. David is a deeply modest man, but the irony of the situation could not have been lost on him. There sat a group of hardworking farmers who'd watched their animals, land, and accounts slide into ruin during the half-century since the USDA declared as its official policy, "Get big or get out."

And there sat their teacher, a farmer who'd stayed small. Small enough, anyway, he would never have to move through his cornfield too quickly to study the soil, or hear the birds answer daylight with their song.

Organically Yours

BY CAMILLE

The word *organic* brings to my mind all the health food stores I've roamed through over the years, which seem to have the same aroma no matter how many miles lie between them: sweet, earthy hints of protein powder, bulk cereal, fresh fruit, and hemp. I guess the word means different things to different people. When applied to food (not a college sophomore's most dreaded chemistry class), "organic" originally described a specific style of agriculture, but now it has come to imply a lifestyle, complete with magazines and brands of clothing. The word has sneaked onto a pretty loose-knit array of food labels too, tiptoeing from "100% organic" over to "contains organic ingredients." Like overused slang, the term has been muddled by rising popularity. It's true, for example, that cookies "made with organic cocoa" have no residue of chemical pesticides or additives in the chocolate powder, but that doesn't vouch for the flour, milk, eggs, and spices that are also in each cookie.

Why should we care which ingredients, or how many, are organic? The reasons go beyond carcinogenic residues. Organic produce actually delivers more nutritional bang for the buck. These fruits and vegetables are tougher creatures than those labeled "conventional," precisely because they've had to fight off predators themselves. Plants live hard lives. They don't have to run around looking for food or building nests to raise their young, but they still have their worries. There's no hiding from predators when you have roots in the ground, and leaves that require direct contact with sunlight. You're stuck, right out in the open. Imagine the Lifetime Original movie: the helpless mother soy plant watching in agony, unable to speak or move, as a loathsome groundhog gobbles down her baby beans one at a time. Starting to tear up yet?

Next plant-kingdom heartache: there is no personal space for the garden vegetable. If you're planted in a row of other tomatoes, there you'll stay for life, watching the neighbors get a nasty case of hornworms, knowing in your heart you're going to get them too. If nobody is spritzing chemicals on the predators, all a plant can do is to toughen up by manufacturing its own disease/pest-fighting compounds. That's why organic produce shows significantly higher levels of antioxidants than conventional—these nutritious compounds evolved in the plant not for *our* health, but for the plant's. Several studies, including research done by Allison Byrum of the American Chemical Society, have shown fruits and vegetables grown without pesticides and herbicides to contain 50 to 60 percent more antioxidants than their sprayed counterparts. The same antioxidants that fight diseases and pests in the plant leaf work similar magic in the human body, protecting us not so much against hornworms as against various diseases, cell aging, and tumor growth. Spending extra money on organic produce buys these extra nutrients, with added environmental benefits for the well-being of future generations (like mine!).

Some of the best-tasting things in life are organic. This dessert (adapted from a Jamie Oliver creation, via our friend Linda) is a great way to use the high-summer abundance of blackberries, which in our part of the country are rain-washed and picked straight from wild fields. The melon salsa will bring one of summer's most luscious orange fruits from the breakfast table to a white tablecloth with candles. It's elegant and delicious over grilled salmon or chicken.

❋ BASIL-BLACKBERRY CRUMBLE

2–3 apples, chopped
2 pints blackberries
2 tablespoons balsamic vinegar
1 large handful of basil leaves, chopped
¼ cup honey—or more, depending on tartness of your berries

Preheat oven to 400°. Combine the above in an ovenproof casserole dish, mix, and set aside.

5 tablespoons flour
3 heaping tablespoons brown sugar
1 stick cold butter

Cut butter into flour and sugar, then rub with your fingers to make a chunky, crumbly mixture (not uniform). Sprinkle it over the top of the fruit, bake 30 minutes until golden and bubbly.

❋ MELON SALSA

(Makes six generous servings.)

1 medium cantaloupe
1 red bell pepper
1 small jalapeño pepper
½ medium red onion
¼ cup fresh mint leaves
1–2 tablespoons honey
2 teaspoons white vinegar

Dice melons and peppers into ¼-inch cubes. Finely mince onion and mint. Toss with honey and vinegar, allow to sit at least one hour before serving over grilled chicken breast or fish filet.

Download these and all other *Animal, Vegetable, Miracle* recipes at www.AnimalVegetableMiracle.com

12 · ZUCCHINI LARCENY

July

The president succumbed to weeds. So did the lost dogs, the want ads, and our county's Miss America hopeful. By the time we returned from vacation at the end of June, our fastidious layers of newspaper mulch were melting into the topsoil. The formerly clean rows between our crops were now smudged everywhere with a hoary green five o'clock shadow. Weeds crowded the necks of the young eggplants and leaned onto the rows of beans. Weeds are job security for the gardener.

Pigweeds, pokeweeds, quackgrass, crabgrass, purslane: we waged war, hoeing and yanking them up until weeds began to twine through our dreams. We steamed and ate some of the purslane. It's not bad. And, we reasoned (with logic typical of those who strategize wars), identifying it as an edible noncombatant helped make it look like we might be winning. *Weed* is, after all, an arbitrary designation—a plant growing where you don't want it. But tasty or not, most of the purslane still had to come out. The agricultural concern with weeds is not aesthetic but functional. Weedy species specialize in disturbed (i.e. newly tilled) soil, and grow so fast they kill the crops if allowed to stay, first through root competition and then by shading.

Conventional farming uses herbicidal chemicals for weed control, but since organic growers don't, it is weeds—even more than insects—that often present the most costly and troublesome challenge. In large operations where a mulching system like ours is impractical, organic farmers often employ three- or four-year crop rotations, using fast-growing cover crops like buckwheat or winter rye to crowd out weeds, then bare-tilling (allowing weeds to germinate, then tilling again to destroy seedlings) before planting the crops. The substitute for chemical-intensive farming is thoughtful management of ecosystems, and that is especially true when it comes to keeping ahead of the weeds. As our Uncle Aubrey says, "Weeds aren't good, but they are smart."

It's not a proud thing to admit, but we were getting outsmarted by the pigweeds. We had gardened this same plot for years, but had surely never had this much of the quackgrass and all its friends. How did they get out of hand this year? Was it weather, fertility imbalance, inopportune tilling times, or the horse manure we'd applied? The heat of composting should destroy weed seeds, but doesn't always. I looked back through my garden journals for some clue. What I found was that virtually every entry, every late June and early July day for the last five years, included the word *weeds:* "Spent morning hoeing and pulling weeds. . . . Started up hand tiller and weeded corn rows. . . . Overcast afternoon, good weeding weather. . . . Tied up grape vines, weeded." And this hopeful entry: "Finished weeding!" (Oh, right.) It's commonly said that humans remember pleasure but forget pain, and that this is the only reason women ever have more than one child. I was thinking now: or more than one garden.

In addition to weeding, we spent the July 4 weekend applying rock lime to the beans and eggplants to discourage beetles, and tying up the waist-high tomato vines to four-foot cages and stakes. In February, each of these plants had been a seed the size of this *o*. In May, we'd set them into the ground as seedlings smaller than my hand. In another month they would be taller than me, doubled back and pouring like Niagara over their cages, loaded down with fifty or more pounds of ripening fruit per plant.

This is why we do it all again every year. It's the visible daily growth, the marvelous and unaccountable accumulation of biomass that makes for the hallelujah of a July garden. Fueled only by the stuff they drink from air and earth, the bush beans fill out their rows, the okra booms, the corn stretches eagerly toward the sky like a toddler reaching up to put on a shirt. Cucumber and melon plants begin their lives with suburban reserve, posted discreetly apart from one another like houses in a new subdivision, but under summer's heat they sprawl from their foundations into disreputable leafy communes. We gardeners are right in the middle of this with our weeding and tying up, our mulching and watering, our trained eyes guarding against bugs, groundhogs, and weather damage. But to be honest, the plants are working harder, doing all the real production. We are management; they're labor.

The days of plenty suddenly fell upon us. On the same July 4 weekend we pulled seventy-four carrots, half a dozen early onions, and the whole garlic crop. (Garlic is fall-planted, braving out the winter under a cover of straw.) We dug our first two pounds of gorgeous new potatoes, red-skinned with yellow flesh. With the very last of the snap peas we gathered the earliest few Silvery Fir Tree and Sophie's Choice tomatoes, followed by ten more the next day. Even more thrilling than the tomatoes were our first precious cucumbers—we'd waited so long for that cool, green crunch. When we swore off transported vegetables, we'd quickly realized this meant life without cukes for most of the year. Their local season is short, and there's no way to keep them around longer except as pickles. So what if they're mostly just water and crunch? I'd missed them. The famine ended July 6 when I harvested six classic dark green Marketmores, two Suyo Longs (an Asian variety that's serpentine and prickly), and twenty-five little Mini Whites, a gourmet cucumber that looks like a fat, snow-white dill pickle. The day after tomorrow we would harvest this many again. And every few days after that, too, for a month or more, if they didn't succumb to wilt and beetles. Cucumbers became our all-day, all-summer snack of choice. We would try to get tired of them before winter.

A pounding all-day rain on the seventh kept me indoors, urging me to get reacquainted with my desk where some deadlines were lurking around. When evening came, for a change, I was not too worn out from garden labor to put time into cooking up a special meal. We used several pounds of cucumbers and tomatoes to make the summer's first gazpacho, our favorite cold soup, spiced with plenty of fresh cilantro. To round out the meal we tossed warm orzo pasta with grated cheese, lots of fresh-picked basil, and several cups of shredded baby squash. After three months of taxing our creativity to feed ourselves locally, we had now walked onto Easy Street.

The squash-orzo combination is one of several "disappearing squash recipes" we would come to depend on later in the season. It's a wonderfully filling dish in which the main ingredient is not really all that evident. Guests and children have eaten it without knowing it contains squash. The importance of this will soon become clear.

⁂

By mid-month we were getting a dozen tomatoes a day, that many cucumbers, our first eggplants, and squash in unmentionable quantities. A friend arrived one morning as I was tag-teaming with myself to lug two full bushel-baskets of produce into the house. He pronounced a biblical benediction: "The harvest is bountiful and the labors few."

I agreed, of course, but the truth is I still had to go back to the garden that morning to pull about two hundred onions—our year's supply. They had bulbed up nicely in the long midsummer days and were now waiting to be tugged out of the ground, cured, and braided into the heavy plaits that would hang from our kitchen mantel and infuse our meals all through the winter. I also needed to pull beets that day, pick about a bushel of green beans, and slip paper plates under two dozen ripening melons to protect their undersides from moisture and sowbugs. In another week we would start harvesting these, along with sweet corn, peppers, and okra. The harvest was bountiful and the labors were blooming endless.

However high the season, it was important for us to remember we were still just gardeners feeding ourselves and occasional friends, not commercial farmers growing food as a livelihood. That is a whole different set of chores and worries. But in our family's "Year of Local," the distinction did blur for us somewhat. We had other jobs, but when we committed to the project of feeding ourselves (and reporting, here, the results), that task became a significant piece of our family livelihood. Instead of the normal modern custom of working for money that is constantly exchanged for food, we worked directly for food, skipping all the middle steps. Basically this was about efficiency, I told myself—and I still do, on days when the work seems as overwhelming as any second job. But most of the time that job provides rewards far beyond the animal-vegetable paycheck. It gets a body outside for some part of every day to work the heart, lungs, and muscles you wouldn't believe existed, providing a healthy balance to desk jobs that might otherwise render us chair potatoes. Instead of needing to drive to the gym, we walk up the hill to do pitchfork free weights, weed-pull yoga, and Hoe Master. No excuses. The weeds could win.

It is also noiseless in the garden: phoneless, meditative, and beautiful. At the end of one of my more ragged afternoons of urgent faxes from magazine editors or translators, copy that must be turned around on a dime, incomprehensible contract questions, and baffling requests from the IRS that are all routine parts of my day job, I relish the short commute to my second shift. Nothing is more therapeutic than to walk up there and disappear into the yellow-green smell of the tomato rows for an hour to address the concerns of quieter, more manageable colleagues. Holding the soft, viny limbs as tender as babies' wrists, I train them to their trellises, tidy the mulch at their feet, inhale the oxygen of their thanks.

Like our friend David who meditates on Creation while cultivating, I feel lucky to do work that lets me listen to distant thunder and watch a nest of baby chickadees fledge from their hole in the fencepost into the cucumber patch. Even the smallest backyard garden offers emotional rewards in the domain of the little miracle. As a hobby, this one could be considered bird-watching with benefits.

Every gardener I know is a junkie for the experience of being out there in the mud and fresh green growth. Why? An astute therapist might diagnose us as codependent and sign us up for Tomato-Anon meetings. We love our gardens so much it hurts. For their sake we'll bend over till our backs ache, yanking out fistfuls of quackgrass by the roots as if we are tearing out the hair of the world. We lead our favorite hoe like a dance partner down one long row and up the next, in a dance marathon that leaves us exhausted. We scrutinize the yellow beetles with black polka dots that have suddenly appeared like chickenpox on the bean leaves. We spend hours bent to our crops as if enslaved, only now and then straightening our backs and wiping a hand across our sweaty brow, leaving it striped with mud like some child's idea of war paint. What is it about gardening that is so addicting?

That longing is probably mixed up with our DNA. Agriculture is the oldest, most continuous livelihood in which humans have engaged. It's the line of work through which we promoted ourselves from just another primate to Animal-in-Chief. It is the basis for successful dispersal from our original home in Africa to every cold, dry, high, low, or clammy region of the globe. Growing food was the first activity that gave us enough prosperity to stay in one place, form complex social groups, tell our stories, and build our cities. Archaeologists have sturdy evidence that plant and animal domestication both go back 14,000 years in some parts of the world—which makes farming substantially older than what we call "civilization" in any place. All the important crops we now eat were already domesticated around five thousand years ago. Early humans independently followed the same impulse wherever they found themselves, creating small agricultural economies based on the domestication of whatever was at hand: wheat, rice, beans, barley, and corn on various continents, along with sheep in Iraq (around 9000 BC), pigs in Thailand (8000 BC), horses in the Ukraine (5000 BC), and ducks in the Americas (pre-Inca). If you want to know which came first, the chicken-in-every-pot or the politician, that's an easy answer.

Hunter-gatherers slowly gained the skills to control and increase their food supply, learned to accumulate surplus to feed family groups through dry or cold seasons, and *then* settled down to build towns, cities, empires, and the like. And when centralization collapses on itself, as it inevitably does, back we go to the family farm. The Roman Empire grew fat on the fruits of huge, corporate, slave-driven agricultural operations, to the near exclusion of any small farms by the end of the era. But when Rome crashed and burned, its urbanized citizenry scurried out to every nook and cranny of Italy's mountains and valleys, returning once again to the work of feeding themselves and their families. They're still doing it, famously, to this day.

Where our modern dependence on corporate agriculture is concerned, some signs suggest we might play out our hand a little smarter than Rome did. Industrialized Europe has lately developed suspicions of the centralized food supply, precipitated by mad cow disease and genetically modified foods. The European Union—through government agencies and enforceable laws—is now working to preserve its farmlands, its local food economies, and the authenticity and survival of its culinary specialties.

Here in the United States we are still, statistically speaking, in the thrall of drive-through dining, but we're not unaware that things have gone wrong with our food and the culture of its production. Sociologists write about "the Disappearing Middle," referring to both middle America and mid-sized operators: whole communities in the heartland left alarmingly empty after a decades-old trend toward fewer, bigger commodity farms. We are quicker to address our problems with regional rather than national solutions. Local agencies throughout the Midwest are devising their own answers, mandating the purchase of locally grown organic food in schools, jails, and other public facilities. Policies in many states aim to bring younger people to farming, a profession whose average age is currently about fifty-five. About 15 percent of U.S. farms are now run by women—up from 5 percent in 1978. The booming organic and market-garden industries suggest that consumers are capable of defying a behemoth industry and embracing change. The direct-sales farming sector is growing. Underneath our stylish clothing it seems we are still animals, retaining some vestigial desire to sniff around the water hole and the food supply.

In the forum of media and commerce, the notion of returning to the land is still reliably stereotyped as a hare-brained hippie enterprise. But image probably doesn't matter much to people who wear coveralls to work and have power meetings with a tractor. In a nation pouring its resources into commodity agriculture—corn and soybeans everywhere and not a speck fit to eat—*back to the land* is an option with a permanent, quiet appeal. The popularity of gardening is evidence of this; so is the huge growth of U.S. agritourism, including U-pick operations, subscription farming, and farm-based restaurants or bed-and-breakfasts. Many of us who aren't farmers or gardeners still have some element of farm nostalgia in our family past, real or imagined: a secret longing for some connection to a life where a rooster crows in the yard.

In summer a young rooster's fancy turns to . . . how can I say this delicately? The most ham-fisted attempts at courtship I've ever had to watch. (And yes, I'm including high school.) As predicted, half of Lily's chick crop was growing up to be male. This was dawning on everyone as the boys began to venture into mating experiments, climbing aboard the ladies sometimes backwards or perfectly sideways. The young hens shrugged them off and went on looking for bugs in the grass. But the three older hens, mature birds we'd had around awhile, did not suffer fools gladly. Emmy, an elderly Jersey Giant, behaved as any sensible grandmother would if a teenager approached her looking for action: she bit him on the head and chased him into a boxwood bush.

These boys had much to learn, and not just the art of love. A mature, skillful rooster takes his job seriously as protector of the flock, using different vocal calls to alert his hens to food, aerial predators, or dangers on the ground. He leads his wives into the coop every evening at dusk. Lacking a proper coop, he'll coax them up onto a tree branch or other safe

Home Grown

Oh sure, Barbara Kingsolver has forty acres and a mule (a donkey, actually). But how can someone like *me* participate in the spirit of growing things, when my apartment overlooks the freeway and other people's windows? Shall I raise a hog in my spare bedroom?

How big *is* that spare bedroom? Just kidding. But even for people who live in urban areas (more than half our population), directly contributing to local food economies isn't out of the question. Container gardening on porches, balconies, back steps, or even a sunny window can yield a surprising amount of sprouts, herbs, and even produce. Just a few tomato plants in big flowerpots can be surprisingly productive.

If you have any yard at all, part of it can become a garden. You can spade up the sunniest part of it for seasonal vegetables, or go for the more understated option of using perennial edibles in your landscaping. Fruit, nut, citrus, or berry plants come in many attractive forms, with appropriate choices for every region of the country.

nighttime roost (hence, his name). The feminist in me balks to admit it, but a flock of free-range hens behaves very differently without a rooster: scattered, vulnerable, a witless wandering of lost souls. Of course, they're chickens. They have bird brains, evolved in polygamous flocks, and have lived for millennia with humans who rewarded docility and egg production. Modern hens of the sturdiest breeds can crank out an egg a day for months at a stretch (until winter days grow too short), and *that* they can do with no need for a fella. Large-scale egg operations keep artificial lights on their hens to extend the laying period, and they don't keep roosters at all. The standard white grocery-store egg is sterile. But in a barnyard where chickens forage and risk predation, flock behavior is more interesting when a guy is ruling the roost.

So Lily wanted one rooster, for flock protection and the chance to watch her hens hatch chicks next year. The position was open for a *good* rooster, not a bad one. Over the years we'd had both. Our historic favorite was Mr. Doodle. If a professional circuit had been open to him, as dogs have their sheepherding trials and such, we could have retired Mr. Doodle for stud. He had a keen eye for hen safety and a heart for justice. I

If you're not a landowner, you can still find in most urban areas some opportunity to garden. Many community-supported agriculture (CSA) operations allow or even require subscribers to participate on their farms; they might even offer a work-for-food arrangement. Most urban areas also host community gardens, using various organizational protocols—a widespread practice in European cities that has taken root here. Some rent garden spaces to the first comers; others provide free space for neighborhood residents. Some are organized and run by volunteers for some specific goal, such as supplying food to a local school, while others accommodate special needs of disabled participants or at-risk youth. Information and locations can be found at the American Community Garden Association site: www.communitygarden.org.

STEVEN L. HOPP

---

saved caterpillars I pulled off my garden so I could throw them into the chicken yard and watch Mr. Doodle run to snatch up each one, cock his head in judgment, and dole one out to each of six hens in turn before he started the next round. Any number of caterpillars not evenly divisible by six would set him into angst; he hated to play favorites.

But that was the ideal husband. The guys we had now were No-Second-Date. They're still young, we allowed. Even a dreamboat has to start somewhere, getting chased into the boxwood a time or two before finding his inner gentleman. We'd be watching our boys closely now as they played a real game of Survivor. All but one would end up on our table, and we couldn't get soft-hearted. Keeping multiple roosters is no kindness. They inevitably engage in a well-known sport that's illegal in forty-eight states.

Who would get to stay? The criteria are strict and varied: good alarm calls, unselfish instincts for foraging and roosting, and a decent demeanor toward humans. Sometimes an otherwise fine rooster will start attacking kids, a capital crime in our barnyard. And finally, our winner would need a good singing voice. We'd be hearing his particular cock-a-doodle for more than a thousand mornings. Chanticleers, as the storybooks call them, are as diversely skilled as opera singers. We wanted a Pavarotti. Crowing skills are mostly genetic, arriving with developing male hormones. So far we'd heard nothing resembling a crow.

And then, one morning, we did. It was in July, soon after my summertime ritual of moving our bedroom outside onto the screened sleeping porch. The summer nights are balmy and marvelous, though it's hard to sleep with so much going on after dark: crickets, katydids, and fireflies fill every visible and aural space. Screech owls send out their love calls. Deer sometimes startle us at close range with the strange nasal whiffle of their alarm call. And in the early hours of one morning, as I watched the forested hillside color itself in slow motion from gray to green, I heard what I thought must be a new Virginian species of frog: "Cro-oak!"

I woke Steven, as wives wake husbands everywhere, to ask: "What's that sound?"

I knew he wouldn't be annoyed, because this was no tedious burglary suspect—it was wildlife. He sat up, attentive. His research interest is bioacoustics: birdsong and other animal communication. He can identify any bird native to the eastern United States by ear, and can nail most insects, mammals, and amphibians at least to category. (Like most mortals, I cannot. I can mistake mammal calls for birds, and certain insects for power tools.) He offered a professional opinion on this pre-dawn croak: "Idunno."

As we listened, it became clear that two of them were having some kind of contest: "Cro-oa-oak!"

(A pause, for formulation of the response.)

"Cri-iggle-ick!"

Steven figured it out way ahead of me. These were our boys of summer. Yikes.

More rooster voices joined the choir as dawn crept over the ridge. Eventually one emerged as something of a leader, to which the others responded together in the call-and-response style of an old-time religious revival.

"Rrrr-arrr-orrrk!"

"Crii-iggle-ick!" "Cro-aok!" "Crr-rdle-rrr!"

We had on our hands what sounded like a newly opened Berlitz School for Rooster, with a teacher hired on a tight budget.

The girls heard us from downstairs, and came up to the sleeping porch to see what was so funny. Soon we were all flopped across the bed laughing after every chorus. Welcome to our funny farm. Did I say we were hoping for a Pavarotti? We had a gang of tone-deaf idol wannabes. For how many weeks would this harrowing audition go on before we could narrow the field of applicants? One outstanding contestant punctuated the end of his croak, every time, with a sort of burp: "Crr-rr-arrrr . . . *bluup!*"

This guy had a future in the culinary arts. Mine.

⸙

Our turkeys were looking gorgeous after their awkward adolescent molt into adult plumage. Bourbon Reds are as handsome as it gets on the turkey runway, with chestnut-red bodies, white wings, and white-tipped tailfeathers. The boys weren't crowing, of course, but this would be their only failing in the department of testosterone. We'd seen that show before. Prior to our move to Virginia I'd raised a few Bourbon Reds as a trial run, to see how we liked the breed before attempting to found a breeding flock. I'd gotten five poults and worried from day one about how I would ever reconcile their darling fuzzy heads with the season of Thanksgiving. But that summer, with the dawning of adolescent hormones, the cuteness problem had resolved itself, *and how*: four of my five birds turned out to be male. They forgot all about me, their former mom, and embarked on a months-long poultry frat party. Picture the classic turkey display, in which the male turkey spreads his colorful tail feathers in an impressive fan. Now picture that times four, continuing nonstop, month after month. The lone female spent the summer probably wishing she'd been born with the type of eyes that can roll. These guys meant to impress her or die trying. They shimmied their wing feathers with a sound like rustling taffeta, stretched their necks high in the air, and belted out a croaky gobble. Over and over and *over*. Our nearest neighbor down the road had called to ask, tentatively, "Um, I don't mean to be nosy, but is your rooster sick?"

Many of us were relieved that year at harvest time, when our first turkey experiment reached its conclusion. By autumn the boys had begun to terrify Lily, who was six that year, by rushing at her gobbling when she entered the poultry yard to feed her chickens. In the beginning she'd lobbied to name the turkeys, which I nixed, but I relented later when I saw what she had in mind. She christened them Mr. Thanksgiving, Mr. Dinner, Mr. Sausage, and—in a wild first-grade culinary stretch—Sushi.

So we knew what we were in for now, as our new flock came of age. By midsummer all our April-born poultry were well settled in. Our poultry house is a century-old, tin-roofed grain barn with wire-screened sides covered by a lattice of weathered wood slats. We had remodeled the building by dividing it into two large rooms, separate nighttime roosting coops for chickens and turkeys (they don't cohabit well), secure from predatory raccoons, possums, coyotes, owls, and large snakes. An entry room at the front of the building, with doors into both the coops, we used for storing grain and supplies. The chicken coop had a whole wall of laying boxes (Lily had high hopes), and a back door that opened directly to the outdoors—the chickens now ranged freely all over our yard during the day, and only had to be shut in at night. The turkey side had a hatch opening into a large, wire-enclosed outdoor run.

The turkey poults had recently figured out how to fly out through this hatch and were enjoying the sunny days, returning to the indoor coop only to roost at night. Although most people think of chickens and turkeys as grain-eaters (and for CAFO birds, grain is the best of what they eat), they consume a lot of grass and leaves when they're allowed to forage. Both chickens and turkeys are also eager carnivores. I've seen many a small life meet its doom at the end of a beak in our yard, not just beetles and worms but salamanders and wild-eyed frogs. (The "free-range vegetarian hens" testimony on an egg-carton label is perjury, unless someone's trained them with little shock collars.) Our Bourbon Reds were skilled foragers, much larger than the chickens now, but a bit slower to mature sexually. It wasn't yet clear how our dozen birds would sort themselves out in that regard. Frankly, I was hoping for girls.

❧

My kids find this hard to believe, but when I was a child I'd never heard of zucchini. We knew of only one kind of summer squash: the yellow crooknecks we grew copiously in our garden. They probably also carried those down at the IGA in summertime, if any unfortunate and friendless soul actually had to buy them. We had three varieties of hard-shelled winter squash: butternuts, pumpkins, and a green-striped giant peculiar to our region called the cushaw, which can weigh as much as a third-grader. We always kept one of these on the cool attic stairs all winter (cushaw, not third-grader) and sawed off a piece every so often for our winter orange vegetable intake. They make delicious pies. And that is the full squash story of my tender youth. Most people might think that was enough.

Not my dad. Always on the lookout for adventure, he went poking into the new Kroger that opened in a town not far from ours when I was in my early teens. Oh, what a brave new world of culinary exotics: they carried actual whole cream pies down there, frozen alive in aluminum plates, and also vegetables of which we were previously unaware. Artichokes, for example. We kids voted for the pies but got overruled; Dad brought home artichokes. Mom dutifully boiled and served them with forks, assuming one would eat the whole thing. We tried *hard.* I didn't touch another artichoke for twenty years.

But our lives changed forever the day he brought home zucchinis. "It's *Italian* food," he explained. We weren't sure how to pronounce it. And while the artichokes had brought us to tears and throat lozenges, we liked these dark green dirigibles a lot. The next year Dad discovered he could order the seeds and grow this foreign food right at home. I was in charge of the squash region of the garden in those days—my brother did the onions—and we were diligent children. I'm pretty sure the point source of the zucchini's introduction into North America was Nicholas County, Kentucky. If not, we did our part, giving them to friends and strangers alike. We ate them steamed, baked, batter-fried, in soup, in summer and also in winter, because my mother developed a knockout zucchini-onion relish recipe that she canned in jars by the score. I come from a proud line of folks who know how to deal with a squash.

So July doesn't scare me. We picked our first baby yellow crooknecks at the beginning of the month, little beauties that looked like fancy restaurant fare when we sautéed them with the blossoms still attached. On July 6 I picked two little pattypans (the white squash that look like flying saucers), four yellow crooknecks, six golden zucchini, and five large Costata Romanescas—a zucchini relative with a beautifully firm texture and a penchant for attaining the size of a baseball bat overnight. I am my father's daughter, always game for the new seed-catalog adventure, and I am still in charge of the squash region of the garden. I can overdo things, but wasn't ready to admit that yet. "I *love* all this squash," I declared, bringing the rainbow of their shapes and colors into the kitchen along with the season's first beans (Purple Romano and Gold of Bacau), Mini White cucumbers, five-color chard, and some Chioggia beets, an Italian heirloom that displays red and white rings like a target when sliced in cross-section. I was still cheerful two days later when I brought in the day's nineteen squash. And then thirty-three more over the next week, including a hefty haul of cubit-long Costatas. Unlike other squash, Costatas are still delicious at this size, though daunting. We split and stuffed them with sautéed onions, bread crumbs, and cheese, and baked them in our outdoor bread oven. All dinner guests were required to eat squash, and then take some home in plastic sacks. We started considering dinner-guest lists, in fact, with an eye toward those who did not have gardens. Our gardening friends knew enough to slam the door if they saw a heavy sack approaching.

Camille gamely did her part. Before her sister's birthday she adapted several different recipes into a genius invention: chocolate chip zucchini cookies. She made a batch of about a hundred, obliterating in the process several green hulks that had been looming in the kitchen. She passed the tray around to Lily's friends at the birthday party, with a sly grin, as they crowded around the kitchen table to watch Lily open her presents. Fourth-graders hate squash. We watched them chew. They asked for seconds. Ha!

Camille dared them to guess the secret ingredient, slanting her eyes suggestively at the dark green blimps that remained (one of them cut in half) on the kitchen counter.

"Cinnamon? Oatmeal! Candy canes??"

We'll never tell. But after the wrapping paper had flown, with all dust settled and the hundred cookies eaten, we still had more of those dirigibles on the counter.

Had we planted too many vines? Should we let the weeds take them early? Oh, constant squash, they never let you down. Early one Saturday morning as I lay sleepless, I whispered to Steven, "We need to get a hog."

"A hog?"

"For the squash."

He knew I couldn't be serious. For one thing, hogs are intelligent enough to become unharvestable. Their eyes communicate an endearing sensibility that poultry eyes don't, even when you've raised them from the darling stage. We didn't need a pet pig.

But we did need something to dispatch all this zucchini—some useful purpose for the pyramid of excess vegetable biomass that was taking over our lives. My family knows I'm congenitally incapable of wasting food. I was raised by frugal parents who themselves grew up in the Depression, when starvation seemed a genuine possibility. I have now, as a grown-up, learned to buy new jeans when mine have patches on the patches, but I have not learned to throw perfectly good food in the garbage. Not even into the compost, unless it has truly gone bad. To me it feels like throwing away a Rolex watch or something. (I'm just guessing on that.) Food was grown by the sweat of someone's brow. It started life as a seed or newborn and beat all the odds. It's intrinsically the most precious product in our lives, from an animal point of view.

But there sat this pile on the kitchen counter, with its relatives jammed into a basket in the mudroom—afloat between garden and kitchen—just waiting for word so they could come in here too: the Boat Zucchinis.

Sometimes I just had to put down my knives and admire their extravagant success. Their hulking, elongated cleverness. Their heft. I tried balancing them on their heads, on their sides: right here in the kitchen we had the beginnings of our own vegetable Stonehenge. Okay, *yes,* I was losing it. I could not stay ahead of this race. If they got a little moldy, then I could compost them. And the really overgrown ones we were cracking open for the chickens to eat—that isn't waste, that's eggs and meat. A hog could really do that for us in spades. . . .

Could they design an automobile engine that runs on zucchini?

It didn't help that other people were trying to give them to *us.* One day we came home from some errands to find a grocery sack of them hanging on our mailbox. The perpetrator, of course, was nowhere in sight.

"Wow," we all said—"*what a good idea!*"

Garrison Keillor says July is the only time of year when country people lock our cars in the church parking lot, so people won't put squash on the front seat. I used to think that was a joke.

I don't want to advertise the presence or absence of security measures in our neighborhood, except to say that in rural areas, generally speaking, people don't lock their doors all that much. The notion of a "gated community" is comprehensible to us only in terms of keeping the livestock out of the crops. It's a relaxed atmosphere in our little town, plus our neighbors keep an eye out and will, if asked, tell us the make and model of every vehicle that ever enters the lane to our farm. So the family was a bit surprised when I started double-checking the security of doors and gates any time we all were about to leave the premises.

"Do I have to explain the obvious?" I asked impatiently. "Somebody might break in and put zucchini in our house."

It was only July. I'd admit no more squash, but I was not ready to admit defeat.

The Spirit of Summer

BY CAMILLE

It's a Saturday afternoon in mid-July, and our farm is overflowing with life. After each trip to the garden we come down the hill bearing armloads of cucumbers, squash, and tomatoes. We're now also harvesting peppers, eggplants, onions, green beans, and chard. In a few hours some friends will be coming over for supper, so my mom and I study our pile of fresh vegetables and begin to prepare. We'll start by making the cucumber soup, which will be served first and needs time to chill. Fourteen small cucumbers go into the blender, one after the other, transformed into bright green, puréed freshness as they meet the whirling blades. Then we stir in the skim-milk yogurt we made yesterday. Finally, we add fresh herbs to the cool, light mixture and wedge the bowl into the refrigerator between gigantic bags of zucchini.

Now it's time for the bread baking. This is the man's job in our household, so Steven gets out his various bags of flour and begins to work his magic. A cup of this and a tablespoon of that fly into the mixer until he's satisfied. Then the machine's bread hook folds it all together and it's left to rise. Later our friends arrive, and Nancy, a true bread artisan, works with Steven to roll out and shape some plump baguettes. Outdoors, a fire has been crackling for hours in the big stone bread oven we built this spring. Nancy has been eager to come over and help try it out. She and Steven set the baguettes on floured pans and slide them into the oven, which has been cleared of coals. The temperature inside is nearly 700 degrees.

Meanwhile Mom and I are working on dessert: cherry sorbet. We picked the cherries from the tree that shades our front porch, teasing us by bearing the most fruit on its highest branches. Every summer Steven and I climb ladders we've set into the back of the pickup truck strategically parked under the tree, while Mom says "Stop! Be careful!" and then finally climbs up there with us. Even so, bushels of shiny, black cherries still stay out of reach for everybody that doesn't have wings. The blue jays get their share, but we still brought in quite a haul this year. Mom pits the cherries, staining her hands the purplish-black color of pen ink, while I heat water and honey on the stove. We mix the fruit and syrup together and let the concoction chill so it will be ready to pour into our ice cream maker after supper.

The baguettes don't take long to bake in the hot stone oven. After about fifteen minutes we take them out and cut them open. The smell is enough to make you give up on cooking a gourmet meal and just eat bread instead. We resist, however, slicing the whole loaves lengthwise and laying the baguettes open-faced on a pan. We stack them with grilled vegetables and cheese, starting with slices of mozzarella we made yesterday with the help of my little sister and our Italian grandmother. Next we pile on slices of nicely browned yellow squash, pattypan squash, green peppers, eggplants, and onions that I've just taken off the grill. The final layer is fresh tomatoes with basil. We stick the pans under the broiler for a few minutes, then sit down to enjoy our meal. I feel like I've been cooking all day, but it's so worth it.

The soup is amazing after its rest in the fridge. It's the perfect finish to a humid afternoon spent chopping vegetables or hovering over a hot grill. We savor every bite, and then bring out the vegetable-loaded baguettes. The broiler has melted the cheese just right, so everything melds together into one extraordinary flavor.

After we sit a while talking, it's time for cherry sorbet. The dessert is almost too purple to be real, and an ideal combination of sweet and sour. Everyone finishes smiling. This has been one of the best meals of my life, not only because it was so delicious, but because all this food came from plants we watched growing from tiny seeds to jungles. We witnessed the moment in the mozzarella's life that the milk turned into curds and whey. We saw the bread go from sandy-colored glop to crusty, golden gorgeous. We had a relationship with this meal.

Here are the recipes that went into our fabulous midsummer menu. I've also included two for secretly serving lots of zucchini. Another *very* easy way to cook squash is to slice it lengthwise, toss in a bowl with olive oil, salt, thyme, and oregano, and slap it on a hot grill alongside burgers or chicken. Whole green beans are also wonderful grilled this way (a grill basket will keep them from leaping into the flames). The squash cookie recipe has passed the ten-year-old test.

❋ CUCUMBER YOGURT SOUP

8 small-medium cucumbers, peeled and chopped
3 cups water
3 cups plain yogurt
2 tablespoons dill
1 tablespoon bottled lemon juice (optional)
1 cup nasturtium leaves and petals (optional)

Combine ingredients in food processor until smooth, chill before serving. Garnish with nasturtium flowers.

❋ GRILLED VEGETABLE PANINI

Summer squash (an assortment)
Eggplant
Onion
Peppers
Olive oil
Rosemary
Oregano
Thyme
Salt and pepper

Slice vegetables lengthwise into strips no thicker than ½ inch. Combine olive oil and spices (be generous with the herbs) and marinate vegetables, making sure all faces of the vegetable slices are covered. Then cook on grill until vegetables are partially blackened; you may want to use grill basket for onions and peppers.

2 loaves French bread (16 to 18 inches)
2 8-ounce balls mozzarella
3 large tomatoes
Basil leaves

Cut loaves of bread lengthwise. Arrange bread on baking sheets and layer with grilled vegetables first, slices of mozzarella next, and slices of tomato last. Drizzle with a little bit of olive oil and place the baking sheets under a broiler until cheese is melted. Garnish with leaves of fresh basil. Cut in pieces to serve.

❋ CHERRY SORBET

2 heaping cups pitted cherries
¾ cup sugar (or honey to taste)
⅔ cup water

While one person pits the cherries, another can combine sugar and water in a saucepan over low heat. Stir until the sugar has dissolved completely (syrup will be clear at this point) and allow the mixture to cool. When cherries are pitted combine them with syrup in a blender. Blend on low until smooth, then refrigerate mixture until you are ready to pour it into an ice cream maker.

❋ DISAPPEARING ZUCCHINI ORZO

¾-pound package orzo pasta (multicolored is fun)

Bring 6 cups water or chicken stock to a boil and add pasta. Cook 8 to 12 minutes.

1 chopped onion, garlic to taste

3 large zucchini
Olive oil for sauté

Use a cheese grater or mandoline to shred zucchini; sauté briefly with chopped onion and garlic until lightly golden.

Thyme
Oregano
¼ cup grated Parmesan or any hard yellow cheese

Add spices to zucchini mixture, stir thoroughly, and then remove mixture from heat.

Combine with cheese and cooked orzo, salt to taste, serve cool or at room temperature.

❋ ZUCCHINI CHOCOLATE CHIP COOKIES

(Makes about two dozen)

1 egg, beaten
½ cup butter, softened
½ cup brown sugar
⅓ cup honey
1 tablespoon vanilla extract

Combine in large bowl.

1 cup white flour
1 cup whole wheat flour
½ teaspoon baking soda
¼ teaspoon salt
¼ teaspoon cinnamon
¼ teaspoon nutmeg

Combine in a separate, small bowl and blend into liquid mixture

1 cup finely shredded zucchini
12 ounces chocolate chips

Stir these into other ingredients, mix well. Drop by spoonful onto greased baking sheet, and flatten with the back of a spoon. Bake at 350°, 10 to 15 minutes.

Don't tell my sister.

Download these and all other *Animal, Vegetable, Miracle* recipes at www.AnimalVegetableMiracle.com

SQUASH-SEASON MEAL PLAN

Sunday ~ Braised chicken with squash, corn, and cilantro

Monday ~ Grilled vegetable panini, served with green salad

Tuesday ~ Sliced cold chicken (cooked Sunday) and zucchini orzo

Wednesday ~ Grilled hamburgers with grilled green beans and squash

Thursday ~ Egg-battered squash blossoms stuffed with cheese, served with salad

Friday ~ Pizza with grilled baby squash, eggplant, caramelized onions, and mozzarella

Saturday ~ Lamb chops and baked stuffed zucchini

13 · LIFE IN A RED STATE

August

I've kept a journal for most of the years I've been gardening. I'm a habitual scribbler, jotting down the triumphs and flops of each season that I always feel pretty sure I'd remember anyway: that the Collective Farm Woman melons were surprisingly prissy; that the Dolly Partons produced such whopping tomatoes, the plants fell over. Who could forget any of that? Me, as it turns out. Come winter when it's time to order seeds again, I always need to go back and check the record. The journal lying open beside my bed also offers a handy incentive at each day's end for making a few notes about the weather, seasonal shifts in bloom and fruiting times, big family events, the day's harvest, or just the minutiae that keep me entertained. The power inside the pea-sized brain of a hummingbird, for example, that repeatedly built her nest near our kitchen door: despite her migrations across continents and the storms of life, her return date every spring was the same, give or take no more than twenty-four hours.

Over years, trends like that show up. Another one is that however jaded I may have become, winter knocks down the hollow stem of my worldliness and I'll start each summer again with expectations as simple as a child's. The first tomato of the season brings me to my knees. Its vital stats are recorded in my journal with the care of a birth announcement: It's an Early Girl! Four ounces! June 16! Blessed event, we've waited so long. Over the next few weeks I note the number, size, and quality of the different tomato varieties as they begin to come in: two Green Zebras, four gorgeous Jaune Flammés, one single half-pound Russian Black. I note that the latter wins our summer's first comparative taste test—a good balance of tart and sweet with strong spicy notes. I describe it in my journal the way an oenophile takes notes on a new wine discovery. On the same day, I report that our neighbor wants to give away all her Russian Blacks on the grounds they are "too ugly to eat." I actually let her give me a couple.

As supply rises, value depreciates. Three weeks after the **First Tomato!** entry in my journal, I've dropped the Blessed Event language and am just putting them down for the count: "10 Romas today, 8 Celebrity, 30 Juliet." I continue keeping track so we'll know eventually which varieties performed best, but by early August I've shifted from numbers to pounds. We bring in each day's harvest in plastic grocery sacks that we heave onto a butcher's scale in our kitchen, jotting down the number on a notepad before moving on to processing.

At this point in the year, we had officially moved beyond hobby scale. My records would show eventually whether we were earning more than minimum wage, but for certain we would answer the question that was largely the point of this exercise: what does it take, literally, to keep a family fed? Organizing the spring planting had been tricky. How many pumpkins does a family eat in a year? How many jars of pickles? My one area of confidence was tomatoes: we couldn't have too many. We loved them fresh, sliced, in soups and salads, as pasta sauces, chutneys, and salsa. I'd put in fifty plants.

In July, all seemed to be going according to plan when we hauled in just over 50 pounds of tomatoes. In August the figure jumped to 302 pounds. In the middle of that month, our neighbor came over while I was canning. I narrowed my eyes and asked her, "Did I let you *give me some tomatoes* a few weeks ago?"

She laughed. She didn't want them back, either.

Just because we're overwhelmed doesn't mean we don't still love them, even after the first thrill wears off. I assure my kids of this, when they point out a similar trend in their baby books: dozens of photos of the first smile, first bath, first steps . . . followed by little evidence that years two and three occurred at all. Tomatoes (like children) never achieve the villainous status of squash—they're too good to wear out their welcome, and if they *nearly* do, our in-town friends are always happy to take them. Fresh garden tomatoes are so unbelievably tasty, they ruin us utterly and forever on the insipid imports available in the grocery. In defiance of my childhood training, I cannot clean my salad plate in a restaurant when it contains one of those anemic wedges that taste like slightly sour water with a mealy texture. I'm amazed those things keep moving through the market, but the world apparently has tomato-eaters for whom "kinda reddish" is qualification enough. A taste for better stuff is cultivated only through experience.

Drowning in good tomatoes is the exclusive privilege of the gardener and farm-market shopper. The domain of excess is rarely the lot of country people, so we'll take this one when we get it. From winter I always look back on a season of bountiful garden tomatoes and never regret having eaten a single one.

⚘

At what point did we realize we were headed for a family tomato harvest of 20 percent of a *ton*? We had a clue when they began to occupy every horizontal surface in our kitchen. By mid-August tomatoes covered the countertops end to end, from the front edge to the backsplash. No place to set down a dirty dish, forget it, and no place to wash it, either. The sink stayed full of red orbs bobbing in their wash water. The stovetop stayed covered with baking sheets of halved tomatoes waiting for their turn in the oven. The cutting board stayed full, the knives kept slicing.

August is all about the tomatoes, every year. That's nothing new. For a serious gardener, the end of summer is when you walk into the kitchen and see red. We roast them in a slow oven, especially the sweet orange Jaune Flammes, which are just the right size to slice in half, sprinkle with salt and thyme, and bake for several hours until they resemble cow flops (the recipe says "shoes," if you prefer). Their slow-roasted, caramelized flavor is great in pizzas and panini, so we freeze hundreds of them in plastic bags. We also slice and slide them into the drawers of the food dryer, which runs 24–7. ("Sun-dried" sounds classy, but Virginia's sun can't compete with our southern humidity; a low-voltage dryer renders an identical product.) We make sauce in huge quantity, packed and processed in canning jars. By season's end our pantry shelves are lined with quarts of whole tomatoes, tomato juice, spaghetti sauce, chutney, several kinds of salsa, and our favorite sweet-sour sauce based on our tomatoes, onions, and apples.

August brings on a surplus of nearly every vegetable we grow, along with the soft summer fruits. Squash are vegetable rabbits in terms of reproductive excess, but right behind them are the green beans, which in high season must be picked every day. They're best when young, slender, and super-fresh, sautéed and served with a dash of balsamic vinegar, but they don't stay young and slender for long. We've found or invented a fair number of disappearing-bean recipes; best is a pureed, bright green dip or spread that's a huge crowd pleaser until you announce that it's green bean paté. It keeps and freezes well, but needs a more cunning title. Our best effort so far is "frijole guacamole," Holy Mole for short.

We process and put up almost every kind of fruit and vegetable in late summer, but somehow it's the tomatoes, with their sunny flavor and short shelf life, that demand the most attention. We wish for them at leisure, and repent in haste. Rare is the August evening when I'm not slicing, canning, roasting, and drying tomatoes—often all at the same time. Tomatoes take over our life. When Lily was too young to help, she had to sit out some of the season at the kitchen table with her crayons while she watched me work. The summer she was five, she wrote and illustrated a small book entitled "Mama the Tomato Queen," which fully exhausted the red spectrum of her Crayola box.

Some moment of every summer finds me all out of canning jars. So I go to town and stand in line at the hardware store carrying one or two boxes of canning jars and lids, renewing my membership in a secret society. Elderly women and some men, too, will smile their approval or ask outright, "What are you canning?" These folks must see me as an anomaly of my generation, an earnest holdout, while the younger clientele see me as a primordial nerdhead, if they even notice. I suppose I'm both. If I even notice.

But canning doesn't deserve its reputation as an archaic enterprise murderous to women's freedom and sanity. It's straightforward, and for tomato and fruit products doesn't require much special equipment. Botulism—the famously deadly bacterium that grows in airless, sealed containers and thus can spoil canned goods—can't grow in a low-pH environment. That means acidic tomatoes, grapes, and tree fruits can safely be canned in a simple boiling water bath. All other vegetables must be processed in a pressure canner that exposes them to higher-than-boiling temperatures; it takes at least 240°F to kill botulism spores. The USDA advises that pH 4.6 is the botulism-safe divide between these two methods. Since 1990, test kitchens have found that some low-acid tomato varieties sit right on the fence, so tomato-canning instructions published years ago may not be safe. Modern recipes advise adding lemon juice or citric acid to water-bath-canned tomatoes. Botulism is one of the most potent neurotoxins on our planet, and not a visitor you want to mess with.

Acidity is the key to safety, so all kinds of pickles preserved in vinegar are fair game. In various parts of the world, pickling is a preservation method of choice for everything from asparagus to zucchini chutney; I have an Indian recipe for cinnamon-spiced pickled peaches. But our Appalachian standard for the noncucumber pickle is the Dilly Bean, essentially dill pickles made of green beans. This year when I was canning them on a July Saturday, Lily and a friend came indoors from playing and marched into the kitchen holding their noses, wanting to know why the whole house smelled like cider vinegar. I pointed my spoon at the cauldron bubbling on the stove and explained I was making pickles. I do wonder what my kids' friends go home to tell their parents about us. This one dubiously surveyed the kitchen: me in my apron, the steaming kettle, the mountain of beans I was trimming to fit into the jars, the corners where my witch's broom might lurk. "I didn't know you could make pickles from beans," she countered. I assured her you could make almost anything into pickles. She came back an hour later when I was cleaning up and my finished jars were cooling on the counter, their mix of green, purple, and yellow beans standing inside like little soldiers in an integrated army. She held her eyes very close to one of the jars and announced, "Nope! They didn't turn into pickles!"

Every year I think about buying a pressure canner and learning to use it, so I could can our beans as beans, but I still haven't. Squash, beans, peas, okra, corn, and basil pesto are easy enough to steam-blanch and put into the freezer in meal-sized bags. But since tomato products represent about half the bulk of our stored garden produce, I'd rather have them on the shelf than using up electricity to stay frozen. (We would also have to buy a bigger freezer.) And besides, all those gorgeous, red-filled jars lining the pantry shelf in September make me happy. They look like early valentines, and they are, for a working mom. I rely on their convenience. I'm not the world's only mother, I'm sure, who frequently plans dinner in the half-hour between work and dinnertime. Thawing takes time. If I think ahead, I can dump bags of frozen or dried vegetables into the Crock-Pot with a frozen block of our chicken or turkey stock, and have a great soup by evening. But if I *didn't* think ahead, a jar of spaghetti sauce, a box of pasta, and a grate of cheese will save us. So will a pint of sweet-and-sour sauce baked over chicken breasts, and a bowl of rice. I think of my canning as fast food, paid for in time up front.

That price isn't the drudgery that many people think. In high season I give over a few Saturdays to canning with family or friends. A steamy canning kitchen full of women discussing our stuff is not so different from your average book group, except that we end up with jars of future meals. Canning is not just for farmers and gardeners, either. Putting up summer produce is a useful option for anyone who can buy local produce from markets, as a way to get these vegetables into a year-round diet. It is also a kindness to the farmers who will have to support their families in December on whatever they sell in August. They can't put their unsold tomatoes in the bank. Buying them now, in quantity, improves the odds of these farmers returning with more next summer.

If canning seems like too much of a stretch, there are other ways to save vegetables purchased in season, in bulk. Twenty pounds of tomatoes will cook down into a pot of tomato sauce that fits into five one-quart freezer boxes, good for one family meal each. (Be warned, the fragrance of your kitchen will cause innocent bystanders to want to marry you.) Tomatoes can even be frozen whole, individually on trays set in the freezer; once they've hardened, you can dump them together into large bags (they'll knock against each other, sounding like croquet balls), and later withdraw a few at a time for winter soups and stews. Having gone nowhere in the interim, they will still be local in February.

⚜

In some supermarket chains in Virginia, North Carolina, and Tennessee, shoppers can find seasonal organic vegetables in packages labeled "Appalachian Harvest." The letters of the brand name arch over a sunny, stylized portrait of plowed fields, a clear blue stream, and the assurance: "Healthy Food, Healthy Farms, Close to Home."

Labels can lie, I am perfectly aware. Plenty of corporations use logo trickery to imply their confined meat or poultry are grown on green pastures, or that their tomatoes are handpicked by happy landowners instead of immigrants earning one cent per pound. But the Appalachian Harvest vegetables really do come from healthy farms, I happen to know, because they're close to *my* home. Brand recognition in mainstream supermarkets is an exciting development for farmers here, in a region that has struggled with chronic environmental problems, double-digit unemployment, and a steady drain of our communities' young people from the farming economy.

But getting some of Appalachia's harvest into those packages has not been simple. Every cellophane-wrapped, organically bar-coded packet of organic produce contains a world of work and specific promises to the consumer. To back them up, farmers need special training, organic certification, reliable markets, and a packaging plant. A model nonprofit called Appalachian Sustainable Development provides all of these in support of profitable, ecologically sound farming enterprises in a ten-county region of Virginia and Tennessee. In 2005, ten years after the program began, participating family farms collectively sold $236,000 worth of organic produce to regional retailers and supermarkets, which those markets, in turn, sold to consumers for nearly $0.3 million.

The Appalachian Harvest packing house lies in a mountain valley near the Virginia-Tennessee border that's every bit as gorgeous as the storybook farm on the product label. In its first year, the resourceful group used a converted wing of an old tobacco barn for its headquarters, using a donated walk-in cooler to hold produce until it could be graded and trucked out to stores. Now the packing plant occupies the whole barn space, complete with truck bays, commercial coolers, and conveyor belts to help wash and grade the produce. Tomatoes are the cash cow of this enterprise, but they are also its prima donnas, losing their flavor in standard refrigeration, but quick to spoil in the sultry heat, so the newest major addition at the packing house is a 100-by-14-foot tomato room where the temperature is held at 56 degrees.

Participating farmers bring vegetables here by the truckload, in special boxes that have never been used for conventional produce. Likewise, the packing facility's equipment is used for organic produce only. Most of the growers have just an acre or two of organic vegetables, among other crops grown conventionally. Those who stick with the program may expand their acreage of organic vegetables, but rarely to more than five, since they're extremely labor-intensive. After planting, weeding, and keeping the crop pest-free all season without chemicals, the final step of picking often begins before dawn. Some farmers have to travel an hour or more to the packing house. In high season they may make three or more trips a week. The largest grower of the group, with fifteen acres in production, last year delivered 200 boxes of peppers and 400 of tomatoes in a single day. Twenty-three crops are now sold under the Appalachian Harvest label, including melons, cucumbers, eggplants, squash, peas, lettuces, and many varieties of tomatoes and peppers.

The packing house manager labels each box as it arrives so the grower's identity will follow the vegetables through washing, grading, and packaging, all the way to their point of wholesale purchase. Farmers are paid after the supermarket issues its check; Appalachian Harvest takes a 25 percent commission, revenue that helps pay for organic training, packing expenses, and organic certification. Cooperating farmers can sell their produce under the umbrella of a group certification, saving them hundreds of dollars in fees and complex bookkeeping, but they still would need individual certification to sell anywhere other than through the Appalachian Harvest label (e.g., a farmers' market). The project's sales have increased dramatically, gaining a few more committed growers each year, even though farmers are notoriously cautious. Many are still on the fence at this point, watching their neighbors to see whether this enormous commitment to new methods will be salvation or disaster. The term "high-value crop" is relative to a dirt-cheap commodity grain like corn; in season, even high quality organic tomatoes will bring the farmer only about 50 to 75 cents per pound. (The lower end, for conventional, is 18 cents.) But that can translate into a cautious living. Participants find the project compelling for many reasons. After learning to grow vegetables organically, many families have been motivated to make their entire farms organic, including hay fields.

The Appalachian Harvest collective pays a full-time marketer named Robin who spends much of her life on the phone, in her vehicle, or pounding the grocery-store pavements, arranging every sale with the supermarkets, one vegetable and one week at a time. As a farmer herself, she knows the stakes. Also on the payroll here are the manager and summer workers who transform truckloads of field-picked vegetables into the clam-shelled or cellophane-wrapped items that ultimately reach the supermarket after produce has been washed and sorted for size and ripeness.

On a midsummer day in the packing house, vegetables roll through the processing line in a quantity that makes the work in my own kitchen look small indeed. Tomatoes bounce down a sorting conveyor, several bushels per minute, dropping through different-sized holes in a vibrating belt. Workers on both sides of the line collect them, check for flaws and ripeness, and package the tomatoes as quickly as their hands can move, finally pressing on the "certified organic" sticker. Watching the operation, I kept thinking of people I know who can hardly even stand to hear that word, because of how *organic* is personified for them. "I'm always afraid I'm going to get the Mr. Natural lecture," one friend confessed to me. "You know, from the slow-moving person with ugly hair, doing back-and-leg stretches while they talk to you . . ." I laughed because, earnest though I am about food, I know this guy too: dreadlocked, Birkenstocked, standing at the checkout with his bottle of Intestinal-Joy Brand wheatgrass juice, edging closer to peer in my cart, reeking faintly of garlic and a keenness to save me from some food-karma error.

For the record, this is what Appalachian Harvest organics look like at the source: Red Wing work boots, barbershop haircuts, Levi's with a little mud on the cuffs, men and women who probably go to church on Sunday but keep their religion to themselves as they bring a day's work to this packing house inside a former tobacco barn. If sanctimony is an additive in their product, it gets added elsewhere.

The tomato room offered a 56-degree respite from the July swelter, but it was all business in there too: full boxes piled on pallets, in columns nearly reaching the ceiling. The stacks on one end of the room were waiting to be processed, while at the other they waited to be trucked out to nearby groceries. Just enough space remained in the center for workers to maneuver, carting out pallets for grading, sorting, and then slapping one of those tedious stickers on every one of the thousands of individual tomatoes that pass through here each day—along with every pepper, cabbage, cucumber, and melon. That's how the cashier ultimately knows which produce is organic.

Supermarkets only accept properly packaged, coded, and labeled produce that conforms to certain standards of color, size, and shape. Melons can have no stem attached, cucumbers must be no less than six inches long, no more than eight. Crooked eggplants need not apply. Every crop yields a significant proportion of perfectly edible but small or oddly shaped vegetables that are "trash" by market standards.

It takes as much work to grow a crooked vegetable as a straight one, and the nutritional properties are identical. Workers at the packing house were as distressed as the farmers to see boxes of these rejects piling up into mountains of wasted food. Poverty and hunger are not abstractions in our part of the world; throwing away good food makes no sense. With the help of several church and social justice groups, Appalachian Harvest arranged to deliver "factory second" vegetables all summer to low-income families. Fresh organic produce entered some of their diets for the first time.

⁂

I grew up among farmers. In my school system we were all born to our rank, as inescapably as Hindus, the castes being only two: "farm" and "town." Though my father worked in town, we did not live there, and so by the numinous but unyielding rules of high school, I was "farmer." It might seem astonishing that a rural-urban distinction like this could be made in a county that boasted, in its entirety, exactly two stoplights, one hardware store, no beer joints (the county was dry), and fewer residents than an average Caribbean cruise ship. After I went away to school, I remained in more or less constant marvel over the fact that my so-called small liberal arts college, with an enrollment of about 2,000, was 25 percent larger than my hometown.

And yet, even in a community as rural as that, we still had our self-identified bourgeoisie, categorically distinguished from our rustics. We of the latter tribe could be identified by our shoes (sometimes muddy, if we had to cover rough country to get to the school bus), our clothes (less frequently updated), or just the bare fact of a Rural Free Delivery mailing address. I spent my childhood in awe of the storybook addresses of some of my classmates, like "14 Locust Street." In retrospect I'm unsure of how fact-based the distinction really was: most of us "farm" kids were well-scrubbed and occasionally even stylish. Nevertheless, the line of apartheid was unimpeachably drawn. Little socializing across this line was allowed except during special events forced on us by adults, such as the French Club Dinner, and mixed-caste dating was unthinkable except to the tragic romantics.

Sustaining the Unsustainable

Doesn't the Federal Farm Bill help out all these poor farmers?

No. It used to, but ever since its inception just after the Depression, the Federal Farm Bill has slowly been altered by agribusiness lobbyists. It is now largely corporate welfare. The formula for subsidies is based on crop type and volume: from 1995 to 2003, three-quarters of all disbursements went to the top-grossing 10 percent of growers. In 1999, over 70 percent of subsidies went for just two commodity crops: corn and soybeans. These supports promote industrial-scale production, not small diversified farms, and in fact create an environment of competition in which subsidized commodity producers get help crowding the little guys out of business. It is this, rather than any improved efficiency or productiveness, that has allowed corporations to take over farming in the United States, leaving fewer than a third of our farms still run by families.

But those family-owned farms are the ones more likely to use sustainable techniques, protect the surrounding environment, maintain green spaces, use crop rotations and management for pest and weed controls, and apply fewer

Why should this have been? How did the leafy, sidewalked blocks behind the newspaper office confer on their residents a different sense of self than did the homes couched among cow pastures and tobacco fields? The townie shine would have dimmed quickly (I now realize) if the merchants' confident offspring were catapulted suddenly into Philadelphia or Louisville. "Urban" is relative. But the bottom line is that it matters. The antipathy in our culture between the urban and nonurban is so durable, it has its own vocabulary: (A) city slicker, tenderfoot; (B) hick, redneck, hayseed, bumpkin, rube, yokel, clodhopper, hoecake, hillbilly, Dogpatch, Daisy Mae, farmer's daughter, from the provinces, something out of *Deliverance.* Maybe you see where I'm going with this. The list is lopsided. I don't think there's much doubt, on either side, as to which class is winning the culture wars.

Most rural people of my acquaintance would not gladly give up their status. Like other minorities, we've managed to turn several of the aforementioned slurs into celebrated cultural identifiers (for use by insiders only). In my own life I've had ample opportunity to reinvent myself as a city person—to *pass,* as it were—but I've remained tacitly rural-identified

---

chemicals. In other words, they're doing exactly what 80 percent of U.S. consumers say we would prefer to support, while our tax dollars do the opposite.

Because of significant protest about this lack of support, Congress included a tiny allotment for local foods in the most recent (2002) Farm Bill: some support for farmers' markets, community food projects, and local foods in schools. But the total of all these programs combined is less than one-half of one percent of the Farm Bill budget, and none of it is for food itself, only the advertising and administration of these programs. Consumers who care about food, health, and the supply of cheap calories drowning our school lunch programs, for example, might want to let their representatives know we're looking for a dramatically restructured Farm Bill. Until then, support for local and sustainable agriculture will have to come directly from motivated customers.

For more information visit www.farmaid.org.

STEVEN L. HOPP

in my psyche, even while living in some of the world's major cities. It's probably this dual citizenship that has sensitized me to my nation's urban-rural antipathy, and how it affects people in both camps. Rural concerns are less covered by the mainstream media, and often considered intrinsically comic. Corruption in city governments is reported as grim news everywhere; from small towns (or Tennessee) it is fodder for talk-show jokes. Thomas Hardy wrote about the sort of people who milked cows, but writers who do so in the modern era will be dismissed as marginal. The policy of our nation is made in cities, controlled largely by urban voters who aren't well informed about the changes on the face of our land, and the men and women who work it.

Those changes can be mapped on worry lines: as the years have gone by, as farms have gone out of business, America has given an ever-smaller cut of each food dollar (now less than 19 percent) to its farmers. The psychic divide between rural and urban people is surely a part of the problem. "Eaters must understand," Wendell Berry writes, "that eating takes place inescapably in the world, that it is inescapably an agricultural act, and that how we eat determines, to a considerable extent, how the world is used." Eaters *must,* he claims, but it sure looks like most eaters *don't.* If they did, how would we frame the sentence suggested by today's food-buying habits, directed toward today's farmers? "Let them eat dirt" is hardly overstating it. The urban U.S. middle class appears more specifically concerned about exploited Asian factory workers.

Symptomatic of this rural-urban identity crisis is our eager embrace of a recently imposed divide: the Red States and the Blue States. That color map comes to us with the suggestion that both coasts are populated by educated civil libertarians, while the vast middle and south are crisscrossed with the studded tracks of ATVs leaving a trail of flying beer cans and rebel yells. Okay, I'm exaggerating a little. But I certainly sense a bit of that when urban friends ask me how I can stand living here, "*so far from everything*?" (When I hear this question over the phone, I'm usually looking out the window at a forest, a running creek, and a vegetable garden, thinking: Define *everything.*) Otherwise sensitive coastal-dwelling folk may refer to the whole chunk of our continent lying between the Cascades and the Hudson River as "the Interior." I gather this is now a common designation. It's hard for me to see the usefulness of lumping Minneapolis, Atlanta, my little hometown in Kentucky, Yellowstone Park, and so forth, into a single category that does not include New York and California. "Going into the Interior" sounds like an endeavor that might require machetes to hack through the tangled vines.

In fact, the politics of rural regions are no more predictable than those in cities. "Conservative" is a reasonable position for a farmer who can lose home and livelihood all in one year by taking a risk on a new crop. But that's *conservative* as in, "eager to conserve what we have, reluctant to change the rules overnight," and unrelated to how the term is currently (often incomprehensibly) applied in party politics. The farm county where I grew up had so few Republicans, they all registered Democrat so they could vote in the only local primary. My earliest understanding of radical, class-conscious politics came from miners' strikes in one of the most rural parts of my state, and of our nation.

The only useful generalization I'd hazard about rural politics is that they tend to break on the line of "insider" vs. "outsider." When my country neighbors sit down with a new social group, the first question they ask one another is not "What do you do?" but rather, "Who are your people?" Commonly we will spend more than the first ten minutes of a new acquaintance tracing how our families might be related. If not by blood, then by marriage. Failing that, by identifying someone significant we have known in common. Only after this ritual of familial placing does the conversation comfortably move on to other subjects. I am blessed with an ancestor who was the physician in this county from about 1910 into the 1940s. From older people I'll often hear of some memorably dire birth or farm accident to which my great-uncle was called; lucky for me he was skilled and Hippocratic. But even a criminal ancestor will get you insider status, among the forgiving. Not so lucky are those who move here with no identifiable family ties. Such a dark horse is likely to remain "the new fellow" for the rest of his natural life, even if he arrived in his prime and lives to be a hundred.

The country tradition of mistrusting outsiders may be unfairly applied, but it's not hard to understand. For much of U.S. history, rural regions have been treated essentially as colonial property of the cities. The carpetbaggers of the reconstruction era were not the first or the last opportunists to capitalize on an extractive economy. When urban-headquartered companies come to the country with a big plan—whether their game is coal, timber, or industrial agriculture—the plan is to take out the good stuff, ship it to the population centers, make a fortune, and leave behind a mess.

Given this history, one might expect the so-called Red States to vote consistently for candidates supporting working-class values. In fact, our nation in almost every region is divided in a near dead heat between two parties that apparently don't distinguish themselves clearly along class lines. If every state were visually represented with the exact blend of red and blue it earned in recent elections, we'd have ourselves a big purple country. The tidy divide is a media just-so story.

Our uneasy relationship between heartland and coasts, farm and factory, country and town, is certainly real. But it is both more rudimentary and more subtle than most political analysts make it out to be. It's about loyalties, perceived communities, and the things each side understands to be important because of the ground, literally, upon which we stand. Wendell Berry summed it up much better than "blue and red" in one line of dialogue from his novel *Jayber Crow,* which is peopled by farmers struggling to survive on what the modern, mostly urban market will pay for food. After watching nearly all the farms in the county go bankrupt, one of these men comments: "I've wished sometimes that the sons of bitches would starve. And now I'm getting afraid they actually will."

⚘

In high summer, about the time I was seeing red in my kitchen, the same thing was happening to some of our county's tomato farmers. They had learned organic methods, put away the chemicals, and done everything right to grow a product consumers claimed to want. They'd waited the three years for certification. They'd watered, weeded, and picked, they'd sorted the round from the misshapen, producing the perfect organic tomatoes ordered by grocery chains. And then suddenly, when the farmers were finally bringing in these tomatoes by the truckload and hoping for a decent payout, some grocery buyers backtracked. "Not this week," one store offered without warning, and then another. Not the next week either, nor the next. A tomato is not a thing that can be put on hold. Mountains of ripe fruits piled up behind the packing house and turned to orange sludge, swarming with clouds of fruit flies.

These tomatoes were perfect, and buyers were hungry. Agreements had been made. But pallets of organic tomatoes from California had begun coming in just a few dollars cheaper. It's hard to believe, given the amount of truck fuel involved, but transportation is tax-deductible for the corporations, so we taxpayers paid for that shipping. The California growers only needed the economics of scale on their side, a cheap army of pickers, and customers who would reliably opt for the lower price.

As simply as that, a year of planning and family labor turned to red mush.

Our growers had been warned that this could happen—market buyers generally don't sign a binding contract. So the farmers took a risk, and took a loss. Some of them will try again next year, though they will likely hedge their bets with Delicata squash and peas as well. Courage, practicality, and making the best of a bad situation are much of what farming is about. Before the tomatoes all rotted away, Appalachian Harvest found a way to donate and distribute the enormous excess of unpurchased produce to needy families. The poor of our county were rich in tomatoes that summer.

"We were glad we could give it away," one of the farmers told me. "We like to be generous and help others, that's fine, that's who we are. But a lot of us are barely making ends meet, ourselves. It seems like it's always the people that have the least who end up giving the most. Why is that?"

In Charlottesville, Asheville, Roanoke, and Knoxville, supermarket shoppers had no way of knowing how much heartache and betrayal might be wrapped up in those cellophane two-packs of California tomatoes. Maybe they noticed the other tomatoes were missing this week, those local ones with the "Healthy Farms, Close to Home" label. Or maybe they just saw "organic tomatoes," picked them up, and dropped them into their carts on top of the cereal boxes and paper towels. *Eaters must understand, how we eat determines how the world is used.*

They will or they won't. And the happy grocery store music plays on.

Canning Season

BY CAMILLE

When I was a kid, summer was as long as a lifetime. A month could pass without me ever knowing what day of the week it was. Time seemed to stretch into one gigantic, lazy day of blackberry picking and crawdad hunting. My friends and I would pretty much spend our lives together, migrating back and forth between the town swimming pool and the woods, where we would pretend to be orphans left to our own devices in the wilderness. School was not on our minds. Our world was green grass, sunshine, and imagination.

Then August would roll around: a tragedy every time. "Already? How can this be?" I would ask, shattered by the terrible truth that I needed a three-ring binder and some #2 pencils. It's not that school was a bad thing. Summer was just so much better.

August is rarely announced to kids by a calendar. For some of my friends it was the shiny floors and fluorescent lights of the department stores with their back-to-school sales that brought the message. For me it was the bubbling canning bath and the smell of tomatoes. In my family the end of summer means the drone of our food-dehydrator is background music, and you can't open the fridge without huge lumpy bags of produce falling out and clobbering your feet. Every spare half-hour goes into cutting up something to be preserved: the beans and corn to be blanched and frozen, the cucumbers sliced and pickled, the squash frozen or dehydrated or pawned off on a friend. And then there are the tomatoes. Pounds of them roll down from the garden each day, staining every one of our kitchen towels with their crimson juices. We slice little ones by the hundred and lay them out on the stackable trays of our food-drier. We can the medium-sized ones, listening afterward for each "ping" that tells us the jar lid has properly sealed. The rest go into big, bubbling pots of tomato sauce.

I'm sure this sounds like a hassle and mess to those who have never done it. But for us it's an important part of summer. Not only because the outcome is great meals for the rest of the year, but because the process is our way of saying good-bye to the sunshine and pace of summer, and reflecting on what the season gave us. August's busy kitchen is our transition from the long, open-ended hours of summer outdoor work to the stricter routines of school and work in the fall. I like to think of it as an end-of-summer meditation.

American culture doesn't allow much room for slow reflection. I watch the working people who are supposed to be my role models getting pushed to go, go, go and take as little vacation time as possible. And then, often, vacations are full of endless activity too, so you might come back from your "break" feeling exhausted. Canning tomato sauce isn't exactly a week at the spa, but it definitely forces a pause in the multitasking whirl of everyday life. It's a "slow down and do one thing at a time" process: now chop vegetables, now stir them until the sauce thickens, now sterilize the jars, make sure each ring is tight. If you're going to do anything else at the same time, it had better just be listening to your own thoughts. Anything else could cause you to blow the entire batch. Canning always puts me in a kind of trance. I reach a point where stirring the bubbling sauce is the world's only task, and I could do it forever. Whether you prefer to sit on a rock in a peaceful place, or take a wooden spoon to a simmering pot, it does the body good to quiet down and tune in.

The basic canning process is as simple as this: (1) tomatoes are dropped into boiling water and peeled, or else cooked down with other ingredients into sauce; (2) they are poured into sterilized mason jars (we take them straight from the dishwasher) and capped with two-part, screw-on lids; and (3) the filled jars are boiled for the number of minutes specified by the recipe, in a big pot of water. We use an enamelware canning kettle that handles ten quarts at a time.

The following recipes are some personal favorites for storing our bounty of summer produce all year long. The family secret in our tomato sauce (which obviously won't be, now) is cinnamon and nutmeg, usually thought of as dessert spices but used in savory tomato dishes in Greek and some Middle Eastern cuisines. The three-sauce recipe is adapted from *The*

Busy Person's Guide to Preserving Food by Janet Chadwick. Our green-bean Holy Mole was inspired by *Recipes from a Kitchen Garden* by Renee Shepherd and Fran Raboff—a helpful book for preparing meals based on fresh garden produce. The recipes are simple but very creative.

❋ FRIJOLE-MOLE

½ pound trimmed green beans

Steam until tender.

1 coarsely chopped onion

1 tablespoon olive oil

Sauté onions over medium heat until they become slightly transparent.

3 hard-boiled eggs

2 cups fresh basil leaves

1 tablespoon lemon juice (optional)

Combine beans, cooked onions, eggs, basil, and lemon juice in food processor and blend into a coarse puree.

Mayonnaise or yogurt

Salt and pepper

Remove puree to a bowl and combine with enough mayonnaise or yogurt to hold mixture together. Add salt and pepper to taste. This spread is fantastic served on crusty bread, crackers, or rice cakes.

❋ FAMILY SECRET TOMATO SAUCE

The point of this recipe is to make a large amount at one time, when tomatoes are in season. *If you're canning it, stick closely to the recipe;* adding additional fresh vegetables will change the pH so it's unsafe for water-bath canning. If you're freezing it, then it's fine to throw in peppers, mushrooms, fresh garlic, whatever you want. This recipe makes 6–7 quarts—you can use a combination of pint and quart canning jars or freezer boxes.

10 quarts tomato puree (about 30 pounds tomatoes)
4 large onions, chopped
1 cup dried basil
½ cup honey
4 tablespoons dried oregano
3 tablespoons salt
2 tablespoons ground dried lemon peel
2 tablespoons thyme
2 tablespoons garlic powder (or more, to taste)
2 tablespoons dried parsley
2 teaspoons pepper
2 teaspoons cinnamon
½ teaspoon nutmeg

Soften onions in a heavy 3-gallon kettle—add a small amount of water if necessary but *no oil if you are canning (very important!).* Add pureed tomatoes and all seasonings, bring to a boil, and simmer on low heat for two to three hours until sauce has thickened to your liking. Stir frequently, especially toward the end, to avoid burning. Meanwhile, heat water in canner bath, sterilize jars in boiling water or dishwasher, and pour boiling water over jar lids.

Bottled lemon juice or citric acid—NOT optional!

Add 2 tablespoons of lemon juice OR ½ teaspoon citric acid to each quart jar (half that much to pint jars). This ensures that the sauce will be safely acidic. When the sauce is ready, ladle it into the jars, leaving ½-inch headspace. Cap jars, lower gently into canner and boil for 35 minutes. Remove, cool, check all seals, label, and store for winter.

❋ RELISH, SAUCE, AND CHUTNEY—ALL IN ONE DAY

If you don't have a garden, you can stock up on tomatoes, peaches, apples, and onions at the end of summer, when your farmers' market will have these at the year's best quality and price. Then, schedule a whole afternoon and a friend for this interesting project that gives you three different, delicious products to eat all winter.

Canning jars and lids: 14 pint jars, 7 half-pint jars

Start with a very large, heavy kettle. You will be adding different ingredients and canning different sauces as you go.

4 quarts tomato puree
24 apples
7 cups chopped onions
2 quarts cider vinegar
6 cups sugar
⅔ cup salt
3 teaspoons ground cloves
3 teaspoons cinnamon
2 teaspoons paprika
2 teaspoons mustard

Puree tomatoes; core and coarsely chop apples; coarsely chop onions. Combine in large pot along with the vinegar, sugar, and seasonings. Bring to a boil and simmer for about 2 hours or until thick. Meanwhile, preheat water in a canner bath and sterilize jars and lids (in boiling water or dishwasher) and keep them hot until use. Fill **7 pint jars** with some of the thickened **Barbecue Relish,** leaving ½ inch headspace in each jar. Put filled jars in canner with lids screwed on tightly and boil for ten minutes. Remove and cool.

2 quarts sliced peaches
6 cups sugar
½ cup water
2 teaspoons garlic powder
1 teaspoon Tabasco sauce

In a separate pan, cook peaches and water for 10 minutes, until soft. Add sugar and bring slowly to boil, stirring until sugar dissolves. Boil until thick (15 minutes), stirring to prevent scorching.

Add peach mixture to the remaining tomato mixture in the kettle and bring back up to a boil to make **Sweet and Sour Sauce.** Fill **7 pint jars,** leaving ½ inch headspace; boil in canner for 10 minutes. Remove and cool.

1 cup raisins
1 cup walnuts

Add these to the kettle, mix well, and bring it back to a boil to make **Chutney.** Fill **7 half-pint jars,** leaving ½ inch headspace. Boil in canner for 10 minutes. Remove.

As all the jars cool, make sure the jar lids pop their seals by creating a vacuum as contents cool. You'll hear them go "ping." To double-check, after they're entirely cooled, push down on each lid's center—it should feel firmly sucked down, not loose. (If a jar didn't seal, refrigerate and use the contents soon.) The ring portion of the lid can be removed before storing; when processed properly, the dome lids will stay securely sealed until you open the jar with a can opener.

Label each product before you forget what's what, and share with the friend who helped. The Barbecue Relish is great on broiled or grilled fish or chicken. The Sweet and Sour Sauce gives an Asian flavor to rice dishes. Chutney can perk up anything.

Download these and all other *Animal, Vegetable, Miracle* recipes at www.AnimalVegetableMiracle.com

TOMATO SEASON MEAL PLAN

Sunday ~ Grilled chicken with tomato salad and corn on the cob
Monday ~ Homemade gazpacho with fresh bread and cream cheese
Tuesday ~ Roasted tomato-eggplant ratatouille with rice (or bread) and grated Parmesan
Wednesday ~ Grilled fish or lamb, served with crusty bread, chutney, and green-bean paté

Thursday ~ Asian vegetable stir fry with soba noodles (or rice) and sweet-and-sour sauce

Friday ~ Pizza with sliced tomatoes, fresh basil, mozzarella, and a drizzle of olive oil

Saturday ~ Pasta with fresh homemade tomato sauce and meatballs

14 · YOU CAN'T RUN AWAY ON HARVEST DAY

September

The Saturday of Labor Day weekend dawned with a sweet, translucent bite, like a Golden Delicious apple. I always seem to harbor a childlike hope through the berry-stained months of June and July that summer will be for keeps. But then a day comes in early fall to remind me why it should end, after all. In September the quality of daylight shifts toward flirtation. The green berries on the spicebush shrubs along our lane begin to blink red, first one and then another, like faltering but resolute holiday lights. The woods fill with the restless singing of migrant birds warming up to the proposition of flying south. The cool air makes us restless too: jeans and sweater weather, perfect for a hike. Steven and I rose early that morning, looked out the window, looked at each other, and started in on the time-honored marital grumble: Was this *your* idea?

We weren't going on a hike today. Nor would we have the postsummer Saturday luxury of sitting on the porch with a cup of coffee and watching the farm wake up. On the docket instead was a hard day of work we could not postpone. The previous morning we'd sequestered half a dozen roosters and as many tom turkeys in a room of the barn we call "death row." We hold poultry there, clean and comfortable with water but no food, for a twenty-four-hour fast prior to harvest. It makes the processing cleaner and seems to calm the animals also. I could tell you it gives them time to get their emotional affairs in order, if that helps. But they have limited emotional affairs, and no idea what's coming.

We had a lot more of both. Our plan for this gorgeous day was the removal of some of our animals from the world of the living into the realm of food. At five months of age our roosters had put on a good harvest weight, and had lately opened rounds of cockfighting, venting their rising hormonal angst against any moving target, including us. When a rooster flies up at you with his spurs, he leaves marks. Lily now had to arm herself with a length of pipe in order to gather the eggs. Our barnyard wasn't big enough for this much machismo. We would certainly take no pleasure in the chore, but it was high time for the testosterone-reduction program. We sighed at the lovely weather and pulled out our old, bloody sneakers for harvest day.

⚘

There was probably a time when I thought it euphemistic to speak of "harvesting" animals. Now I don't. We calculate "months to harvest" when planning for the right time to start poultry. We invite friends to "harvest parties," whether we'll be gleaning vegetable or animal. A harvest implies planning, respect, and effort. With animals, both the planning and physical effort are often greater, and respect for the enterprise is substantially more complex. It's a lot less fun than spending an autumn day picking apples off trees, but it's a similar operation on principle and the same word.

Killing is a culturally loaded term, for most of us inextricably tied up with some version of a command that begins, "Thou shalt not." Every faith has it. And for all but perhaps the Jainists of India, that command is absolutely conditional. We know it does not refer to mosquitoes. Who among us has never killed living creatures on purpose? When a child is sick with an infection we rush for the medicine spoon, committing an eager and purposeful streptococcus massacre. We sprinkle boric acid or grab a spray can to rid our kitchens of cockroaches. What we mean by "killing" is to take a life cruelly, as in murder—or else more accidentally, as in "Oops, looks like I killed my African violet." Though the results are incomparable, what these different "killings" have in common is needless waste and some presumed measure of regret.

Most of us, if we know even a little about where our food comes from, understand that every bite put into our mouths since infancy (barring the odd rock or marble) was formerly alive. The blunt biological truth is that we animals can only remain alive by eating other life. Plants are inherently more blameless, having been born with the talent of whipping up their own food, peacefully and without noise, out of sunshine, water, and the odd mineral ingredient sucked up through their toes. Strangely enough, it's the animals to which we've assigned some rights, while the saintly plants we maim and behead with moral impunity. Who thinks to beg forgiveness while mowing the lawn?

The moral rules of destroying our fellow biota get even more tangled, the deeper we go. If we draw the okay-to-kill line between "animal" and "plant," and thus exclude meat, fowl, and fish from our diet on moral grounds, we still must live with the fact that every sack of flour and every soybean-based block of tofu came from a field where countless winged and furry lives were extinguished in the plowing, cultivating, and harvest. An estimated 67 million birds die each year from pesticide exposure on U.S. farms. Butterflies, too, are universally killed on contact in larval form by the genetically modified pollen contained in most U.S. corn. Foxes, rabbits, and bobolinks are starved out of their homes or dismembered by the sickle mower. Insects are "controlled" even by organic pesticides; earthworms are cut in half by the plow. Contrary to lore, they won't grow into two; both halves die.

To believe we can live without taking life is delusional. Humans may only cultivate nonviolence in our diets by degree. I've heard a Buddhist monk suggest the *number* of food-caused deaths is minimized in steak dinners, which share one death over many meals, whereas the equation is reversed for a bowl of clams. Others of us have lost heart for eating any steak dinner that's been shoved through the assembly line of feedlot life—however broadly we might share that responsibility. I take my gospel from Wendell Berry, who writes in *What Are People For,* "I dislike the thought that some animal has been made miserable in order to feed me.

If I am going to eat meat, I want it to be from an animal that has lived a pleasant, uncrowded life outdoors, on bountiful pasture, with good water nearby and trees for shade. And I am getting almost as fussy about food plants."

I find myself fundamentally allied with a vegetarian position in every way except one: however selectively, I eat meat. I'm unimpressed by arguments that condemn animal harvest while ignoring, wholesale, the animal killing that underwrites vegetal foods. Uncountable deaths by pesticide and habitat removal—the beetles and bunnies that die collaterally for our bread and veggie-burgers—are lives plumb wasted. Animal harvest is at least not gratuitous, as part of a plan involving labor and recompense. We raise these creatures for a reason. Such premeditation may be presumed unkind, but without it our gentle domestic beasts in their picturesque shapes, colors, and finely tuned purposes would never have had the distinction of existing. To envision a vegan version of civilization, start by erasing from all time the Three Little Pigs, the boy who cried wolf, *Charlotte's Web,* the golden calf, *Tess of the d'Urbervilles.* Next, erase civilization, brought to you by the people who learned to domesticate animals. Finally, rewrite our evolutionary history, since *Homo sapiens* became the species we are by means of regular binges of carnivory.

Most confounding of all, in the vegan revision, are the chapters addressing the future. If farm animals have civil rights, what aspect of their bondage to humans shall they overcome? Most wouldn't last two days without it. Recently while I was cooking eggs, my kids sat at the kitchen table entertaining me with readings from a magazine profile of a famous, rather young vegan movie star. Her dream was to create a safe-haven ranch where the cows and chickens could live free, happy lives and die natural deaths. "Wait till those cows start bawling to be milked," I warned. Having nursed and weaned my own young, I can tell you there is no pain to compare with an overfilled udder. We wondered what the starlet might do for those bursting Jerseys, not to mention the eggs the chickens would keep dropping everywhere. What a life's work for that poor gal: traipsing about the farm in her strappy heels, weaving among the cow flops, bending gracefully to pick up eggs and stick them in an incubator where they would maddeningly *hatch,* and grow up bent on laying *more* eggs. It's dirty work, trying to save an endless chain of uneaten lives. Realistically, my kids observed, she'd hire somebody.

Forgive us. We know she meant well, and as fantasies of the super-rich go, it's more inspired than most. It's just the high-mindedness that rankles; when moral superiority combines with billowing ignorance, they fill up a hot-air balloon that's awfully hard not to poke. The farm-liberation fantasy simply reflects a modern cultural confusion about farm animals. They're human property, not just legally but biologically. Over the millennia of our clever history, we created from wild progenitors whole new classes of beasts whose sole purpose was to feed us. If turned loose in the wild, they would haplessly starve, succumb to predation, and destroy the habitats and lives of most or all natural things. If housed at the public expense they would pose a more immense civic burden than our public schools and prisons combined. No thoughtful person really wants those things to happen. But living at a remove from the actual workings of a farm, most humans no longer learn appropriate modes of thinking about animal harvest. Knowing that our family raises meat animals, many friends have told us—not judgmentally, just confessionally—"I don't think I could kill an animal myself." I find myself explaining: It's not what you think. It's nothing like putting down your dog.

Most nonfarmers are intimate with animal life in only three categories: people; pets (i.e., junior people); and wildlife (as seen on nature shows, presumed beautiful and rare). Purposely beheading any of the above is unthinkable, for obvious reasons. No other categories present themselves at close range for consideration. So I understand why it's hard to think about harvest, a categorical act that includes cutting the heads off living lettuces, extended to crops that blink their beady eyes. On our farm we don't especially enjoy processing our animals, but we do value it, as an important ritual for ourselves and any friends adventurous enough to come and help, because of what we learn from it. We reconnect with the purpose for which these animals were bred. We dispense with all delusions about who put the *live* in livestock, and who must take it away.

A friend from whom we buy pasture-grazed lamb and poultry has concurred with us on this point. Kirsty Zahnke grew up in the U.K., and observes that American attitudes toward life and death probably add to the misgivings. "People in this country do everything to cheat death, it seems. Instead of being happy with each moment, they worry so much about what comes next. I think this gets transposed to animals—the preoccupation with 'taking a life.' My animals have all had a good life, with death as its natural end. It's not without thought and gratitude that I slaughter my animals, it is a hard thing to do. It's taken me time to be able to eat my own lambs that I had played with. But I always think of Kahlil Gibran's words:

> *" 'When you kill a beast, say to him in your heart:*
> *By the same power that slays you, I too am slain, and I too shall be consumed.*
> *For the law that delivers you into my hand shall deliver me into a mightier hand.*
> *Your blood and my blood is naught but the sap that feeds the tree of heaven.' "*

Kirsty works with a local environmental organization and frequently hosts its out-of-town volunteers, who camp at her farm while working in the area. Many of these activists had not eaten meat for many years before arriving on the Zahnkes' meat farm—a formula not for disaster, she notes, but for education. "If one gets to know the mantras of the farm owners, it can change one's viewpoint. I would venture to say that seventy-five percent of the vegans and vegetarians who stayed at least a week here began to eat our meat or animal products, simply because they see what I am doing as right—for the animals, for the environment, for humans."

I respect every diner who makes morally motivated choices about consumption. And I stand with nonviolence, as one of those extremist moms who doesn't let kids at her house pretend to shoot each other, *ever,* or make any game out of human murder. But I've come to different conclusions about livestock. The ve-vangelical pamphlets showing jam-packed chickens and sick downer-cows usually declare, as their first principle, that all meat is factory-farmed. That is false, and an affront to those of us who work to raise animals humanely, or who support such practices with our buying power. I don't want to cause any creature misery, so I won't knowingly eat anything that has stood belly deep in its own poop wishing it was dead until *bam,* one day it was. (In restaurants I go for the fish, or the vegetarian option.)

But meat, poultry, and eggs from animals raised on open pasture are the traditional winter fare of my grandparents, and they serve us well here in the months when it would cost a lot of fossil fuels to keep us in tofu. Should I overlook the suffering of victims of hurricanes, famines, and wars brought on this world by profligate fuel consumption? Bananas that cost a rain forest, refrigerator-trucked soy milk, and prewashed spinach shipped two thousand miles in plastic containers do not seem cruelty-free, in this context. A hundred different paths may lighten the world's load of suffering. Giving up meat is one path; giving up bananas is another. The more we know about our food system, the more we are called into complex choices. It seems facile to declare one single forbidden fruit, when humans live under so many different kinds of trees.

To breed fewer meat animals in the future is possible; phasing out those types destined for confinement lots is a plan I'm assisting myself, by raising heirloom breeds. Most humans could well consume more vegetable foods, and less meat. But globally speaking, the vegetarian option is a luxury. The oft-cited energetic argument for vegetarianism, that it takes ten times as much land to make a pound of meat as a pound of grain, only applies to the kind of land where rain falls abundantly on rich topsoil. Many of the world's poor live in marginal lands that can't support plant-based agriculture. Those not blessed with the fruited plain and amber waves of grain must make do with woody tree pods, tough-leaved shrubs, or sparse grasses. Camels, reindeer, sheep, goats, cattle, and other ruminants are uniquely adapted to transform all those types of indigestible cellulose into edible milk and meat. The fringes of desert, tundra, and marginal grasslands on every continent—coastal Peru, the southwestern United States, the Kalahari, the Gobi, the Australian outback, northern Scandinavia—are inhabited by herders. The Navajo, Mongols, Lapps, Masai, and countless other resourceful tribes would starve without their animals.

Domestic herds can also carry problems into these habitats. Overgrazing has damaged plenty of the world's landscapes, as has clearing rain forests to make way for cattle ranches. But well-managed grazing can actually benefit natural habitats where native grazers exist or formerly existed. Environmental research in North and South American deserts has shown that careful introduction of cattle, sheep, or goats into some grasslands helps return the balance of their native vegetation, especially mesquite trees and their kin, which coevolved for millennia with large grazing mammals (mastodons and camels) that are now extinct. Mesquite seeds germinate best after passing through the stomach of a ruminant. Then the habitat also needs the return of fire, and prairie dog predation on the mesquite seedlings—granted, it's complicated. But grazers do belong.

In northwestern Peru, in the extremely arid, deforested region of Piura, an innovative project is using a four-legged tool for widespread reforestation: goats. This grassless place lost most of its native mesquite forests to human refugees who were pushed out of greener places, settled here, and cut down most of the trees for firewood. Goats can subsist on the seedpods of the remaining mesquites (without damaging the thorny trees) and spread the seeds, depositing them across the land inside neat fertilizer pellets. The goats also provide their keepers with meat and milk, in a place where rainfall is so scarce (zero, in some years), it's impossible to subsist on vegetable crops. The herds forage freely when mesquite beans are in season, and live the rest of the year on pods stored in cement-block granaries. These low-maintenance animals also reproduce themselves free of charge, so the project broadens its reforesting and hunger-relief capacities throughout the region, year by year.

We cranky environmentalists tend to nurture a hunch that humans and our food systems are always dangerous to the earth. But when I visited Piura to study the mesquite-goat project, I could not name any measure by which the project was anything but successful. The "before" scenario involved malnourished families in a desiccated brown landscape. Within a few years after receiving goats, the families still lived in simple mud-and-lath homes, but their villages were shaded by green oases of fast-growing native vegetation. They milked the goats, made cheese, burned mesquite pods for cooking fuel, and looked forward to eating meat several times each month. Small, irrigated gardens of beans and leafy greens provided supplementary nutrition, but in this climate it's animal products that can offer the prospect of ending malnutrition. Each goat-owning family makes an agreement with the donor organizations (Heifer International and the local group ACBIODESA) to give the first female offspring to another family, thus moving their own status from "poor" to "benefactor"—a powerfully important distinction in terms of local decision-making and further stewardship of the land. For the same money, a shipment of donated wheat, rice, or corn would only have maintained the region's widespread poverty through another few months, and deepened its environmental crisis. Between vegetable or animal solutions to that region's problems, my vote goes to the goats.

The mountainous part of the United States where I live, though neither destitute nor desiccated, has its own challenges. The farms here are small and steep. Using diesel tractors to turn the earth every spring (where that's even possible) sends our topsoil downhill into the creeks with every rain, creating many problems at once. One of the region's best options for feeding ourselves and our city neighbors may be pasture-based hoof stock and poultry. Cattle, goats, sheep, turkeys, and chickens all have their own efficient ways of turning steep, grass-covered hillsides into food, while fertilizing the land discreetly with their manure. They do it without drinking a drop of gasoline.

Managed grazing is healthier for most landscapes, in fact, than annual tilling and planting, and far more fuel-efficient. Grass is a solar-powered, infinitely renewable resource. As consumers discover the health benefits of grass-based meat, more farmers may stop plowing land and let animals go to work on it instead. A crucial part of this enterprise involves recovering the heritage cattle, poultry, and other livestock that can fatten on pasture grass. It's news to most people that chickens, turkeys, and pigs can eat foliage at all, since we're used to seeing them captive and fed. Even cattle are doing less and less grass-eating, since twentieth-century breeding programs gave us animals that tolerate (barely) a grain-based diet for weight gain during their final eight months in confinement. For decades, the public has demanded no meat animals but these.

More lately, though, conditions inside CAFOs have been exposed by voices as diverse as talk-show host Oprah Winfrey and *Fast Food Nation* author Eric Schlosser. In an essay titled "Food with a Face," journalist Michael Pollan wrote: "More than any other institution, the American industrial animal farm offers a nightmarish glimpse of what Capitalism can look like in the absence of moral or regulatory constraint. Here, in these places, life itself is redefined—as protein production—and with it, suffering. That venerable word becomes 'stress,' an economic problem in search of a cost-effective solution. . . . The industrialization—and dehumanization—of American animal farming is a relatively new, evitable, and local phenomenon: no other country raises and slaughters its food animals quite as intensively or as brutally as we do." U.S. consumers may take our pick of reasons to be wary of the resulting product: growth hormones, antibiotic-resistant bacteria, unhealthy cholesterol composition, deadly *E. coli* strains, fuel consumption, concentration of manure into toxic waste lagoons, and the turpitude of keeping confined creatures at the limits of their physiological and psychological endurance.

It's that last one that finally ended it for me. Yes, I am a person who raises some animals for the purpose of whacking them into cuts of meat to feed my family. But this work has made me more sympathetic, not less, toward the poor wretches that have to live shoulder-to-shoulder with their brethren waiting for the next meal of stomach-corroding porridge. In '97, when our family gave up meat from CAFOs, that choice was synonymous with becoming a vegetarian. No real alternatives existed. Now they do. Pasture-based chicken and turkey are available in whole food stores and many mainstream supermarkets. Farmers' markets are a likely source for free-range eggs, poultry, beef, lamb, and pork. Farmers who raise animals on pasture have to charge more, of course, than factories that cut every corner on animal soundness. Some consumers will feel they have to buy the cheaper product. Others will eat meat less often and pay the higher price. As demand rises, and more farmers can opt out of the industrial system, the cost structure will shift.

After many meatless years it felt strange to us to break the taboo, but over time our family has come back to carnivory. I like listening to a roasting bird in the oven on a Sunday afternoon, following Julia Child's advice to "regulate the chicken so it makes quiet cooking noises" as its schmaltzy aroma fills the house. When a friend began raising beef cattle entirely on pasture (rather than sending them to a CAFO as six-month-olds, as most cattle farmers do), we were born again to the idea of hamburger. We can go visit his animals if we need to be reassured of the merciful cowness of their lives.

As meat farmers ourselves we are learning as we go, raising heritage breeds: the thrifty antiques that know how to stand in the sunshine, gaze upon a meadow, and munch. (Even mate without help!) We're grateful these old breeds weren't consigned to extinction during the past century, though it nearly did happen. Were it not for these animals that can thrive outdoors, and the healthy farms that maintain them, I would have stuck with tofu-burgers indefinitely. That wasn't a bad life, but we're also enjoying this one.

⁂

Believing in the righteousness of a piece of work, alas, is not what gets it done. On harvest day we pulled on our stained shoes, sharpened our knives, lit a fire under the big kettle, and set ourselves to the whole show: mud, blood, and lots of little feathers. There are some things about a chicken harvest that are irrepressibly funny, and one of them is the feathers: in your hair, on the backs of your hands, dangling behind your left shoe the way toilet paper does in slapstick movies. Feathery little white tags end up stuck all over the chopping block and the butchering table like Post-it notes from the chicken hereafter. Sometimes we get through the awful parts on the strength of black comedy, joking about the feathers or our barn's death row and the "dead roosters walking."

But today was not one of those times. Some friends had come over to help us, including a family that had recently lost their teenage son in a drowning accident. Their surviving younger children, Abby and Eli, were among Lily's closest friends. The kids were understandably solemn and the adults measured all our words under the immense weight of grief as we set to work. Lily and Abby went to get the first rooster from the barn while I laid out the knives and spread plastic sheets over our butchering table on the back patio. The guys stoked a fire under our fifty-gallon kettle, an antique brass instrument Steven and I scored at a farm auction.

Really, We're Not Mad

Cows must have some friends in high places. If a shipment of ground beef somehow gets contaminated with pathogens, our federal government does not have authority to recall the beef, only to request that the company issue a recall. When the voluntary recall is initiated, the federal government does not release information on where the contaminated beef is being sold, considering that information proprietary. Apparently it is more important to protect the cows than the people eating them. Now I need to be careful where I go next, because (for their own protection) there are laws in thirteen states that make it illegal to say anything bad about cows.

One serious disease related to our friends the cows has emerged in the past twenty years: bovine spongiform encephalopathy (BSE), or so-called mad cow disease. Mad cow, and its human variant, Creutzfeldt-Jakob disease, are invariably fatal for both cows and humans. Unfortunately, tracking mad cow is complicated by the fact that it frequently incubates for years in the victim. The disease became infamous during the 1980s outbreak in England, where more than 150 humans died from eating BSE beef, and thousands of cattle were destroyed. A tiny malformed protein called a prion is the BSE culprit. The prions cause other proteins in the victim to rearrange into their unusual shape, and destroy tissue. Prions confine their activities to the nervous system, where they cause death. How do cows contract prions? Apparently from eating other cows. What? Yes, dead cow meat gets mixed into their feed, imposing cannibalism onto their lifestyle. It's a way to get a little more mileage from the byproducts of the slaughterhouse.

An appropriate response would be to stop this, which the British did. They also began testing 100 percent of cows over two years old at slaughter for BSE, and removing all "downer" cows (cows unable to walk on their own) from the food supply. As a result, the U.K. virtually eradicated BSE in two years. Reasonably enough, Japan implemented the same policies.

In the United States, the response has been somewhat different. U.S. policies restrict feeding cow tissue directly to other cows, but still allow cows to be fed to other animals (like chickens) and the waste from the chickens to be fed back to the cows. Since prions aren't destroyed by extreme heat or any known drug, they readily survive this food-chain loop-de-loop. Cow blood (yum) may also be dinner for other cows and calves, and restaurant plate wastes can also be served.

After the first detected case of U.S. mad cow disease, fifty-two countries banned U.S. beef. The USDA then required 2 percent of all the downer cows to be tested, and 1 percent of all cows that were slaughtered. After that, the number of downer cows reported in the United States decreased by 20 percent (did I mention it was voluntary reporting?), and only two more cases of BSE were detected. In May 2006, the USDA decided the threat was so low that only one-tenth of one percent of all slaughtered cows needed to be tested. Jean Halloran, the food policy initiatives director at Consumers Union, responded, "It approaches a policy of don't look, don't find."

How can consumers respond to this? Can we seek out beef tested as BSE-free by the meat packers? No. One company tried to test all its beef, but the USDA declared that illegal (possibly to protect any BSE cows from embarrassment). Would I suggest a beef boycott? Heavens, no; cows are our friends (plus, I believe that would be illegal). But it might be worth remembering this: not a single case of BSE, anywhere, has ever turned up in cattle that were raised and finished on pasture grass or organic feed. As for the other 99 percent of beef in the United States, my recommendation would be to consider the words of Gary Weber, the National Cattlemen's Beef Association head of regulatory affairs: "The consumers we've done focus groups with are comfortable that this is a very rare disease."

For more information visit www.organicconsumers.org/madcow.html.

STEVEN L. HOPP

The girls returned carrying rooster #1 upside down, by the legs. Inversion has the immediate effect of lulling a chicken to sleep, or something near to it. What comes next is quick and final. We set the rooster gently across our big chopping block (a legendary fixture of our backyard, whose bloodstains hold visiting children in thrall), and down comes the ax. All sensation ends with that quick stroke. He must then be held by the legs over a large plastic bucket until all the blood has run out. Farmers who regularly process poultry have more equipment, including banks of "killing cones" or inverted funnels that contain the birds while the processor pierces each neck with a sharp knife, cutting two major arteries and ending brain function. We're not pros, so we have a more rudimentary setup. By lulling and swiftly decapitating my animal, I can make sure my relatively unpracticed handling won't draw out the procedure or cause pain.

What you've heard is true: the rooster will flap his wings hard during this part. If you drop him he'll thrash right across the yard, unpleasantly spewing blood all around, though the body doesn't *run*—it's nothing that well coordinated. His newly detached head silently opens and closes its mouth, down in the bottom of the gut bucket, a world apart from the ruckus. The cause of all these actions is an explosion of massively firing neurons without a brain to supervise them. Most people who claim to be running around like a chicken with its head cut off, really, are not even close. The nearest thing might be the final convulsive seconds of an All-Star wrestling match.

For Rooster #1 it was over, and into the big kettle for a quick scald. After a one-minute immersion in 145-degree water, the muscle tissue releases the feathers so they're easier to pluck. "Easier" is relative—every last feather still has to be pulled, carefully enough to avoid tearing the skin. The downy breast feathers come out by handfuls, while the long wing and tail feathers sometimes must be removed individually with pliers. If we were pros we would have an electric scalder and automatic plucker, a fascinating bucket full of rotating rubber fingers that does the job in no time flat. For future harvests we might borrow a friend's equipment, but for today we had a pulley on a tree limb so we could hoist the scalded carcass to shoulder level, suspending it there from a rope so several of us could pluck at once. Lily, Abby, and Eli pulled neck and breast feathers, making necessary observations such as "Gag, look where his head came off," and "Wonder which one of these tube thingies was his windpipe." Most kids need only about ninety seconds to get from *eeew gross* to solid science. A few weeks later Abby would give an award-winning, fully illustrated 4-H presentation entitled "You Can't Run Away on Harvest Day."

Laura and Becky and I answered the kids' questions, and also talked about Mom things while working on back and wing feathers. (Our husbands were on to the next beheading.) Laura and I compared notes on our teenage daughters—relatively new drivers on the narrow country roads between their jobs, friends, and home—and the worries that come with that territory. I was painfully conscious of Becky's quiet, her ache for a teenage son who never even got to acquire a driver's license. The accident that killed Larry could not have been avoided through any amount of worry. We all cultivate illusions of safety that could fall away in the knife edge of one second.

I wondered how we would get through this afternoon, how *she* would get through months and years of living with impossible loss. I wondered if I'd been tactless, inviting these dear friends to an afternoon of ending lives. And then felt stupid for that thought. People who are grieving walk with death, every waking moment. When the rest of us dread that we'll somehow remind them of death's existence, we are missing their reality. Harvesting turkeys—which this family would soon do on their own farm—was just another kind of work. A rendezvous with death, for them, was waking up each morning without their brother and son.

By early afternoon six roosters had lost their heads, feathers, and viscera, and were chilling on ice. We had six turkeys to go, the hardest piece of our work simply because the animals are larger and heavier. Some of these birds were close to twenty pounds. They would take center stage on our holiday table and those of some of our friends. At least one would be charcuterie—in the garden I had sage, rosemary, garlic, onions, everything we needed for turkey sausage. And the first two roosters we'd harvested would be going on the rotisserie later that afternoon.

We allowed ourselves a break before the challenge of hoisting, plucking, and dressing the turkeys. While Lily and her friends constructed feather crowns and ran for the poultry house to check in with the living, the adults cracked open beers and stretched out in lawn chairs in the September sun. Our conversation turned quickly to the national preoccupation of that autumn: Katrina, the hurricane that had just hit southern Louisiana and Mississippi. We were horrified by the news that was beginning to filter out of that flooded darkness, the children stranded on rooftops, the bereaved and bewildered families slogging through streets waist-deep in water, breaking plate glass windows to get bottles of water. People drowning and dying of thirst at the same time.

It was already clear this would be an epic disaster. New Orleans and countless other towns across southern Louisiana and Mississippi were being evacuated and left for dead. The news cameras had focused solely on urban losses, sending images of flooded streets, people on rooftops, broken storefronts, and the desperate crises of people in the city with no resources for relocating or evacuating. I had not seen one photograph from the countryside—a wrecked golf course was the closest thing to it. I wondered about the farmers whose year of work still lay in the fields, just weeks or days away from harvest, when the flood took it all. I still can't say whether the rural victims of Katrina found their support systems more resilient, or if their hardships simply went unreported.

The disaster reached into the rest of the country with unexpected tentacles. Our town and schools were already taking in people who had lost everything. The office where I'd just sent my passport for renewal was now underwater. Gasoline had passed $3 a gallon, here and elsewhere, leaving our nation in sticker shock. U.S. citizens were making outlandish declarations about staying home. Climate scientists were saying, "If you warm up the globe, you eventually pay for it." Economists were eyeing our budget deficits and predicting collapse, mayhem, infrastructure breakdown. In so many ways, disaster makes us take stock. For me it had inspired powerful cravings about living within our means. I wasn't thinking so much of my household budget or the national one but the *big* budget, the one that involves consuming approximately the same things we produce. Taking a symbolic cue from my presumed-soggy passport, I suddenly felt like sticking very close to home, with a hand on my family's production, even when it wasn't all that easy or fun—like today.

Analysts of current events were mostly looking to blame administrators. Fair enough, but there were also, it seemed, obvious vulnerabilities here—whole populations depending on everyday, long-distance lifelines, supplies of food and water and fuel and everything else that are acutely centralized. That's what we consider normal life. Now nature had written a hugely abnormal question across the bottom of our map. I wondered what our answers might be.

⁂

Our mood stayed solemn until Eli introduced the comedy show of poultry parts. He applied his artistry and grossout-proof ingenuity to raw materials retrieved from the gut bucket. While the rest of us merely labored, Eli acted, directed, and produced. He invented the turkey-foot backscratcher, the inflated turkey-crop balloon. Children—even when they have endured the unthinkable—have a gift for divining the moment when the grown-ups really need to lighten up. We got a little slap-happy egging on the two turkey heads that moved their mouths to Eli's words, starring in a mock TV talk show. As I gutted the last bird of the day, I began thinking twice about what props I was tossing into the gut bucket. I was not sure I wanted to see what an eight-year-old boy could do with twelve feet of intestine.

The good news was that we were nearly done. I encouraged the rest of the adults to go ahead and wash up, I had things in hand. They changed out of the T-shirts that made them look like *Braveheart* extras. The girls persuaded Eli to retire the talking heads and submit to a hosing-down. Our conversation finally relaxed fully into personal news, the trivial gripes and celebrations for which friends count on one another: what was impossible these days at work. How the children were faring with various teachers and 4-H projects. How I felt about having been put on *the list*.

That question referred to a book that had been released that summer, alerting our nation to the dangers of one hundred people who are Destroying America. It was popular for nearly a week and a half, so I'd received a heads-up about my being the seventy-fourth most dangerous person in America. It gave a certain pizzazz to my days, I thought, as I went about canning tomatoes, doing laundry, meeting the school bus, and here and there writing a novel or essay or whatever, knowing full well that kind of thing only leads to trouble. My thrilling new status had no impact on my household position: I still had to wait till the comics were read to get the Sudoku puzzle, and the dog ignored me as usual. Some of my heroes had turned up much higher on the list. Jimmy Carter was number 6.

"When you're seventy-four, you try harder," I now informed my friends, as I reached high up into the turkey's chest cavity from the, um, lower end. I was trying to wedge my fingers between the lungs and ribs to pull out the whole package of viscera in one clean motion. It takes practice, dexterity, and a real flair for menace to disembowel a deceased turkey. "Bond. *James* Bond," a person might say by way of introduction, in many situations of this type. My friends watched me, openly expressing doubts as to my actual dangerousness. They didn't think I even deserved to be number 74.

"Hey," I said, pretty sure I now had the gizzard in hand, "don't distract me. I'm on the job here. Destroying America is not the walk in the park you clearly think it is."

Someone had sent me a copy of this book, presumably to protect me from myself. A couple of people now went into the house to fetch it so they could stage dramatic readings from the back jacket. These friends I've known for years uncovered the secrets they'd never known about me, President Carter, and our ilk: "*These,*" the book warned, "are the cultural elites who look down their snobby noses at 'ordinary' Americans. . . ."

All eyes turned fearfully to me. My "Kentucky NCAA Champions" shirt was by now so bloodstained, you would think I had worn it to a North Carolina game. Also, I had feathers sticking to my hair. I was crouched in something of an inharmonious yoga pose with both my arms up a turkey's hind end, more than elbow deep.

With a sudden sucking sound the viscera let go and I staggered back, trailing intestines. My compatriots laughed very hard. *With* me, not *at* me, I'm sure.

And that was the end of a day's work. I hosed down the butcher shop and changed into more civilized attire (happy to see my wedding ring was still on) while everybody else set the big picnic table on our patio with plates and glasses and all the food in the fridge we'd prepared ahead. The meat on the rotisserie smelled really good, helping to move our party's mindset toward the end stages of the "cooking from scratch" proposition. Steven brushed the chicken skin with our house-specialty sweet-and-sour sauce and we uncorked the wine. At dusk we finally sat down to feast on cold bean salad, sliced tomatoes with basil, blue potato salad, and meat that had met this day's dawn by crowing.

We felt tired to our bones but anointed by life in a durable, companionable way, for at least the present moment. We the living take every step in tandem with death, *naught but the sap that feeds the tree of heaven,* whether we can see that or not. We bear it by the grace of friendship, good meals, and if we need them, talking turkey heads.

Carnivory

BY CAMILLE

The summer I was eleven, our family took a detour through the Midwest on our annual drive back from our farm in Virginia to Tucson. We passed by one feedlot after another. The odor was horrifying to me, and the sight of the animals was haunting: cows standing on mountains of their own excrement, packed so tightly together they had no room to walk. All they could do was wearily moo and munch on grain mixed with the cow pies under their feet.

Looking out the window at these creatures made my heart sink and my stomach lose all interest. The outdoor part of the operation seemed crueler than anything that might go on inside a slaughterhouse. Whether or not it was scheduled to die, no living thing, I felt, should have to spend its life the way those cows were. When we got home I told my parents I would never eat beef from a feedlot again. Surprisingly, they agreed and took the same vow.

I had another eye-opening experience that fall, in my junior high cafeteria: most people, I learned, really don't want to know what their hamburger lived through before it got to the bun. Some of the girls at my usual lunch table stopped sitting with me because they didn't like the reasons I gave them for not eating the ground-beef spaghetti sauce or taco salad the lunch ladies were serving. I couldn't imagine my friends would care so little about something that seemed so important. To my shock, they expressed no intention of changing their ways, and got mad at *me* for making them feel badly about their choice. A very important lesson for me.

Nobody (including me) wants to be told what church to attend or how to dress, and people don't like being told what to eat either. Food is one of our most intensely personal systems of preference, so obviously it's a touchy subject for public debate. Eight years after my cafeteria drama, I can see plainly now I was wrong to try to impose my food ethics on others, even friends. I was recently annoyed when somebody told me I should not eat yogurt because "If I were a cow, how would I like to be milked?" At the same time, we create our personal and moral standards based on the information we have, and most of us (beyond grade seven) want to make informed choices.

Egg and meat industries in the United States take some care not to publicize specifics about how they raise animals. Phrases like "all natural" on packaged meat in supermarkets don't necessarily mean the cow or chicken agrees. Animals in CAFOs live under enormous physiological stress. Cows that are fed grain diets in confinement are universally plagued with gastric ailments, most commonly subacute acidosis, which leads to ulceration of the stomach and eventually death, though the cattle don't usually live long enough to die of it. Most cattle raised in this country begin their lives on pasture but are sent to feedlots to fatten up during the last half of life. Factory-farmed chickens and turkeys often spend their entire lives without seeing sunlight.

On the other hand, if cattle remain on pasture right to the end, that kind of beef is called "grass finished." The differences between this and CAFO beef are not just relevant to how kindly you feel about animals: meat and eggs of pastured animals also have a measurably different nutrient composition. A lot of recent research has been published on this subject, which is slowly reaching the public. USDA studies found much lower levels of saturated fats and higher vitamin E, beta-carotene, and omega-3 levels in meat from cattle fattened on pasture grasses (their natural diet), compared with CAFO animals. In a direct approach, *Mother Earth News* hosts a "Chicken and Egg Page" on its Web site, inviting farmers to send eggs from all over the country into a laboratory for nutritional analyses, and posting the results. The verdict confirms research published fifteen years ago in the *New England Journal of Medicine:* eggs from chickens that ranged freely on grass have about half the cholesterol of factory-farmed eggs, and it's mostly HDL, the cholesterol that's good for you. They also have more vitamin E, beta-carotene and omega-3 fatty acids than their cooped-up counterparts. The more pasture time a chicken is allowed, the greater these differences.

As with the chickens, the nutritional benefits in beef are directly proportional to the fraction of the steer's life it spent at home on the range eating grass instead of grain-gruel. Free-range beef also has less danger of bacterial contamination because feeding on grass maintains normal levels of acidity in the animal's stomach. At the risk of making you not want to sit at my table, I should tell you that the high-acid stomachs of grain-fed cattle commonly harbor acid-resistant strains of *E. coli* that are very dangerous to humans. Because CAFOs are so widespread in our country, this particular strain of deadly bacteria is starting to turn up more and more commonly in soil, water, and even other animals, causing contamination incidents like the nationwide outbreak of spinach-related illnesses and deaths in 2006. Free-range grazing is not just kinder to the animals and the surrounding environment; it produces an entirely different product. With that said, I leave the decision to you.

Pasture-finished meat is increasingly available, and free-range eggs are now sold almost everywhere. Here is the recipe for one of my family's standard, easy egg-based meals. If you feel more adventurous, you can get some free-range turkey meat and freak out your kids' friends with my parents' sausage recipe.

❋ VEGGIE FRITTATA

Olive oil for pan
8 eggs
½ cup milk

Beat eggs and milk together, then pour into oiled, oven-proof skillet over medium heat.

Chopped kale, broccoli, asparagus, or spinach, depending on the season
Salt and pepper to taste
Feta or other cheese (optional)

Promptly add vegetables and stir evenly into egg mixture. At this point you can also add feta or other cheeses. Cook on low without stirring until eggs are mostly set, then transfer to oven and broil 2–4 minutes, until lightly golden on top. Cool to set before serving.

❋ SPICY TURKEY SAUSAGE

2½ pounds raw turkey meat, diced, including dark meat and fat
½ cup chopped onion
¼ cup chopped garlic
½ tablespoon paprika
1½ teaspoons ground cumin
2 teaspoons fresh oregano (or 1 teaspoon dry)
2 teaspoons fresh thyme (or 1 teaspoon dry)
1 teaspoon ground black pepper
2 teaspoons cayenne (optional)
Hog casings (ask your butcher, optional)

Combine seasonings in a large bowl and mix well. Toss with turkey meat until thoroughly coated. If the meat is very lean, you may need to add olive oil to moisten. Cover and refrigerate overnight. Then grind the mixture in a meat grinder or food processor. You can make patties, or stuff casings to make sausage links. An inexpensive sausage-stuffing attachment is available for KitchenAid and other grinders; your butcher may know a source for organic hog casings.

Download these and all other *Animal, Vegetable, Miracle* recipes at www.AnimalVegetableMiracle.com

15 · WHERE FISH WEAR CROWNS

September

Steven came downstairs with the suitcases and found me in the kitchen studying a box full of papery bulbs. My mail-order seed garlic had just arrived.

His face fell. "You're going to plant those *now*?"

In two hours we were taking off on our first real vacation without kids since our honeymoon—a trip to Italy we'd dreamed of for nearly a decade. My new passport had escaped, by one day, the hurricane that destroyed the New Orleans office of its issue. We had scrupulously organized child care for Lily, backup child care, backup-backup plus the animal chores and so forth. We'd put the garden away for the season, cleaned the house, and finally were really going to do this: the romantic dinners alfresco, the Tuscan sun. The second-honeymoon bride reeking of garlic . . .

"Sorry," I said. I put the bulbs back in the box.

I confess to a ludicrous flair for last-minute projects before big events. I moved nine cubic yards of topsoil the day before going into labor with my first child. (She was overdue, so yes, I was trying.) On the evening of my once-in-a-lifetime dinner at the White House with President and Mrs. Clinton, my hands were stained slightly purple because I'd been canning olives the day before. I have hoed, planted, and even butchered poultry in the hours before stepping onstage for a fund-raising gala. Some divas get a manicure before a performance; I just try to make sure there's nothing real scary under my fingernails. My mother raised children who feel we need to earn what this world means to give us. When I sat back and relaxed on the flight to Rome, I left behind a spit-shined kitchen, a year's harvest put away, and some unplanted garlic. I'd live with it.

With the runway of the Leonardo da Vinci airport finally in sight and our hearts all set for *andiamo,* at the last possible moment the pilot aborted our landing. Wind shear, he announced succinctly. We circled Rome, flying low over ruddy September fields, tile-roofed farmhouses, and paddocks enclosed by low stone walls. The overnight flight had gone smoothly, but now I had ten extra minutes to examine my second thoughts. Would this trip be everything we'd waited for? Could I forget about work and the kids, indulging in the luxury of hotels and meals prepared by someone else?

Finally the nose cone tipped down and our 767 roared low over a plowed field next to the airport. Drifting in the interzone between waiting and beginning, suspended by modern aerodynamics over an ancient field of pebbled black soil, I found myself studying freshly turned furrows and then the farmer himself. A stone's throw from the bustle of Rome's international airport, this elderly farmer was plowing with harnessed draft horses. For reasons I didn't really understand yet, I thought: I've come home.

⚜

I am Italian by marriage: both Steven's maternal grandparents were born there, emigrating as young adults. His mother and aunts grew up in an Italian-speaking home, deeply identified with the foodways and all other ways of the mother country. Steven has ancestors from other parts of the world too, but we don't know much about them. It's my observation that when Italian genes are present, all others duck and cover. His daughter looks like the apple that fell not very far from the olive tree; when asked, Lily identifies herself as American and invariably adds, "but really I'm Italian."

After arriving on the ancestral soil I figured out pretty quickly why that heritage swamps all competition. It's a culture that sweeps you in, sits you down in the kitchen, and feeds you so well you really don't want to leave.

In the whole of Italy we could not find a bad meal. Not that we were looking. But a spontaneous traveler inevitably will end up with the tummy gauge suddenly on empty, in some place where cuisine is not really the point: a museum cafeteria, or late-night snack bar across from the concert hall.

Eating establishments where cuisine isn't the point—is that a strange notion? Maybe, but in the United States we have them galore: fast-food joints where "fast" is the point; cafeterias where it's all about efficient caloric load; sports bars where the purported agenda is "sports" and the real one is to close down the arteries to the diameter of a pin. In most airport restaurants the premise is "captive starving audience." In our country it's a reasonable presumption that unless you have gone out of your way to find good food, you'll be settling for mediocre at best.

What we discovered in Italy was that if an establishment serves food, then *food is the point.* Museum cafeterias offer crusty panini and homemade desserts; any simple diner serving the lunch crowd is likely to roll and cut its own pasta, served up with truffles or special house combinations. Pizzerias smother their pizzas with fresh local ingredients in widely recognized combinations with evocative names. I took to reading these aloud from the menu. Most of the named meals I'd ever known about had butch monikers like Whopper, Monster, and Gulp. I was enchanted with the idea of a lunch named Margherita, Capricciosa, or Quattro Stagioni.

Reading the menus was reliable entertainment for other reasons too. More Italians were going to chef school, apparently, than translator school. This is not a complaint; it's my belief that when in Rome, you speak the best darn Italian you can muster. So we mustered. I speak some languages, but that isn't one of them. Steven's Italian consisted of only the endearments and swear words he grew up hearing from his Nonnie. I knew the Italian vocabulary of classical music, plus that one song from *Lady and the Tramp.* But still, I'd be darned if I was going to be one of those Americans who stomp around Italy barking commands in ever-louder English. *I* was going to be one of those Americans who traversed Italy with my forehead knit in concentration, divining words from their Latin roots and answering by wedging French cognates into Italian pronunciations spliced onto a standard Spanish verb conjugation.

To my astonishment, this technique served really well about 80 percent of the time. Italians are a deeply forgiving people. Or else they are polite, and still laughing. *Va bene.* With a dictionary and grammar book in hand, learning a little more actual Italian each day, we traveled in our rental car from Rome up the winding mountain roads to Steven's grandmother's hometown in Abruzzi, then north through the farmsteads of Umbria and Tuscany, and finally by train to Venice, having fascinating conversations along the way with people who did not speak English. I've always depended on the kindness of strangers. In this case they were kind enough to dumb down their explanations and patiently unscramble a romance language omelet.

So we didn't expect English translations on the menu. No problem. Often there was no menu at all, just the meal of the day in a couple of variations. But restaurants with printed menus generally offered some translation, especially around cities and tourist destinations. I felt less abashed about my own wacky patois as I puzzled through entries such as "Nose Fish," "Pizza with fungus," and the even less appetizing "Polyps, baked or grilled." It seemed "Porky mushrooms" were in season everywhere, along with the perennial favorite (but biologically challenging) "bull mozzarella."

The fun didn't stop with printed menus: an impressive sculpture in the Vatican Museum was identified as the "Patron Genius of Childbirth." (So *that's* who thought it up.) A National Park brochure advised us about hiking preparedness, closing with this helpful tip: "Be sure you have the necessary equipments to make funny outings in respects of nature!" One morning after breakfast we found a polite little sign in our hotel room that warned: "Due to general works in the village, no water or electricity 8:30 to 11:00. Thank you for your comprehension."

Comprehension is just what was called for in these situations. Sooner or later we always figured out the menus, though we remained permanently mystified by a recurring item called "oven-baked rhombus." We were tempted to order it just to put the question to rest, but never did. Too square, I guess.

Italian food is not delicious for its fussiness or complexity, but for the opposite reason: it's simple. And it's an obsession. For a while I thought I was making this up, an outsider's exaggerated sensitivity to a new cultural expression. But I really wasn't. In the famous Siena cathedral I used my binoculars to study the marble carvings over the entry door (positioned higher than the Donatello frescoes), discovering these icons to be eggplants, tomatoes, cabbages, and zucchini. In sidewalk cafés and trattoria with checkered tablecloths, we eavesdropped on Italians at other tables engaging in spirited arguments, with lots of hand gestures. Gradually we were able to understand they were disagreeing over not politics, but olive oils or the best wines. (Or soccer teams.) In small towns the restaurant staff always urged us to try the local oil, and then told us in confidence that the olive oil from the next town over was terrible. Really, worse than terrible: (*sotto voce*) it was *mierda!* Restaurateurs in the next town over, naturally, would repeat the same story in reverse. We always agreed. Everything was the best.

Simple cuisine does not mean *spare,* however. An Italian meal is like a play with many acts, except if you don't watch it you'll be stuffed to the gills before intermission. It took us a while to learn to pace ourselves. First comes the antipasto—in September this was thinly sliced prosciutto and fresh melon, or a crostini of toasted bread with ripe tomatoes and olive oil. That, for me, could be lunch. But it's not. Next comes the pasta, usually handmade, in-house, the same day, served with a sauce of truffles or a grate of pecorino cheese and chopped *pomodoro.* And *that,* for me, could be supper. But it's not, we're still at the lunch table. Next comes the *secondo* (actually the third), a meat or fish course. In the mountains, in autumn, it was often rabbit stewed "hunter's style" or wild boar sausage served with porcini mushrooms; near the coast it was eels, crayfish, anchovies, or some other fresh catch sautéed with lemon juice and fresh olive oil.

With all this under the belt, the diner comes into the home stretch with the salad or *contorno*—a dish of roasted red peppers, eggplants, or sliced tomatoes with basil. Finally—in case you've just escaped from a kidnapping ordeal and find you are still hungry—comes the option of dessert, the only course that can be turned down with impunity. I tried politely declining other courses, but this could generate consternations over why we disliked the food, whether the damage could somehow be repaired, until I was left wondering what part of "No, grazie" was an insult to the cook. Once when I really insisted on skipping the pasta, our server consented only on the condition that he bring us, instead, the house antipasto, which turned out to be a platter of prosciutto, mixed cheeses, pickled vegetables, stuffed mushrooms, fried zucchini flowers stuffed with ham, and several kinds of meat pastries. (The *secondo* was still coming.) Also nearly obligatory are the postprandial coffee and liqueur: grappa, limoncello, meloncello (made from cantaloupes), or some other potent regional specialty.

I was not a complete stranger to meals served in this way. But prior to our trip I'd expected to encounter such cuisine only in fancy, expensive restaurants. Silly me. Whether it's in the country or the town, frequented by tourists or office workers or garage workers or wedding guests, a sit-down restaurant in Italy aims for you to sit down *and stay there.* Steven and I immediately began to wonder if we would fit into the airplane seats we had booked for our return in two weeks. How is it possible that every citizen of Italy doesn't weigh three hundred pounds? They don't, I can tell you that.

By observing our neighbors we learned to get through the marathon of lunch (followed by the saga of dinner) by accepting each course as a morsel. City dining is often more formal, but the rural places we preferred generally served family style, allowing us to take just a little from the offered tray. If a particular course was a favorite it was fine to take more, but in most cases a few bites seemed to be the norm. Then slow chewing, and joy. Watching Italians eat (especially men, I have to say) is a form of tourism the books don't tell you about. They close their eyes, raise their eyebrows into accent marks, and make sounds of acute appreciation. It's fairly sexy. Of course I don't know how these men behave at home, if they help with the cooking or are vain and boorish and mistreat their wives. I realize Mediterranean cultures have their issues. Fine, don't burst my bubble. I didn't want to marry these guys, I just wanted to watch.

The point of eating one course at a time, rather than mixing them all on a single loaded plate, seems to be the opportunity to concentrate one's attention on each flavor, each perfect ingredient, one uncluttered recipe at a time. A consumer trained to such mindful ingestion would not darken the door of a sports bar serving deep-fried indigestibles. And consumption controls the market, or so the economists tell us. That's why it's hard to find a bad meal in Italy. When McDonald's opened in Rome, chefs and consumers together staged a gastronomic protest on the Spanish Steps that led to the founding of Slow Food International.

We did have some close gastronomic shaves in our travels, or so we thought anyway, until the meal came. Early in the trip when we were still jet-lagged and forgetting to eat at proper mealtimes, we found ourselves one afternoon on a remote rural road, suddenly ravenous. Somehow we'd missed breakfast and then lunchtime, by a wide margin. The map showed no towns within an hour's reach. As my blood sugar dipped past grouchy into the zone of stupefaction, Steven made the promise we have all, at some point, made and regretted: he'd stop the car at *the very next place that looked open.*

We rejoiced when a hotel-restaurant materialized at a motorway crossroads, but to be honest, we were also a little disappointed. How quickly the saved can get picky! It looked generic: a budget chain hotel of the type that would, in the United States, serve steam-table food from SYSCO. We resigned ourselves to a ho-hum lunch.

In the parking lot, every member of a rambunctious bridal party was busy taking snapshots of all the others with raised champagne glasses. We tiptoed past them, to be met at the restaurant entrance by a worried-looking hostess. "Mi dispiace!" she cried, truly distressed. The whole dining room was booked all afternoon for a late wedding luncheon. While trying not to sink to my knees, I tried to convey our desperation. The words *affamatto* and *affogato* blurred in my mind. (One means "hungry" and the other is, I think, a poached egg.) The hostess let us in, determined in her soul to find a spot for these weary pilgrims from Esperanto. She seated us near the kitchen, literally behind a potted palm. It was perfect. From this secret vantage point we could be wedding crashers, spies, even poached eggs if that was our personal preference, *and* we could eat lunch.

The hostess scurried to bring us antipasto, then some of the best pasta

I've ever tasted. We didn't poach on the wedding banquet, just the three ordinary courses they'd whipped up to feed the staff. While we ate and recovered our senses we watched the banquet pass by, one ornate entry after another. Forget all previous remarks about simplicity being the soul of Italian cuisine, this was an edible Rose Bowl parade. The climax was the Coronated Swordfish: an entire sea creature, at least four feet long from snout to tail, stuffed and baked and presented in a semi-lounging "S" shape on its own rolling cart. It seemed to be smiling as it reclined in languid, fishy glory on a bed of colorful autumn vegetables, all cupped delicately in a nest of cabbage leaves. Upon its head, set at a rakish angle, the fish wore a crown carved from a huge red bell pepper. Its sharp nose poked out over the edge of the cart, just at eye level to all the bambini running around, so in the interest of public safety the tip of His Majesty's sword was discreetly capped with a lemon cut into the shape of a tulip.

I imagined the kitchen employees who carved this pepper crown and lemon tulip, arranging this fish on his throne. No hash slingers here, but food poets, even in an ordinary budget roadside hotel. We'd come in expecting steam-table food, and instead we found cabbages and kings.

⚜

The roads of Abruzzi, Umbria, and Tuscany led us through one spectacular agrarian landscape after another. On the outskirts of large cities, most of the green space between apartment buildings was cordoned into numerous tidy vegetable gardens and family-sized vineyards. Growing your own, even bottling wine on a personal scale, were not eccentric notions here. I've seen these cozy, packed-in personal gardens in blocks surrounding European cities everywhere: Frankfurt, London, every province of France. After the abrupt dissolution of the Soviet Union's food infrastructure, community gardeners rallied to produce a majority of the fruits and vegetables for city populations that otherwise might have starved.

Traversing the Italian countryside, all of which looked ridiculously perfect, we corroborated still another cliché: all roads actually do lead to Rome. Every crossroads gave us a choice of blue arrows pointing in both directions, for ROMA. Beyond the cities, the wide valleys between medieval hilltop towns were occupied by small farms, each with its own modest olive grove, vineyard, a few fig or apple trees (both were ripe in September), and a dozen or so tomato plants loaded with fruit. Each household also had its own pumpkin patch and several rows of broccoli, lettuces, and beans. Passing by one little stuccoed farmhouse we noticed a pile of enormous, yellowed, overmature zucchini. I made Steven take a

"Dig! Dig! Dig! And Your Muscles Will Grow Big"

On July 9, 2006, in Edinburgh, Scotland, the world lost one of its most successful local-foods advocates of all time: John Raeburn. At the beginning of World War II when Germany vowed to starve the U.K. by blocking food imports with U-boats, Raeburn, an agricultural economist, organized the "Dig for Victory" campaign. British citizens rallied, planting crops in backyards, parks, golf courses, vacant lots, schoolyards, and even the moat of the Tower of London. These urban gardens quickly produced twice the tonnage of food previously imported, about 40 percent of the nation's food supply, and inspired the "Victory Garden" campaign in the United States. When duty called, these city farmers produced.

A similar sense of necessity is driving a current worldwide growth of urban-centered food production. In developing countries where numbers of urban poor are growing, spontaneous gardening on available land is providing substantial food: In Shanghai over 600,000 garden acres are tucked into the margins of the city. In Moscow, two-thirds of families grow food. In Havana, Cuba, over 80 percent of produce consumed in the city comes from urban gardens.

In addition to providing fresh local produce, gardens like these serve as air filters, help recycle wastes, absorb rainfall, present pleasing green spaces, alleviate loss of land to development, provide food security, reduce fossil fuel consumption, provide jobs, educate kids, and revitalize communities. Urban areas cover 2 percent of the earth's surface but consume 75 percent of its resources. Urban gardens can help reduce these flat-footed ecological footprints. Now we just need promotional jingles as good as the ones for John Raeburn's campaign: *"Dig! Dig! Dig! And your muscles will grow big."*

For more information visit www.cityfarmer.org or www.urbangardeninghelp.com.

STEVEN L. HOPP

picture, as proof of some universal fact of life: they couldn't give all theirs away either. At home we would have considered these "heavers" (that's what we do with them, over the back fence into the woods). But these were carefully stacked against the back wall of the house like a miniature cord of firewood, presumably as winter fuel for a pig or chickens. The garden's *secondos* would be next year's prosciutto.

On a rural road near Lake Trasimeno we stopped at a roadside stand selling produce. We explained that we weren't real shoppers, just tourists with a fondness for vegetables. The proprietor, Amadeo, seemed thrilled to talk with us anyway (slowly, for the sake of our comprehension) about his life's work and passion. He was adamantly organic, a proud founding member of Italy's society of organic agriculture.

His autumn display was anchored by melons, colorful gourds, and enough varieties of pumpkin to fill a seed catalog for specialists. I was particularly enchanted with one he had stacked into pyramids all around his stand. It was unglamorous by conventional standards: dark blue-green, smaller than the average jack-o'-lantern, a bit squat, and covered over 100 percent of its body with bluish warts. He identified it as Zucche de Chioggia. We took photos of it, chatted a bit more, and then moved on, accepting the Italian tourist's obligation to visit more of the world's masterpieces than the warty pumpkin-pyramids of Amadeo.

At day's end we were headed back, after having taken full advantage of an olive oil museum, a farmers' market, two castles, a Museum of Fishing, and a peace demonstration sponsored by the Italian government. We passed by the same vegetable stand on our return trip and couldn't resist stopping back in to say hello. Amadeo recognized us as the tourists with no vegetable purchasing power, but was as hospitable as ever. He'd had a fine day, he said, though his pyramids had not exactly been ransacked. I admired that pumpkin, asking its name again (writing it down this time), and whether it was edible. Amadeo sighed patiently. *Edible,* signora? He gave me to know this wart-covered cucurbit I held in my hand was the most delicious vegetable known to humankind. If I was any kind of cook, any kind of gardener, I needed to grow and eat them myself.

I asked if he had any seeds, glancing around for one of those racks. He leaned toward me indulgently, summoning the disposition that all good people of the world maintain toward the earnest dimwitted: the *seeds,* he explained, are *inside the pumpkin.*

Oh. Yes, right. But . . . I struggled (in the style to which I'd become accustomed) to explain our predicament, gesturing toward our rental car. We were just passing through, in possession of no knives, no kitchen, no means of getting the seeds out of the pumpkin. Amadeo suffered our helplessness patiently. We would be going to a hotel? he asked. A hotel with . . . *a kitchen*? They could cook this pumpkin for us any number of ways, baked, sautéed, turned into soup, after setting aside the seeds for us.

I frankly could not imagine sallying into the kitchen of our hotel and asking anyone to carve up a pumpkin, but we were in so deep by now I figured I'd just buy the darn thing and leave it in a ditch somewhere. Or maybe, somehow, figure out how to extract its seeds. But I had to ask one more question. A pumpkin grown in a field with other kinds of winter squash would be cross-pollinated by bees. The seeds would sprout into all sorts of interesting combinations, none of which you'd ever really want. I asked about this in my halting Franco-Spitalian.

The light dawned over his face. He understood perfectly, and began talking a mile a minute. He gestured toward a basket of assorted bright, oddly shaped gourds, and told me *those* were allowed to cross with each other freely, with the obvious sordid results. He wanted to make sure I understood. "Signora, it would be as if you had not married an Italian. Your children could be anything at all!"

Not married an Italian—Mama mia! I shuddered, to make my sentiments clear. Amadeo then seemed satisfied that he could continue the genetics lesson. On his farm, Zucche de Chioggia was prized above all other pumpkins, and thus was raised in a "seminario" where the seeds would breed true to type.

A *seminary*? I pondered the word, struggling for cognates, only able to picture a classroom of pious young pumpkins devoting themselves to Bible study. Then I chuckled, realizing there must be a common root somewhere—the defining condition having to do with these chaste fellows all keeping their genes to themselves. How could we not buy such a well-qualified vegetable? Off we drove with our precious cargo, warts and all.

Back in our hotel room, I paced around staring at it, trying to summon the courage to take it down to the dining room. "*You* take it," I prodded Steven.

"No way," was his helpful reply.

We're too American. We lost our nerve. We dined well, but no seminary-trained pumpkin met its maker that night.

⚜

A type of tourist establishment exists in Italy that does not easily translate: categorically called *agriturismo,* it's a guest accommodation on a working family farm. The rooms tend to be few in number, charmingly furnished, in a picturesque setting, similar to a bed and breakfast with the addition of lunch and dinner, plus the opportunity to help hoe the turnips and harvest the grapes if a guest is so inclined. The main point of the visit for the guest, however, is dinner, usually served family-style at a long wooden table adjacent to the kitchen. Virtually everything set down upon that table, from the wine, olive oil, and cheeses to the after-dinner liqueur, will have been grown and proudly fabricated on the premises. The growers and fabricators will be on hand to accept the diners' queries and appreciation. The host family will likely join the guests at the table, discussing the meal's preparation while enjoying it. By law, this type of accommodation must be run by farmers whose principal income derives from farming rather than tourism. The guest rooms must be converted from farm buildings; all food served must be the farm's own. Fakes are not tolerated.

This hospitality tradition is big business in Italy, with 9,000 establishments hosting more than 10 million bed-nights in a typical recent year, turning over nearly 500 million euros. The notion of agri-vacationing originated in the days (not so long ago) when urban Italians routinely made trips to the countryside to visit relatives and friends who were still on the farm. Any farmstead with a little extra in the storehouse could hang a leafy bough out on the public highway, announcing that travelers were welcome to stop in, sample, and purchase some of the local bounty to take home. It was customary for city-dwelling Italians to spend a few nights out in the country, whenever they could get away, tasting regional specialties at their freshest and best.

It still is customary. The farmhouse holiday business attracts some outsiders, but during our foray through Italian agritourism we met few other foreigners, mostly from elsewhere in Europe. The great majority of our companions at the farm table had traveled less than 100 kilometers. Whether old or young, from Rome or Perugia, their common purpose was to remind themselves of the best flavors their region had to offer. We chatted with elderly couples who were nostalgic for the tastes of their rural childhoods. One young couple, busy working parents, had looked forward to this as their first romantic getaway since the birth of their twins two years earlier. Most guests were urban professionals whose hectic lives were calmed by farm weekends when they could exchange the cell phone's electronic jingle for a rooster's wake-up call and the gentle mooing of Chianina cattle. And more to the point, *eating* the aforementioned beasts.

These Italian agri-tourists were lovely dinner companions who met the arrival of each course with intense interest, questions, and sometimes applause. My customary instincts about rural-urban antipathy ran aground here where our farmer hosts, wearing aprons tied over their work clothes, were the stars of the evening, basking in the glow of their city guests' reverent appreciation.

The farm hotel often has the word *fattoria* in its name. It sounds like a place designed to make you fat, and I can't argue with that, but it means "farm," deriving from the same root as *factory*—a place where things get made. Our favorite *fattoria* was in Tuscany, not far from Siena, where many things were getting made on the day we arrived, including wine. We watched the grapes go through the crusher and into giant stainless steel fermenting tanks in a barn near our guest room. Some guests had brought work gloves to help pick grapes the next morning. We were on vacation from farm work, thanks, but walked around the property to investigate the gardens and cattle paddocks. A specialty of the house here was the beef of Chianina, the world's largest and oldest breed of cattle, dating back to Etruscan times. Snow white, standing six feet at the shoulder, they are gentle by reputation but I found them as intimidating as bison.

In a vegetable garden near the main house an elderly farm worker walked slowly along on his knees, planting lettuces. When he finished his row he came over to talk with us, making agreeable use of his one word of English, "Yes!" Nevertheless we managed to spend an hour asking him questions; he was patient with the linguistically challenged, and his knowledge was encyclopedic. It takes 100 kilograms of olives to make 14 kilos of oil. The pH is extremely important. The quality of olives grown on this farm were (naturally) the best, with some of the lowest acid levels in all the world's olive oil. This farm also produced a nationally famous label of wine, along with beef and other products. I was curious about wintertime temperatures here, which he said rarely dropped lower than freezing, although in '87 a Siberian wind brought temperatures of –9°F, killing olive trees all over central Italy. On this farm they'd had no harvest for six years afterward, but because they had a very old, established orchard, it recovered.

All the olives here were harvested by hand. Elsewhere, in much of Italy, older trees have been replaced in the last two decades by younger orchards trimmed into small, neat box shapes for machine harvest. Mechanization obviously increases productivity, but the Italian government is now making an effort to preserve old olive orchards like the ones on this farm, considering their twisted trunks and spreading crowns to be a classic part of the nation's cultural heritage.

Our new tour guide was named Amico. We seemed to be working our way through the alphabet of colorful Italian patriarchs. And if Amadeo was poetically devout, Amico really was Friendship personified. He showed us the field of saffron crocuses, an ancient and nearly extinct Tuscan crop that is recently being revived. Next came the vineyards. I asked how they protected the grapes from birds. He answered, by protecting their predators: civet cats, falcons, and owls patrol the farm by turns, day and night. Amico was also a big fan of bats, which keep down the insects, and of beneficial ladybugs. And he *really* loved the swallows that build their mud nests in the barns. They are marvelous birds, he insisted, rolling his eyes heavenward and cupping a hand over his heart. His other deepest passion appeared to be Cavolo Nero de Toscana—Tuscan black kale. He gave us an envelope of its tiny seeds, along with careful instructions for growing and cooking it. We were going to have Tuscan kale soup for supper that very night, a secret he disclosed with one of those long Italian-guy sighs.

At dinnertime, we were surprised when Amico took his place at the head of the table. He introduced himself to the rest of the guests, opening his arms wide to declare, "*Amico de tutti!*" This kind old man in dirty jeans we had taken that afternoon for a field hand was in fact the owner of this substantial estate. I took stock of my assumptions about farmers and landowners, modesty and self-importance. I'm accustomed to a culture in which farmers are either invisible, or a joke. From the moment I'd spied draft horses turning the soil of a glamorous city's outskirts, Italy just kept surprising me.

That does it, I told Steven later that night, retrieving our well-traveled Zucche de Chioggia from the car. They're not going to kick us out of this place for possession of pumpkin with agricultural intent. At worst they might tell us it's doo-doo because it was grown in a different province. But no, the housekeeper admired it the next morning, and the kitchen staff without hesitation handed over an enormous knife and spoon. They suggested we butcher it down by the chicken and pig pens, where any fallout would be appreciated.

Indeed, the enormous pink pigs sniffed the air, snorted, and squealed as we whacked open our prize. Its thick flesh was custardy yellow, with a small cavity in the center packed with huge white seeds. We took pity on the wailing hogs and chopped up the pumpkin flesh for their dinner rather than ours, distributing the pieces carefully among the pens to avoid a porcine riot. Back in our room, I washed the sticky pulp from the seeds as best I could in our bathroom sink, but really needed a colander, which the kitchen staff also happily supplied. For the rest of our stay we spread out the seeds on towels in the sun, but they weren't completely dried by the time we had to depart from the *fattoria.* They would mildew and become inviable if we left them packed in a suitcase, so the rest of our Italian vacation (in the rain, on the train) was in some way organized around opportunities to spread out the seeds for further drying. Most challenging was a fancy hotel room in Venice, where I put them in a heavy, hand-blown glass ashtray on the dresser. We were going out for the whole day so I left a note to housekeeping, whom I feared would throw them away. I gave it my best, something like: "Please not to disturb the seminary importants, Thanking You!"

But most of our days were spent in bucolic places where we and our seeds could all bask in the Tuscan sunshine, surrounded by views too charming to be believed. The landscape of rural central Italy is nothing like the celebrated tourist sights of North America: neither the uninhabited wildness of the Grand Tetons, nor the constructed grandeur of the Manhattan skyline. Tuscany is just farms, like my home county. Its beauty is a harmonious blend of the natural and the domestic: rolling hills quilted with yellowy-green vineyard rows running along one contour, silvery-green circles of olive trees dotting another. The fields of alfalfa, sunflowers, and vegetables form a patchwork of shapes in shades of yellow and green, all set at different angles, with dark triangles of fencerows and woodlots between them.

The effect is both domestic and wild, equal parts geometric and chaotic. It's the visual signature of small, diversified farms that creates the picture-postcard landscape here, along with its celebrated gastronomic one. Couldn't Americans learn to love landscapes like these around our cities, treasuring them not just gastronomically but aesthetically, instead of giving everything over to suburban development? Can we only love agriculture on postcards?

Tuscans and Umbrians have had a lot more time than we have, of course, to recognize the end of the frontier when they see it, and make peace with their place. They were living on and eating from this carefully honed human landscape more than a thousand years before the Pilgrims learned to bury a fish head under each corn plant in the New World. They have chosen to retain in their food one central compelling value: that it's fresh from the ground beneath the diners' feet. The simple pastas still taste of sunshine and grain; the tomatoes dressed with fruity olive oil capture the sugars and heat of late summer; the leaf lettuce and red chicory have the specific mineral tang of their soil; the black kale soup tastes of a humus-rich garden.

On the road to the *fattoria* we'd passed a modest billboard that seemed to epitomize the untranslatable difference between Italy's food culture and our own. It's a statement you just don't hear from the tourist boards of America. It promised, simply:

Nostro terra . . .
E suo sapore.

I'm no expert, but I see what it means: You can taste our dirt.

16 · SMASHING PUMPKINS

October

Driving through our little town in late fall, still a bit love-struck for Tuscany's charm, I began to see my home through new eyes. We don't have medieval hilltop towns here, but we do have bucolic seasonal decor and we are not afraid to use it. "Look," I cried to my family, "we live in *Pleasantville.*" They were forced to agree. Every store window had its own cheerful autumnal arrangement to celebrate the season. The lampposts on Main Street had corn shocks tied around them with bright orange ribbons. The police station had a scarecrow out front.

As I have mentioned, yard art is an earnest form of self-expression here. Autumn, with its blended undertones of "joyful harvest" and "Trick-or-Treat kitsch," brings out the best and worst on the front lawns: colorful displays of chrysanthemums and gourds. A large round hay bale with someone's legs hanging out of its middle. (A pair of jeans and boots stuffed with newspaper, I can only hope; we'll call it a farm safety reminder.) One common theme runs through all these dioramas, and that is the venerable pumpkin. They were lined up in rows, burnished and proud and conspicuous, the big brass buttons on the uniform of our village. On the drive home from our morning's errands we even passed a pumpkin field where an old man and a younger one worked together to harvest their crop, passing up the orange globes and stacking them on the truck bed to haul to market. We'd driven right into a Norman Rockwell painting.

Every dog has its day, and even the lowly squash finally gets its month. We may revile zucchini in July, but in October we crown its portly orange cousin the King Cucurbit and Doorstop Supreme. In Italy I had nursed a growing dread that my own country's food lore had gone over entirely to the cellophane side. Now my heart was buoyed. Here was an actual, healthy, native North American vegetable, non-shrink-wrapped, locally grown, and in season, sitting in state on everybody's porch.

The little devil on my shoulder whispered, "*Oh yeah?* You think people actually know it's edible?"

The angel on the other shoulder declared "*Yeah*" (too smugly for an angel, probably), the very next morning. For I opened our local paper to the food section and found a colorful two-page spread under the headline "Pumpkin Possibilities." Pumpkin Curry Soup, Pumpkin Satay! The food writer urged us to think past pie and really dig into this vitamin-rich vegetable. I was excited. We'd grown three kinds of pumpkins that were now lodged in our root cellar and piled on the back steps. I was planning a special meal for a family gathering on the weekend. I turned a page to find the recipes.

As I looked them over, Devil sneered at Angel and kicked butt. Every single recipe started with the same ingredient: "1 can (15 oz) pumpkin."

I could see the shopping lists now:

1 can pumpkin (for curry soup)
1 of those big orangey things (for doorstep).

Come *on,* people. Doesn't anybody remember how to take a big old knife, whack open a pumpkin, scrape out the seeds, and bake it? We can carve a face onto it, but can't draw and quarter it? Are we not a nation known worldwide for our cultural zest for blowing up flesh, on movie and video screens and/or armed conflict? Are we in actual fact too squeamish to stab a large knife into a pumpkin? Wait till our enemies find out.

Two days later my mother walked in the kitchen door, catching me in the act of just such a murder, and declared "Barbara! That looks dangerous."

I studied my situation objectively: the pumpkin was bluish (not from asphyxiation) but clinging tenaciously to life. I was using a truly enormous butcher knife, but keeping my fingers out of harm's way. "Mom, this is safe," I insisted. "I have good knife skills."

Where her direct descendants are concerned, my mother's opinion is that a person is never too old to lose digits or eyesight in the normal course of events.

Dad is another story. To be frank, he's the reason Mom's worry-skills remain acutely honed. Naturally he wanted to get in on the act here. "Step right up!" I said. My father's career, before he retired, covered most kinds of emergency surgery imagined in the twentieth century, carried out in operating rooms that occasionally did not come equipped with electricity. I wasn't going to argue with his knife skills.

The pumpkin kept the two of us sawing and sweating for a good thirty minutes while we made no appreciable progress. Our victim was a really large pumpkin of the variety called Queensland Blue. The seed catalog had lured me in with testimonials about its handsome, broad-shouldered Aussie physique and tasty yellow flesh. Next year, Amadeo's warty Zucche de Chioggia would get to vie for the World Cup title of our pumpkin patch, but for now the Queensland Blue ruled. And this one was not yielding. My surgical assistant and I sawed some more, taking frequent breaks to review and strategize. I lusted in my heart after one of those "1 can (15 oz) pumpkin" recipes.

Cooking pumpkin from scratch may not be for the fainthearted, but it's generally not *that* hard. I wasn't merely trying to hack it to pieces—if that had been our goal, Dad would have dispatched it in no time flat. But I was being girly, insisting on cutting open the top as neatly as possible to scoop out the seeds and turn the whole thing into a presentable tureen in which to bake pumpkin soup. I have two different cookbooks that feature this as a special-guest recipe. Meanwhile we received plenty of advice from the bystanders (what are families for?), all boiling down to the opinion that it would taste exactly the same if we just smashed it. But this was a special dinner and I was doing it up right. "You'll see, go away," I said sweetly, waving my knife. I never went to chef school but I know what they say: presentation, presentation, presentation.

Our occasion was Thanksgiving declared a month early, since all of us wouldn't be able to convene on the actual day. Turkeys galore were now wedged into our freezer, postharvest, but today's family gathering included some vegetarians who would not enjoy a big dead bird on the table, however happily it might have lived its life. Meatless cooking is normal to me; I was glad to make a vegetarian feast. But today, as hostess, I felt oddly pressed by tradition to have some kind of large, autumnally harvested *being,* not just ingredients, as the centerpiece for our meal. A hearty pumpkin soup baked in its own gorgeous body would be just the thing for a turkeyless Thanksgiving.

Or so I'd thought, before I knew this one would not go gentle into that good night (as Dylan Thomas advised). Even our meanest roosters hadn't raged this hard against the dying of the light. At length, Dad and I ascertained the problem was our Queensland Blue, which was almost solid flesh, lacking the large open cavity in its center that makes the standard jack-o'-lantern relatively easy to open up. We finally performed something

Trading Fair and Square

The local food movement addresses many important, interconnected food issues, including environmental responsibility, agricultural sustainability, and fair wages to those who grow our food. Buying directly from small farmers serves all these purposes, but what about things like our pumpkin pie spices, or our *coffee,* that don't grow where we live?

We can apply most of the same positive food standards, minus the local connection, to some imported products. Coffee, tea, and spices are grown in environmentally responsible ways by some small-scale growers, mostly in the developing world. We can encourage these good practices by offering a fair wage for their efforts. This approach, termed fair trade, has grown into an impressive international effort to counter the growing exploitation of farmers in these same countries. Consumer support for conscientious small growers helps counter the corporate advantage, and sustains their livelihoods, environments, and communities.

Coffee is an example of how fair trade can work to the advantage of the grower, consumer, and environment. As an understory plant, coffee was traditionally grown under a shaded mixture of fruit, nut, and timber trees. Large-scale modern production turned it into a monoculture, replacing wild forests with

like trephination on our client, creating a battle-weary but still reasonably presentable hollowed-out tureen. I rubbed the cavity with sea salt and poured in milk I'd heated with plenty of sage and roasted garlic. (Regular, lactose-free, or soy milk work equally well.) I set it carefully into the oven to bake. According to the recipe, after an hour or so of baking I could use a large spoon to scrape gently at the inside of the tureen, stirring the soft, baked pumpkin flesh into the soup.

I'll now add my own cautions to this recipe: don't scrape too hard, and don't overbake. I confess I may have hoped for a modest flutter of kiss-the-cook applause as I set our regal centerpiece on the table, instead of the round of yelping and dashes for the kitchen towels that actually ensued. The whole thing collapsed. Fortunately, I'd baked it in a big crockery pie dish. We saved the tablecloth, and nine-tenths of the soup.

My loved ones have eaten my successes and failures since my very first rice pudding, at age nine, for which I followed the recipe to the letter

single-crop fields, utterly useless as wildlife habitat, doused heavily with fertilizers and pesticides. This approach is highly productive in the short term, but causes soil erosion and kills tropical biodiversity—including the migratory birds that used to return to our backyards in summer. Not to mention residual chemicals in your coffee. In contrast, farmers using traditional growing methods rely on forest diversity to fertilize the crop (from leaf litter) and help control coffee pests (from the pest predators that are maintained). Although their yields are lower, the shade-grown method sustains itself and supports local forest wildlife. Selecting shade-grown and fair-trade coffee allows these small-farm growers a chance to compete with larger monocrop production, and helps maintain wildlife habitat.

Independent certification agencies (similar to those that oversee organic agriculture) ensure that fair trade standards are maintained. As demand grows, the variety of products available has also grown to include chocolate, nuts, oils, dried fruits, and even hand-manufactured goods. For more information see www.transfairusa.org, www.fairtrade.net or www.ifat.org.

STEVEN L. HOPP

but didn't understand the "1 cup rice" had to be cooked first. Compared with that tooth-cracking concoction, everyone now agreed, my pumpkin soup was great. Really it was, by any standards except *presentation* (which I *flunked flunked flunked*), and besides, what is the point of a family gathering if nobody makes herself the goat of a story to be told at countless future family gatherings? Along with Camille's chard lasagna, our own fresh mozzarella, and the season's last sliced tomatoes, we made a fine feast of our battered centerpiece. If anyone held Dad and me under suspicion of vegetable cruelty, they didn't report us. We remain at large.

A pumpkin is the largest vegetable we consume. Hard-shelled, firm-fleshed, with fully ripened seeds, it's the caboose of the garden train. From the first shoots, leaves, and broccoli buds of early spring through summer's small soft tomato, pepper, and eggplant fruits, then the larger melons, and finally the mature, hard seeds such as dry beans and peanuts, it's a long and remarkable parade. We had recently pulled up our very last crop to mature: peanuts, which open their orange, pealike flowers in midsummer, pollinate and set their seeds, and then grow weirdly long, down-curved stems that nosedive the seed pods earthward, drilling them several inches into the soil around the base of the plant. Many people do realize peanuts are an underground crop (their widespread African name is "ground nut"), although few ever pause to ruminate on how a *seed,* product of a *flower,* gets under the dirt. In case you did, wonder no more. Peanuts are the dogged overachievers of the plant kingdom, determined to plant their own seeds without help. It takes them forever, though. Our last ritual act before frost comes is to pull up the peanut bushes and shake the dirt from this botany-freak-show snack food.

We were ready now for frost to fall on our pastures. We had eaten one entire season of botanical development, in the correct order. Months would pass before any new leaf poked up out of the ground. So . . . now what?

Another whole category of vegetable used to carry people through winter, before grocery chains erased the concept of season. This variant on vegetable growth is not an exception but is auxiliary to the leaf-bud-flower-fruit-seed botanical time lapse I initially posted as the "vegetannual rule" for thinking about what's in season. A handful of food plants are not annuals, but biennials. Their plan is to grow all summer from a seed, lay low through one winter, then burst into flower the following spring. To do it, they frugally store the sugars they've manufactured all summer in a bulky tuber or bulb that hides underground waiting for spring, after their leaves have died back.

Humans thwart them opportunistically, murdering the plant and robbing its savings account just when the balance is fattest. Carrots, beets, turnips, garlic, onions, and potatoes are all the hard-earned storage units of a plant that intended to live another season in order to fulfill its sexual destiny. As a thrifty person myself, raised to trust hard work, I feel like such a cheat when I dig the root crops. If I put emotion in charge of my diet I would not only be a vegetarian, I'd end up living on air and noodles like a three-year-old because I also feel sorry for the plants. In virtuous green silence they work as hard as any chicken or cow. They don't bleat or wail as we behead them, rip them from their roots, pull their children from their embrace. We allow them no tender mercies.

But heaven help me, I eat them like nobody's business. Root crops are the deliverance of the home-food devotee. Along with dry beans and grains, they bring vegetable nutrition into months when nothing else fresh is handy. Because they store well, it's easy enough for gardeners to produce a year's worth in the growing season. Some of my neighbors grumble about the trouble of growing potatoes when a giant bag at the store costs less than a Sunday newspaper. And still, every spring, we are all out there fighting with the cold, mucky late-winter soil, trying to get our potatoes in on schedule. We're not doing it for the dimes we'll save. We know the fifty-pound bag from the store *tastes* about like a Sunday newspaper, compared with what we can grow. A batch of tender new Carolas or Red Golds freshly dug in early summer is its own vegetable: waxy, nutty, and sweet. Peruvian Blues, Russian Banana fingerlings, Yukon Golds: the waxy ones hold together when boiled and cut up for potato salad; others get fluffy and buttery-colored when baked; still others are ideal for oven-roasting. A potatophile needs them all.

The standard advice on potato-planting time is the same as for onions and peas: "as early as the soil can be worked." That is a subjective date, directly related to impatience. I always get stirred up around Saint Patrick's Day and go through my annual ritual of trooping out to the potato bed with a shovel, sticking it in the ground, and scientifically discerning that it's still a half-frozen swamp. You don't need a groundhog for that one: wait a few more weeks. We generally get them in around the first of April.

Potato plants don't mind cool weather, as long as they're not drowning. They were bred from wild ancestors in the cool, dry equatorial Andean highlands where days and nights are equal in length, year-round. They don't respond to changes in day length to control their maturity. Other root crops are triggered by summer's long days to start banking starch, preparing for the winter ahead. In fact, onions are so sensitive to day length, onion growers must choose their varieties with a latitude map.

Temperatures are not a reliable cue—they can rise and fall capriciously during a season, giving us dogwood winter, Indian summer, and all the other folklorically named false seasons. But no fickle wind messes with the track of the sun. It's a crucial decision for a living thing: When, exactly, to shut down leaf growth and pull all resources down into the roots to stock up for winter? A mistake will cost a plant the chance to pass on its genes. So in temperate climates, evolution has tied such life-or-death decisions to day length. Animals use it also, to trigger mating, nesting, egg-laying, and migration.

But potatoes, owing to their origin in the summerless, winterless, high-altitude tropics, evolved without day-length cues. Instead they have a built-in rest period that is calendar-neutral, and until it's over the tubers won't sprout, period. I learned this the hard way, early in my gardening career, when I planted some store-bought potatoes. I waited as the weather grew warm, but no sprouts emerged. Potatoes are often treated with chemicals to keep them dormant, but I'd planted organic ones with no such excuse for sloth. After about a month I dug them up to see what in the heck was going on with the lazy things. (I'm much more relaxed with my children, I swear.) My potatoes were still asleep. Not one eye was open, not a bump, shoot, or bud.

Now I know: potatoes have a preprogrammed naptime which cannot for any reason be disturbed. Seed potatoes aren't ready to plant until after they've spent their allotted months in cool storage. I had assumed a spring potato was a spring potato, but these I'd bought from the grocery in March must have been harvested recently in some distant place where March was not the end of winter. The befuddlements of a seasonless vegetable universe are truly boundless.

If the potatoes in the produce section are already sprouting, on the other hand, it means they're ready to get up. They're edible in that condition, as long as they haven't been exposed to light and developed a green cast to the skin. Contrary to childhood lore these photosynthesizing potatoes won't kill you, but like all the nightshades—including tomatoes and eggplants—all green parts of the plant contain unfriendly toxins and mutagens. Sprouting and a tad wrinkly, though, they're still okay to eat. Rolled in a sturdy paper bag in the bottom drawer of the refrigerator they'll keep six months and more. But when I open our labeled bags of seed potatoes in April I always find a leggy mess, the whole clump spiderwebbed together with long sprouts.

That means the wake-up call has come. We toss them in the ground and hill them up. From each small potato grows a low, bushy plant with root nodules that will grow into eight or more new potatoes. (Fingerling potatoes produce up to twenty per hill, though they're smaller.) The plants have soft leaves and a flower ranging from white to pink or lavender, depending on the variety. In the Peruvian Andes, where farmers still grow many more kinds of potatoes than most of us can imagine, I've seen fields of purple-flowered potatoes as striking in their way as a Dutch tulip farm in bloom.

The little spuds in the roots continue to gain size until the plant gets tired and dies down, four to five months after planting. Few garden chores are more fun for kids than recovering this buried treasure at the end of the season. We plant eight different kinds, plus a mongrel bag that Lily calls the "Easter egg hunt" when we dig them: I turn up each hill with a pitchfork and she dives in after the red, blue, golden, and white tubers. For my part it's a cross between treasure-hunt and *ER,* as I have to shout *"Clear!"* every time I dig, to prevent an unfortunate intersection of pitchfork and fingers.

Even though the big score comes at summer's end, we'd already been sneaking our hands down into the soil under the bushy plants all summer, to cop out little round baby spuds. This feels a bit like adolescent necking antics, but is a time-honored practice described by a proper verb: grabbling. (Magazine editors want to change it to "grabbing." Writers try to forgive them.) The proper time for doing it is just before dinner: like corn, new potatoes are sweetest if you essentially boil them alive.

In late summer we'd finished digging the Yukon Golds, All-Blues, and other big storage potatoes, best for baking. In September we'd harvested the fingerlings—yellow, finger-shaped gourmet potatoes that U.S. consumers have discovered fairly recently. I found them in a seed catalog, ordered some on a whim, and got hooked. The most productive potato in our garden is a fingerling adored by the French, called by a not-so-romantic name, La Ratte. (Means what it sounds like.) The name baffled me, until I grew them. When a mature hill is uncovered it looks like a nest of fifteen suckling baby rodents, all oriented with their blunt noses together in the center and their long tails pointed out. By virtue of incredible flavor and vigor, these little rats take over a larger proportion of my potato patch every year.

By October all our potatoes were in the root cellar, along with the beets, carrots, and sweet potatoes (a tropical vine botanically unrelated to potatoes). Our onions were harvested too. We pull those up in late summer after their tall strappy tops begin to fall over, giving the patch an "off the shoulder" look with the bulbs bulging sensuously above the soil line. Comparisons with a Wonderbra are impossible to avoid. By the time they're fully mature, onions are pushing themselves out of the ground. It's an easy task on a dry, late summer day to walk down the row tugging them up, leaving them to cure in the sun for the afternoon, then laying them on newspapers for a couple of weeks in a shed or garage with plenty of air circulation.

Garlic gets a similar treatment. No matter how well it's ventilated, the aroma of the curing shed gets intense; in earlier days children would have been made to sit in there to cure the category of ailment my grandmother called "the epizootie." She also used to speak of children wearing "asafetidy bags" around their necks to prevent colds. Genuine asafoetida is a

European plant in the parsley family, but the root word is *fetid*. Garlic obviously worked as well, the medicinal property being that nobody would get close enough to your children to cough on them. As they used to say in New York, "A nickel will get you on the subway, but garlic will get you a seat." In fairness to its devotees, I should point out that real medicinal value has been attributed to garlic, stemming from its antibacterial sulfur compounds and its capacity to break down fibrin and thin the blood. The Prophet Muhammad recommended it for snakebite, Eleanor Roosevelt took it in chocolate-covered pills to improve her memory, and Pliny the Elder claimed it was good for your sex life. I wouldn't bet on that last one.

We stick to culinary uses. When the tops of the cured garlic and onions have faded from green to brown but are still pliable, and their odor has tamed, I braid them into the heavy skeins that adorn our kitchen all winter. The bulbs keep best when they're hung in open air at room temperature, and we clip off the heads as we need them, working from the top down. Garlic is the spice of life in our kitchen: spaghetti sauce, lasagna, chicken soup, just about everything short of apple pie begins with some minced cloves of garlic sautéed in olive oil. I spend it as the currency of our culinary happiness, cutting off the heads week by week, working the braids slowly down to their bottom ends, watching them like the balance of a bank account. With good management we'll reach the end in midsummer, just as the new crop gets harvested. But we always come down the home stretch on empty.

Garlic, like the potato, is a more subtle vegetable than most people know, since most groceries carry only one silverskin variety that keeps like Egyptian royalty. Garlic connoisseurs know the rest of the story. Seed Savers Exchange lists hundreds of varieties, each prized for its own qualities of culture and flavor. They fall into two basic categories: the lusty, primitive hardnecks have a "scape"—a flower stalk that shoots up from the center of the bulb in early spring, striking an incomprehensibly circular path, growing itself into something like an overhand knot before it blooms. The more domestic softnecks are better for braiding and storage. Beyond this, garlics are as hard to categorize as red and white wines, with equally enthusiastic legions of fans. Inchelium Red has taken first place in taste-tests on several continents. Red Toch (according to my seed catalog) has "a multidimensional quality, a spicy fragrance, and consummate flavor." Persian Star, obtained from a bazaar in Samarkand, Uzbekistan, has "a mild spicy zing." Brown Tempest starts hot and intense, then simmers down to a warm pleasant finish; Music is rich and pungent; Chesnok Red is the best for baking, "very aromatic with an abiding flavor." Elephant is just plain huge.

Who can decide? A sucker for seed-catalog prose, I ordered six varieties and planted about a dozen cloves of each. Buried under the soil with a blanket of straw mulch on top, each clove would spend the winter putting out small, fibrous roots. The plant begins working on top growth in early spring, as soon as air temperatures rise above freezing. By midsummer it will have from eight to a dozen long leaves, with one clove at the base of each leaf nestled into the tight knot of a new head. I pull them in late June, and tend to think of that moment as the end of some sort of garden fiscal year. It never really stops, this business of growing things—garlic goes into the ground again in October, just as other frost-killed crops are getting piled onto the compost heap. Food is not a product but a process, and it never sleeps. It just goes underground for a while.

⚘

October ceded to us the unexpected gifts of a late first frost: a few more weeks of tomatoes, eggplants, peppers, and basil. Usually it's late September when the freeze comes to knock these tender plants down to black stalks—a predictable enough event that still somehow takes us by surprise every time. The late evening newscast predicting it will create a mild hubbub in the neighborhood, sending all the gardeners out in the chill to scoop up green tomatoes and pull peppers in the dark.

Both peppers and eggplants are tropical by nature, waking slowly to summer and just getting into full swing in late September. We were happy for the October bonus, bringing in bushels of blocky orange, red, and yellow peppers at our leisure instead of under siege. In the evenings we built fires in the patio fire pit and lined the peppers up shoulder-to-shoulder on the grill like sunbathers on a crowded beach. The heat makes their skins sear and bubble. Afterward we freeze them by the bagful, so their smoky flavor can lively up our homemade pizzas all winter. If it's the scent of burning leaves that invokes autumn for most people, for us it's roasting peppers.

My most nostalgic harvest rituals from childhood are of the apple variety. My parents took us on annual excursions to a family-owned orchard in the next county where we could watch cider being pressed, and climb into the trees to pick fruit that the clay-footed adults couldn't reach. We ate criminal quantities of apples up in the boughs. Autumn weather still brings that crisp greenish taste to the roof of my mouth. I realized other members of my family must share this olfactory remembrance of things past, when our October gathering spontaneously rallied into a visit to an apple orchard near our farm. We bought bushels, inspired to go home and put up juice and apple butter.

I don't know what rituals my kids will carry into adulthood, whether they'll grow up attached to homemade pizza on Friday nights, or the scent of peppers roasting over a fire, or what. I do know that flavors work their own ways under the skin, into the heart of longing. Where my kids are concerned I find myself hoping for the simplest things: that if someday they crave orchards where their kids can climb into the branches and steal apples, the world will have trees enough with arms to receive them.

Splendid Spuds

BY CAMILLE

Dietitians seem to love or hate potatoes. This vegetable is a great source of quality carbohydrates that will keep you feeling energized and satisfied. Or else, it should be avoided because its sinful starchiness will make you gain weight like crazy.

So which is it? First, it's not safe to generalize. The Idaho spuds at grocery stores are just the tip of the iceberg. I've eaten potatoes ranging from gumball to guinea-pig size, golden, brick red, or even brilliant blue. When you dig those out of the ground they look almost black, but after a good scrub they have a deep purplish shine you would expect in a jewelry store, not under layers of caked mud. I'm always amazed to slide my knife into the first Peruvian Blue of the season, finding its insides as vivid as a ripe blueberry. As it happens, the phytochemical responsible for the gorgeous color of blueberries and blue potatoes is the same one—a powerful antioxidant. Some antiaging facial products on the market feature mashed blueberries as a main ingredient. Nobody in my household has tried rubbing potatoes on her face to preserve a youthful complexion, but we take them internally, appreciating all the nutrients in our potato rainbow. Even white potatoes, eaten with their skins, give us vitamins C and B6, and nearly twice the potassium per serving as a banana.

The potato's bad rep comes from its glycemic index (GI), a tool used to rate the increase in blood sugar caused by eating particular foods. Foods with a high GI (like sugar and corn sweetener, also) cause a sharp rise in blood glucose that stimulates high levels of blood insulin. It's hard on the body when that happens frequently, and has been linked to Type II diabetes. But if potatoes are eaten along with foods containing some fiber, protein, and fat, those lower the body's glycemic response so that insulin levels stay calm.

A potato's nutritional value is a package best delivered in its own wrapper. Unfortunately, eating potato skins is not always safe. Because they grow underground, conventional potatoes are among the most pesticide-contaminated vegetables. Potatoes in the United States commonly contain chlorinated hydrocarbon insecticides and sometimes even residues of DDT, dieldrin, and chlordane, extremely hazardous chemicals that have been banned since 1978 but linger in the soil. Conventionally grown potatoes are so contaminated, the Environmental Working Group warns parents not to feed them to infants and toddlers unless they're thoroughly peeled and boiled. This makes a strong case for buying organic potatoes from trustworthy growers who know the history of the land where their produce is raised.

We're lucky enough to have home-grown. Potato salad is one dish our household never goes without, although it changes with the seasons. We eat many other root vegetables too, and plenty of sweet potatoes and winter squash in the autumn months. I've included Mom's infamous pumpkin soup recipe here, and dinner menus for a typical autumn week at our house.

❋ FOUR SEASONS OF POTATO SALAD

WINTER

4 cups large storage potatoes, coarsely diced and boiled until firmly tender
3 hard-boiled eggs, peeled and coarsely chopped
¾ cup last summer's dill pickles, finely chopped
2 tablespoons dill
Salt to taste
Mayonnaise—a few tablespoons

Combine potatoes, eggs, and pickles, being careful not to mash anything. Add dill and salt to mixture and combine thoroughly. Add just enough mayonnaise to hold the salad together.

SPRING

4 cups storage potatoes, coarsely diced
⅓ cup fresh mint leaves
1–2 cups new peas
1 cup crumbled feta
¼ cup extra-virgin olive oil

Boil diced potatoes as above. Combine ingredients.

SUMMER

2 pounds red or golden new potatoes, cut in 1-inch chunks
3 tablespoons olive oil
Coarse salt
2 yellow or red bell peppers, cut in chunks
2 cups green beans (stringed and broken in 1-inch lengths)
1–2 ears sweet corn on cob

Toss potatoes with salt and oil and spread on baking sheet. Roast in 450° oven until tender (20–30 minutes). Place ears of corn, lightly oiled, with the potatoes. Add peppers and green beans to roast for last 10 minutes. When done, loosen the vegetables with a spatula, cut corn kernels off cob, and combine in a large, shallow bowl.

2 cups tomatoes cut in wedges
½ cup fresh basil
¼ cup olive oil whipped together with 1 tablespoon balsamic or other mellow vinegar

Toss tomatoes, basil, and dressing with roasted vegetables; salt to taste.

FALL

2 pounds fingerling potatoes (such as Russian Banana, Rose Finn, La Ratte)
Seasonal vegetables
4 tablespoons dried basil
¼ cup olive oil whipped with 1 tablespoon balsamic vinegar

Prepare roasted potatoes as in "Summer" recipe, combining with late-season peppers and green beans, shelly beans, or limas, roasted along with the potatoes. Toss with fresh tomato wedges, basil, and the dressing. As the season progresses and different things become available, you can mix and match other vegetables with the potatoes to your heart's content, keeping proportions roughly the same. Cubed winter squash and sweet potatoes are wonderful in this roasted dish, requiring about the same amount of time in the oven. Don't hesitate to combine sweet and regular potatoes—they are unrelated, and marry well!

❋ PUMPKIN SOUP IN ITS OWN SHELL

1 five-pound pumpkin (if smaller or larger, adjust the amount of liquid)

Cut a lid off the top, scoop out the seeds and stringy parts, and rub the inside flesh with salt. Set the pumpkin in a large roasting pan or deep pie dish.

1 quart chicken or vegetable stock
1 quart milk or soy milk
½ cup fresh sage leaves (use less if dried)
3 garlic cloves
2 teaspoons sea salt
Pepper to taste

Roast garlic cloves whole in oven or covered pan on low heat, until soft. Combine with liquid and spices in a large pot, mashing the cloves and heating carefully so as not to burn the milk. Fill the pumpkin with the liquid and replace the lid, putting a sheet of foil between the pumpkin and its top so it doesn't fall in. (If you accidentally destroyed the lid while hollowing the pumpkin, just cover with foil.) Bake the filled pumpkin at 375° for 1–2 hours, depending on the thickness of your pumpkin. Occasionally open lid and check with a spoon, carefully scraping some inside flesh into the hot liquid. If the pumpkin collapses or if the flesh is stringy, remove liquid and flesh to a blender and puree. With luck, you can serve the soup in the pumpkin tureen.

Download these and all other *Animal, Vegetable, Miracle* recipes at www.AnimalVegetableMiracle.com

AUTUMN MEAL PLAN

Sunday ~ Grilled steak and fall potato salad

Monday ~ Sweet potato and chard quesadilla

Tuesday ~ Twice-baked potatoes with cheese and late-season broccoli

Wednesday~ Spanish tortilla (potato and onion frittata)

Thursday ~ Three-bean soup with fresh bread

Friday ~ Pizza with tomato sauce, turkey sausage, roasted onions, and mozzarella

Saturday ~ Cheese and squash quiche

17 · CELEBRATION DAYS

November–December

The closing-down season of the year set us to dragging out storm windows and draining outdoor pipes, but Lily had a whole different agenda: her egg enterprise opened for business. Her April chicks had matured into laying hens, surprising us with their first eggs in late October. Winter is the slow season for egg-laying, with many breeds ceasing production altogether when days are less than thirteen hours long. We'd counseled Lily not to expect much from her flock until next spring.

Never underestimate the value of motivational speeches from the boss. Lily shot out of bed extra early every morning so she'd have time to spend in the chicken coop before the school bus came. Her hens have special nest boxes that open from outside the chickens' roosting quarters, so it's possible to stand (in clean shoes) in the front room of the poultry barn and reach through to collect the eggs. Or in Lily's case, to stand for hours peering in, supervising the hens at their labors. She actually has watched eggs exiting the hens' oviducts—a sight few people on earth have yet checked off their to-do lists, I imagine. When planning this flock she had chosen antique, heavy-bodied breeds with good reputations for laying right through cold weather. By mid-November she was bringing in as many as a dozen eggs a day from her nineteen layers.

Lily apparently knew all along that her workforce could actualize its potential. She had also been working her customer base for months, taking phone numbers in advance. A CEO wears many hats—accountant, supervisor, egg scrubber—but this company's special strength was public relations. Advance planning had taken into account not just winter production, but also egg color. The products from her different breeds of hens crossed a palette from soft green to pink, tan, and chocolate brown. Lily arranged them so every carton contained a rainbow, and printed out her own label, "Lily's Lovely Layers," with a photo of herself holding one of the lovelies. She pasted this over the Brand X names on the recycled cartons friends had saved for her. By the time she made her first sale, customers were practically lined up in the driveway.

Lily was beside herself, dancing around the kitchen with her first dollars. Seeing my young entrepreneur realize her dream made me feel proud too, and also mystified, in the way of all parents who watch their kids acquire skills beyond our ken. When I was that age, the prospect of selling even a Girl Scout cookie mortified me to tears. Now I watched my nine-year-old stand a couple of inches taller each time she picked up the phone to arrange an egg pickup, always remembering first to ask, "How are you today, and how's your family?" In the evenings she sat down at the kitchen table with the account book I'd helped her set up to keep track of customer information, inventory, and expenses. Finally she was entering numbers in the "Income" column.

I soon wondered if I'd have to walk down the driveway and get in line myself. I reminded Lily that our family still needed eggs too. We'd stayed well supplied for the past year from her three old pet hens, which I had presumed were not going to go on payroll. But now their eggs went straight into the Lovely Layer cartons with all the rest. They could be mine, I learned, for $2.50 a dozen. Taking into account the cost of feed, this price gave Lily a small profit margin and still pleased her customers.

I, however, balked at it. Of course I didn't mind rewarding my hard-working daughter, that wasn't the problem. She had been diligent about caring for her hens, closing them safely into their coop every night, even cracking ice off their water bowl on cold mornings. She kept her ears permanently tuned to the chicken voices outside, so knew immediately when a coyote had crept into the yard, and barreled screaming for the front door before the rest of us had a clue. (I don't know about the coyote, but I nearly needed CPR.) These hens owed their lives and eggs to Lily, there was no question.

But since she was taking her business so seriously, I wanted her to understand it genuinely. Businesses have start-up costs. I reminded her that I'd paid for the chicks, and also the feed they'd eaten for six months before they started laying—grazing hens still need supplementary protein, calcium, and other nutrients. I explained to Lily about capitalization, credit, and investors. She listened with interest. "I'll pay you back," she said immediately. "I want the business to be really mine, not just some little kid thing."

She sat down with her ledger to figure the size of the zero-interest loan I'd fronted in venture capital. We had the receipts. We have to buy organic feed in bulk, so we'd already purchased all she'd need until next spring. What she had to calculate was the cost of thirty-two mail-order chicks and the edible wages required to keep the layers producing one full year. (Roosters had been dispatched.) I wondered myself what the figure would be. She bent over her ledger for a long time, pigtail ends brushing the table as her pencil scratched, erased, and scratched some more. Finally she spoke: "Two hundred and eighty-five dollars!"

She rolled out of her chair, flopping dramatically onto the floor, eyes squeezed shut and tongue thrust out to convey either despair or grave fiscal alarm. As I said, a CEO wears many hats.

"Don't panic," I said, sitting down beside her on the floor. "Let's talk about this."

She opened one eye. "Mom, I won't make that much till I'm fifty or something."

"Trust me, you will. It's not as much as it sounds. Plus, if we're keeping track of everything, I owe you for the roosters we ate and all the eggs we've used since April. I forgot about that. Add those up and we'll see where we are."

Where we were, at the beginning of November, was just under $155 in debt. Lily and I made a deal. She would give me all the eggs I wanted, subtracting $2.50 per carton from her debt. I wouldn't charge interest, but I would ask to be considered a priority customer. No standing in line. At two cartons a week, she'd be debt-free in about thirty weeks. To a fourth-grader that sounds roughly the same as life without parole.

"But you'll still be earning real money from all your other customers," I pointed out. "You'll be opening a bank account before you know it. And everybody's going to need extra eggs for our baking, with the holidays coming up."

Cheered by the prospect of holiday baking, the Corporate Executive Officer took the situation in hand. As far as I know, the workforce was never apprised of the crisis.

Of all holidays we celebrate in the United States, few come with food traditions that are really our own. Most of the holy days and bank holidays on our calendar have come from other cultures, some of them ancient, others too modern to have settled yet into having their own menus. The only red-blooded American holiday food customs, it seems to me, arrive on the Fourth of July and Thanksgiving.

They couldn't be more different. The first is all about charring things on a grill, burgers and hot dogs and the like, washed down with plenty of beer or soda, the purpose being to stay outdoors for a long afternoon culminating at dusk in elaborate explosions of gunpowder. Aside from the flagpole that may be somewhere in the scene, there is nothing about this picnic that's really rooted in our land. The pyrotechnics are Chinese, technically, and the rest of the deal is as packaged as food can be. That might even be what's most American about it. At the end of a Fourth of July party, if asked to name the sources of what we'd consumed, we'd be hard pressed to muster an ingredient list.

The other holiday is all about what North America has to offer at the end of a good growing season. Thanksgiving is my favorite, and always has been, I suppose because as a child of the farmlands I appreciate how it honestly belongs to us. On Saint Patrick's Day every beer-drinking soul and his brother is suddenly Irish. Christmas music fills our ears with tales of a Palestinian miracle birth, a generous Turkish saint whom the Dutch dressed in a red suit, and a Druid ceremonial tree . . . I think. But Turkey Day belongs to *my* people. Turkeys have walked wild on this continent since the last ice age, whereas Old Europe was quite turkeyless. (That fact alone scored them nearly enough votes to become our national bird, but in the end, I guess, looks do matter.) Corn pudding may be the oldest New World comfort food; pumpkins and cranberries, too, are exclusively ours. It's all American, the right stuff at the right time.

To this tasty native assembly add a cohort of female relatives sharing work and gossip in the kitchen, kids flopped on the living room floor watching behemoth cartoon characters float down a New York thoroughfare on TV, and men out in the yard pretending they still have the upper-body strength for lateral passes, and that is a perfect American day. If we need a better excuse to focus a whole day on preparing one meal, eating it, then groaning about it with smiles on our faces, just add a dash of humility and hallelujah. Praise the harvest. We made it through one more turn of the seasons.

In modern times it's mostly pageantry, of course, this rejoicing over harvest and having made it to winter's doorstep with enough food. But at our house this year, the harvest was real and the relief literal. Also, for the first time since we'd begun our local-food experiment, we approached a big dining event for which the script was already written. Local turkey? We had some whose lives began in the palms of our hands and ended twelve steps from the back door. Pumpkin pies, mashed potatoes, corn pudding, sweet potatoes, green beans, celery and chestnuts for the stuffing—how could it be this easy? On our continent, this party plans itself.

I had no complaint about celebrating Thanksgiving twice, this time carnivorously. Any excuse to spend a day with friends and my husband and kids is good with me, and I'm partial to the traditional menu. I love carving up Tom on the table, and then revisiting him throughout the following weeks in sandwiches, soups, and casseroles. I have such a fondness for the stuffing, my post-Thanksgiving bliss in childhood was to make stuffing sandwiches. (Dr. Atkins, roll over.) Our recipe starts with a skillet full of sautéed onions, garlic, home-grown celery, and chestnuts from our Chinese chestnut, tossed with a whole loaf of Steven's wheat bread torn to pieces, softened with stock, and spiced with loads of sage and thyme.

We started the evening before, baking several loaves of bread and checking the progress of our thawing bird. I also hacked a Queensland

Blue handily to pieces (ten minutes flat—revenge is sweet) to cook down for pumpkin pies. Lily helped roll out the dough. Both girls have always helped with Thanksgiving dinner, since they were tall enough to stand on a chair and mix the stuffing with their hands like a splendid mud pie. One year earlier, at ages eight and seventeen, they had taken responsibility for an entire holiday meal when I was sidelined with a broken leg. With some heavy-lifting help from Steven they pulled it off beautifully: turkey, pies and all. Cooking is 80 percent confidence, a skill best acquired starting from when the apron strings wrap around you twice.

On Thursday morning we baked the pies. In the afternoon we roasted sweet potatoes, braised winter squash, sautéed green beans with chestnuts, boiled and mashed the potatoes, all while keeping a faithful eye on Mr. T. By herself, Lily cracked half a dozen eggs into a bowl (subtract $1.25 from the I.O. Mama column) and made the corn pudding, using corn we'd cut from the cob and frozen in summer. Our garden provided everything, with one exception. Cranberries mostly grow farther north. I'd planted a small experimental cranberry patch but had nothing yet to show for my efforts. We discussed a cranberryless Thanksgiving, and agreed that would be like kissing through a screen door. Who needs it?

Did *we* need it—was it essential that this feast be 100 percent pure Hoppsolver-grown? Personal quests do have a way of taking on lives of their own, even when nobody else knows or cares: recreational runners push themselves another mile, Scrabblers keep making bigger words. Our locavore project nudged us constantly toward new personal bests. But it always remained fascination, not fanaticism. We still ate out at restaurants with friends sometimes, and happily accepted invitations to dine at their homes. People who knew about our project would get flustered sometimes about inviting us, or when seeing us in a restaurant would behave as if they'd caught the cat eating the canary. We always explained, "We're converts in progress, not preachers. No stone tablets." Our Thanksgiving dinner would include a little California olive oil, a pinch of African nutmeg, and some Virginia flour that likely contained wheat from Pennsylvania and points north. Heeding the imperatives of tradition, we also bought a bag of lipstick-colored organic cranberries from Wisconsin. As the first store-bought fruit or vegetable to enter our house in many months, they looked wildly exotic lounging on our counter, dressed in their revealing cellophane bag. All of us, I think, secretly fondled them before Camille cooked them into a gingery sauce.

Our guests came over in the afternoon to hang out in the kitchen and inhale while everything roasted and braised. We had given away some of our harvested Bourbon Reds, but Mr. Thanksgiving had been chosen while still on his feet, headed all his life for this appointment with our table. He weighed eighteen pounds, even without the mega-breast the Broad-Breasted Whites push around. Historic breeds tend to have proportionally more dark meat, and true to type, this guy made up for his narrower chest with a lot of leg, thigh, and overall heft. His color and texture were so different from the standard turkey, it's hard to compare them. He was a pleasure to cook, remaining exceptionally moist and tender. Thanks to a thin layer of egg-yolk-colored fat under the skin of the breast, the meat seemed to baste itself and gained a delicate aroma and flavor reminiscent, I swear, of lobster. The prices these birds earn on the specialty market are deserved. Taste is the reason for the success of Slow Food USA's turkey project, through which customers sign up in springtime for heritage-breed turkeys delivered by farmers at Thanksgiving. This Bourbon Red, we and our friends agreed when we bit into him, was the richest, most complex-flavored turkey we had eaten.

But a perfect turkey is no more important than any other part of this ritual: lighting the candles, passing the gravy, telling some of the same stories every time. Eating until you swear you are miserable, and then happily eating dessert too. In addition to the pumpkin pies, our friend Maruśka brought a strudel she had learned to make as a young girl in Slovenia. She apologized for its shape, explaining that for holidays it was supposed to be turned like a horseshoe but she didn't have the right pan. Everyone told her, of course, that it didn't matter, it would taste the same. But I understood her longing to re-create in every detail a comestible memory. It's why we'd needed the cranberries for the tart, pinky tinge that oozes into the gravy and makes everything Thanksgiving. It's why we go to the trouble to make a meal with more vegetable courses than many people consume in a week, and way more bird than anyone needs at one sitting.

Having more than enough, whether it came from the garden or the grocery, is the agenda of this holiday. In most cases it may only be a pageant, but holidays are symbolic anyway, providing the dotted lines on the social-contract treasure map we've drawn up for our families and nations. As pageantry goes, what could go more to the heart of things than this story of need, a dread of starvation, and salvation arriving through the unexpected blessing of harvest? Even feigning surprise, pretending it was unexpected and saying a ritual thanks, is surely wiser than just expecting everything so carelessly. Wake up now, look alive, for here is a day off work just to praise Creation: the turkey, the squash, and the corn, these things that ate and drank sunshine, grass, mud, and rain, and then in the shortening days laid down their lives for our welfare and onward resolve. There's the miracle for you, the absolute sacrifice that still holds back seeds: a germ of promise to do the whole thing again, another time.

Oh, yes, I know the Squanto story, we replayed it to death in our primitive grade-school pageantry ("Pilgrim friends! Bury one fish beneath each corn plant!"). But that hopeful affiliation ended so badly, I hate to keep bringing it up. Bygones are what they are. In my household credo, Thanksgiving is Creation's birthday party. Praise harvest, a pause and sigh on the breath of immortality.

⁂

Snow fell on our garden in December, leaving the dried corn stalks and withered tomato vines standing black on white like a pen-and-ink drawing titled *Rest.* I postponed looking at seed catalogs for awhile. Those of us who give body and soul to projects that never seem to end—child rearing, housecleaning, gardening—know the value of the occasional closed door. We need our moments of declared truce.

The farmers' market closed for the year. We paid our last call to the vendors there, taking phone numbers and promising to keep in touch for all kinds of reasons: we would miss our regular chats; we would need advice about the Icelandic sheep we were getting in the spring; we might drive out sometimes to get winter greens from their cold frames. We stocked up on enough frozen meat to see us through winter, including a hefty leg of lamb for one of our holiday dinners.

The tunnel of winter had settled over our lives, ushered in by that great official Hoodwink, the end of daylight saving time. Personally I would vote for one *more* hour of light on winter evenings instead of the sudden, extra-early blackout. Whose idea was it to jilt us this way, leaving us in cold November with our unsaved remnants of daylight petering out before the workday ends? In my childhood, as early as that, I remember observing the same despair every autumn: the feeling that sunshine, summertime, and probably life itself had passed me by before I'd even finished a halfway decent tree fort. But mine is not to question those who command the springing forward and the falling back. I only vow each winter to try harder to live like a potato, with its tacit understanding that time is time, no matter what any clock might say. I get through the hibernation months by hovering as close as possible to the woodstove without actual self-immolation, and catching up on my reading, cheered at regular intervals by the excess of holidays that collect in a festive logjam at the outflow end of our calendar.

We are a household of mixed spiritual backgrounds, and some of the major holidays are not ours, including any that commands its faithful to buy stuff nobody needs. But we celebrate plenty. We give away our salsas and chutneys as gifts, and make special meals for family and friends: turkey and stuffing. Leg of lamb with mint jelly and roasted root vegetables tossed with rosemary and olive oil. For New Year's Day, the traditional southern black-eyed peas and rice, for good luck. Always in the background, not waiting for a special occasion, is the businesslike whir of the bread-machine paddles followed by the aroma of Steven's bread-of-the-day filling the whole house. We have our ways of making these indoor months a more agreeable internment.

When a brand-new organic corn chip factory opened its doors twenty miles from our house, at least one member of our family took it as a sign that wishes do come true. But finding wheat flour for our bread continued to be our most frustrating pursuit. A historic mill five miles from our house processes corn and other specialty flours, but not whole wheat. So we were excited to discover a wheat-flour mill about an hour's drive away, a family operation we were happy to support. But the product, frankly, wasn't what we wanted: bromide-bleached white flour. They also sold a biscuit mix fortified with MSG. We asked if they could process batches of whole wheat or unbleached white flour for us, but we were just one family without enough influence to change even a small company's program. We needed fellow locavores to add clout to our quest, and in time we'll have them. For the time being we liberally supplemented the local product with an organic brand made from wheat grown in Vermont. We sometimes made our own pasta, but more commonly were buying that, too, from outside our state. Ditto for breakfast cereals, though the motherlode was a large package of David and Elsie's amazing oatmeal they sent us as a gift. Some things followed us home from Italy, too, including permanently influenced tastes in wine. But we stuck by our commitment to local meats and produce. In the realm of processed foods, we'd mostly forgotten what's out there.

We'd long since said good-bye to summer's fruits, in exchange for some that are bountiful in December: antique apples, whose flavor improves with cold weather; native persimmons, which aren't edible until after frost hits the tree. It's also the season of citrus in the Deep South, and if you don't live there, the transfer of oranges across a few states from Flor-

How to Impress your Wife, Using a Machine

I know you've got one around somewhere: maybe in the closet. Or on the kitchen counter, so dusty nobody remembers it's there. A bread machine. You can actually use that thing to make some gourmet bread for about 50 cents a loaf, also becoming a hero to your loved ones.

First, get the machine out of the closet (or the box, if it's still in there). Second, I'm sorry, but you'll have to read enough of the manual to know how to put together a basic loaf. Then do exactly that: find a basic recipe for the white or whole wheat loaf and make it a few times, to get a feel for it. Use fresh ingredients; throw out that old flour and yeast and start with new flour milled specifically for bread, preferably organic.

Now comes the creative part. Visit your local health food store or grocery and find the flour section. Most will stock what might be called alternative flours; these are the key to your gourmet bread. Among these are wheat varieties like kamut, pumpernickel, durum, and other grains such as spelt, oats, or rye. Other flours are made from rice, soy, buckwheat, millet, corn, potato, and barley. Beida or Texas still seems more reasonable than some fruit on walkabout from another hemisphere. Our holiday food splurge was a small crate of tangerines, which we found ridiculously thrilling after an eight-month abstinence from citrus. No matter where I was in the house, that vividly resinous orangey scent woke up my nose whenever anyone peeled one in the kitchen. Lily hugged each one to her chest before undressing it as gently as a doll. Watching her do that as she sat cross-legged on the floor one morning in pink pajamas, with bliss lighting her cheeks, I thought: Lucky is the world, to receive this grateful child. Value is not made of money, but a tender balance of expectation and longing.

On Boxing Day we had friends over for a no-holds-barred Italian dinner made from our own garden goods (chestnut and winter-squash-stuffed ravioli) combined with some special things we'd brought back from Tuscany: truffles, olive oil, lupini beans. Starting with rolling out and stuffing the ravioli, we proceeded through the cheeses and bread brought by our guests, a stunning bottle of Bordeaux that was a gift from a French colleague, antipasto of our dried Principe Borghese tomatoes, salad greens from generous friends with a greenhouse, and several other

---

yond these are less familiar (and less appreciated) grains such as teff, amaranth, or quinoa, tubers like yams or arrowroot ground into flours, and meals made from nuts or seeds: chickpeas, flaxseeds, or almonds. In some regions you may find mesquite or malanga. Pick a few of these you'd like to try, and stock them with your other ingredients.

When you put together your next loaf of bread, substitute some of your alternatives for the regular flour. Be experimental, but use only a little at first, just ¼ to ⅓ cup—too much nonwheat flour can compromise the texture or rise of the loaf. Flaxseed meal and buckwheat are especially healthy and successful additions. With practice, you'll find desirable blends, and might be tempted to try out a loaf in your oven. Even the failures will be fresh, warm, and make the house smell great. The successes will become indispensable additions to your good local meals.

STEVEN L. HOPP

courses culminating in a dessert of homemade yogurt, gingered figs, and local honey. We managed to stretch dinner into a five-hour-long social engagement in the Mediterranean fashion. It took ten years for Steven and me to work ourselves up to a vacation in Italy, but from there we were quick studies on how to have dinner.

For most people everywhere, surely, food anchors holiday traditions. I probably spent some years denying the good in that, mostly subconsciously—devoutly refusing the Thanksgiving pie, accepting the stigma my culture has attached to celebrating food, especially for women my age. Because of the inscriptions written on our bodies by the children we've borne, the slowing of metabolisms and inevitable shape-shifting, we are supposed to pretend if we are strong-willed that food is not all that important. Eat now and pay later, we're warned. Stand on the scale, roll your eyes, and on New Year's Day resolve to become a moral person again.

But most of America's excess pounds were not gained on national holidays. After a certain age we can't make a habit of pie, certainly, but it's a soul-killing dogma that says we have to snub it even on Thanksgiving. Good people eat. So do bad people, skinny people, fat people, tall and short ones. Heaven help us, we will never master photosynthesis. Planning complex, beautiful meals and investing one's heart and time in their preparation is the opposite of self-indulgence. Kitchen-based family gatherings are process-oriented, cooperative, and in the best of worlds, nourishing and soulful. A lot of calories get used up before anyone sits down to consume. But more importantly, a lot of talk happens first, news exchanged, secrets revealed across generations, paths cleared with a touch on the arm. I have given and received some of my life's most important hugs with those big oven-mitt potholders on both hands.

Holiday gatherings provide a category of cheer I especially need in winter after the depressing Daylight Robbery incident. Fortunately, the first one follows right on the heels of the clock fall-back, at the beginning of November: Dia de los Muertos. I learned to celebrate the Mexican Day of the Dead during many years of living among Mexican-American friends, and brought it with me to a surprisingly receptive community in southwestern Virginia. It seemed too important to leave behind.

The celebration has its roots in Aztec culture, whose Micteca-cihuatl—"Lady of the Dead"—presided over rituals that welcomed dead friends and ancestors back among the living. Spanish priests arriving among the Aztecs were alarmed to find people dancing around with skeletons, making flowery altars, and generally making whoopee with the memory of their deceased. This would never do. The priests tried moving it from midsummer to November 1 and cloaking it in the Roman Catholic aegis of All Saints Day. Surely everyone would get more from this jolly pagan hootenanny if it were renamed and observed with droning in Latin about an endless list of dead saints.

The date is the only part of that plan that stuck. Dia de los Muertos is still an entirely happy ritual of remembering one's departed loved ones, welcoming them into the living room by means of altars covered with photographs and other treasured things that bring memory into the present. Families also visit cemeteries to dress up the graves. I've seen plots adorned not just with flowers but also seashells, coins, toys, the Blessed Virgin, cigarettes, and tequila bottles. (To get everybody back, you do what you have to do.) Then the family members set out a picnic, often directly on top of a grave, and share reminiscences about the full cast of beloved dead, whether lured in by the flowers or the tequila, and it's the best party of the year. Food is the center of this occasion, especially aromatic dishes that are felt to nourish spiritual presence. The one indispensable food is *pan de muerto,* bread of the dead, a wonderfully sweet, full-of-eggs concoction that Frida Kahlo raised to an art form. For our own Dia de los Muertos celebration this year we cracked enough eggs to make *pan de muerto* for thirty. Thus Frida took a personal hand in lifting Lily's debt.

Anthropologists who write about this holiday always seem surprised by how pleasant the festivals are, despite the obvious connections with morbidity. Most modern lives include very few days penciled onto the calendar for talking and thinking about people we miss because they've died. Death is a gulf we rarely broach, much less celebrate joyfully. By coincidence (or actually, because of those priests again), a different, ancient non-Christian holiday from northern Europe is also celebrated at the same time of year. That one is called Hallowe'en and reinforces an opposite tradition, characterizing death as horrifying and grotesque. Far be it from me to critique an opportunity to dress up and beg free candy, but I prefer Dia de los Muertos. It's not at all spooky. It's funny and friendly.

Most of what's known about religious practices in pre-Hispanic Mexico has come to us through a Catholic parish priest named Hernando Ruiz de Alarcón, one of the few who ever became fluent in the Nahuatl language. He spent the 1620s writing his *Treatise on the Superstitions and Heathen Customs that Today Live Among the Indians Native to This New Spain.* He'd originally meant it to be something of a "field guide to the heathens" to help priests recognize and exterminate indigenous religious rites and their practitioners. In the process of his documentation, though, it's clear from his writings that Father Ruiz de Alarcón grew sympathetic. He was particularly fascinated with how Nahuatl people celebrated the sacred in ordinary objects, and encouraged living and spirit realities to meet up in the here and now. He noted that the concept of "death" as an *ending* did not exactly exist for them. When Aztec people left their bodies, they were presumed to be on an exciting trip through the ether. It wasn't something to cry about, except that the living still wanted to visit with them. People's sadness was not for the departed, but for themselves, and could be addressed through ritual visiting called Xantolo, an ordinary communion between the dead and the living. Mexican tradition still holds that Xantolo is always present in certain places and activities, including wild marigold fields, the cultivation of corn, the preparation of tamales and *pan de muerto.* Interestingly, farmers' markets are said to be loaded with Xantolo.

I'm drawn to this celebration, I'm sure, because I live in a culture that allows almost no room for dead people. I celebrated Dia de los Muertos in the homes of friends from a different background, with *their* deceased relatives, for years before I caught on. But I think I understand now. When I cultivate my garden I'm spending time with my grandfather, sometimes recalling deeply buried memories of him, decades after his death. While shaking beans from an envelope I have been overwhelmed by a vision of my Pappaw's speckled beans and flat corn seeds in peanut butter jars in his garage, lined up in rows, curated as carefully as a museum collection. That's Xantolo, a memory space opened before my eyes, which has no name in my language.

When I'm cooking, I find myself inhabiting the emotional companionship of the person who taught me how to make a particular dish, or with whom I used to cook it. Slamming a door on food-rich holidays, declaring food an enemy, sends all the grandparents and great aunts to a lonely place. I have been so relieved lately to welcome them back: my tiny great-aunt Lena who served huge, elaborate meals at her table but would never sit down there with us herself, insisting on eating alone in the kitchen instead. My grandmother Kingsolver, who started every meal plan with dessert. My other grandmother, who made perfect rolls and gravy. My Henry grandfather, who used a cool attic room to cure the dark hams and fragrant cloth-wrapped sausages he made from his own hogs. My father, who first took me mushroom hunting and taught me to love wild asparagus. My mother, whose special way of beating eggs makes them fly in an ellipse in the bowl.

Here I stand in the consecrated presence of all they have wished for me, and cooked for me. Right here, canning tomatoes with Camille, making egg bread with Lily. Come back, I find myself begging every memory. Come back for a potholder hug.

Food Fright

BY CAMILLE

When I travel on airplanes I often indulge in one of my favorite guilty pleasures: trashy magazines. Nothing makes the time fly like most-embarrassing-first-date stories and completely impractical fashion advice. And of course, always, the diet dos and don'ts. Which ten foods you should eat to melt fat *and* have more energy. On a recent trip I came across an article warning about the Danger Foods for Dieters: the hazards of hidden calories and craving triggers, revealed in a tone I'd thought was reserved for shows like *Unsolved Mysteries.* Would I even be able to sleep that night for fear of an 800-calorie smoothie (disguised as a healthy fruit drink) jumping out from under my bed and pouring itself down my esophagus? Yikes!!

Can we really be this afraid of the stuff that sustains human life? Of where our food comes from, and what it might do to us? We can, we are. TV dinners and neon blue Jell-O *are* unsolved mysteries. As far removed as most of us are from the processes of growing and preparing our food, it makes a certain kind of sense to see food as the enemy. It's very natural to fear the unknown.

The first step toward valuing and trusting food is probably eating food that has some integrity. People who hold their traditions of food preparation and presentation in high regard don't tend to bargain-shop for cheap calories. Associating food with emotional comfort can lead to a life of scary habits and pitfalls, if the training ground is candy bars for good report cards and suckers for bravery during a booster shot. But there are other ways to go. Some of my happiest family memories involve making and eating elaborate meals for special occasions. Food turns events into celebrations. It's not just about the food, but the experience of creating and then consuming it. People need families and communities for this kind of experience. Kids need parents, or some kind of guide, to lead them toward the food routines our bodies need. Becoming familiar with the process of food production generates both respect and a greater sense of calm about the whole idea of dinner.

This Thanksgiving corn pudding goes with one of our roasted turkeys, baked sweet potatoes, steamed Brussels sprouts or braised winter squash, and more than enough stuffing. Here, also, is a recipe for *pan de muerto*, the traditional food for Day of the Dead celebrations. Finally, I've included some recipes we developed for our preserved tomatoes. Dried tomatoes are one of many foods that are ridiculously expensive to buy, inexpensive to make. If you have (or can borrow) a food dryer, you can save a hundred dollars fast by purchasing five extra pounds of small tomatoes every time you visit the market and dehydrating them. In winter we toss them into soups and stews as well as making this antipasto and pesto, which we often pack in fancy anchovy jars for holiday gifts.

❋ HOLIDAY CORN PUDDING A NINE-YEAR-OLD CAN MAKE

3 cups corn kernels
2 eggs, beaten
1 cup milk
1 cup grated Gouda or Jack cheese
2 tablespoons parsley (dried)
1 tablespoon marjoram (dried)
Salt and pepper to taste

Combine ingredients and pour into greased baking dish. Bake at 350° for 45 minutes or until top is puffy and golden.

❋ FRIDA KAHLO'S *PAN DE MUERTO*

This recipe makes 30 small breads. The hard part is making them look like she did: shaped like skulls and dancing whirligig bones. Just making it tasty is not complicated, but you do have to start the dough the evening before your party is scheduled, then bake them just beforehand.

7½ cups white flour, sifted
2 cups sugar (or 1½ cups honey)
1¼ cups butter
2 packages active dry yeast dissolved in 5 tablespoons warm milk
12 eggs
2 teaspoons cinnamon
2 teaspoons vanilla extract

Put flour into a large bowl, cut in the butter, make a well in the center, and pour in the yeast and milk, eggs, sugar or honey, cinnamon, and vanilla. Work it with a spoon, then your hands, until it pulls away from the sides of the bowl. If dough is too soft, knead in more flour. If using honey, more flour will be necessary. Shape into a ball, grease and flour it lightly, and let stand in a warm place for 2½ hours, until doubled. Refrigerate overnight. Shape chilled dough into balls the size of a peach. Then shape or decorate them in any way that makes you think of your deceased ancestors. Place on greased baking sheets and let rise until doubled, about 1½ hours. Dust with powdered sugar and cinnamon, and bake at 350° for 30 minutes, until the bottoms sound hollow when tapped.

❋ ANTIPASTO TOMATOES

Lots of tomatoes

This step involves thinking ahead. Small tomatoes work best for drying—Juliets, Principe Borgheses, Sun Golds, or cherry types. Cut in half and arrange skin-side-down on trays in a food dehydrator, or the sun if you live in a dry climate. Dry until they feel between leathery and brittle.

Vinegar
Dried thyme
Capers
Olive oil

Place dried tomatoes in a bowl. Heat vinegar in a saucepan or microwave, then pour enough into the bowl to cover the tomatoes. Soak for 10 minutes, then pour it off and save (it makes a great vinaigrette). Press off excess vinegar with the back of a wooden spoon. Then toss the damp tomatoes with thyme, or other spices that appeal to you. Pack loosely in glass jars with capers and enough olive oil to cover. They will keep on the shelf this way for several weeks, but taste so good they probably won't last that long.

❋ DRIED TOMATO PESTO

2 cups dried tomatoes
1 cup coarsely chopped walnuts
¾ cup olive oil
⅓ cup grated Parmesan
¼ cup dried basil
4 cloves garlic
2 tablespoons balsamic or other good vinegar
½ teaspoon salt

Puree all ingredients in a food processor until smooth. Add a little water if it seems too sticky, but it should remain thick enough to spread on a slice of bread.

Download these and all other *Animal, Vegetable, Miracle* recipes at www.AnimalVegetableMiracle.com

18 · WHAT DO YOU EAT IN JANUARY?

"January brings the snow . . . ," began the well-thumbed, illustrated children's book about the seasons that my children cleaved to as gospel, while growing up in a place where January did nothing of the kind. Our sunny Arizona winters might bring a rim of ice on the birdbath at dawn, but by midafternoon it would likely be warm enough to throw open the school bus windows. Tucson households are systematically emptied of all sweatshirts and jackets in January, as kids wear them out the door in the morning and forget all about them by noon, piling up derelict sweatshirt mountains in the classroom corners.

Nevertheless, in every winter of the world, Arizona schoolchildren fold and snip paper snowflakes to tape around the blackboard. In October they cut out orange paper leaves, and tulips in spring, just as colonial American and Australian schoolchildren once memorized poems about British skylarks while the blue jays or cockatoos (according to continent) squawked outside, utterly ignored. The dominant culture has a way of becoming more real than the stuff at hand.

Now, at our farm, when the fully predicted snow fell from the sky, or the leaves changed, or tulips popped out of the ground, we felt a shock of thrill. For the kids it seemed like living in storybook land; for Steven and me it was a more normal return to childhood, the old days, the way things ought to be. If we remembered the snow being deeper, the walks to school harder and longer, we refrained from mentioning that to any young person. But the seasons held me in thrall.

And so those words from the Sara Coleridge poem, "January brings the snow," were singing a loop in my head as I sat at the kitchen table watching the flakes blow around in one of those featherweight boxing-match snowstorms. It was starting to drift at bizarre angles, in very odd places, such as inside the eaves of the woodshed. The school bus would likely bring Lily home early if this kept up, but at the moment I had the house to myself. My sole companion was the crackling woodstove that warms our kitchen: talkative, but easy to ignore. I was deeply enjoying my solitary lunch break, a full sucker for the romance of winter, eating a warmed-up bowl of potato-leek soup and watching the snow. Soon I meant to go outside for a load of firewood, but found it easy to procrastinate. I perused the newspaper instead.

Half the front page (above the fold) was covered by a photo of a cocker spaniel with an arrow running entirely through his poor fuzzy torso. The headline—*A MIRACLE: UNHARMED!*—stood in 48-point type, a letter size that big-city newspapers probably reserve for special occasions such as Armageddon. Out here in the heartland, we are not waiting that long. Our local paper's stance on the great big headline letters is: You got 'em, you use 'em.

The rest of it reads about like any local daily in the land, with breaking news, features, and op-eds from the very same wire services and syndicates that fill the city papers I also read. What sets our newspaper apart from yours, wherever you live, is our astounding front-page scoop—the unharmed transpierced dog, the burned-to-the-ground chicken house, the discovery of an unauthorized garbage dump. That, plus our own obituaries and a festive, locally produced lifestyle section.

We newspaper readers all have our pet vexations. Somewhere in one of those sections is the column we anxiously turn to for the sole purpose of disagreeing with the columnist. Volubly. Until family members, rolling their eyes, remind us it's a free country and you don't have to read it *every time.* My own nemesis is not in the World or Op-Ed sections; it's the food column. While I am sick to death of war, corporate crime, and science writers who can't understand the difference between correlation and causation, I try to be open-minded. And yet this food writer has less sense than God gave a goose about where food comes from.

I'd worked on our relationship, moving through the stages of bafflement, denial, and asking this guy out loud, "Where do you live, the *moon*?" I knew the answer: he didn't. He was a local fellow writing just for our region of bountiful gardens and farms, doing his best I'm sure. But no one was ever keener on outsourcing the ingredients. The pumpkins of his world all grow in cans, it goes without saying. If it's fresh ingredients you need, you can be sure the combinations he calls for won't inhabit the same continent or season as one another, or you. On this cozy winter day when I was grooving on the snow that stuck in little triangles on my windowpanes, he wanted to talk pesto.

To lively up anything from pasta to chicken, he said, I should think about fresh basil pesto this week. How do I make it? Easy! I should select only the youngest, mildest flavored leaves, bruising them between my fingers to release the oils before dumping them in my blender with olive oil to make a zingy accompaniment to my meal.

Excuse me? The basil leaves of our continent's temperate zones had now been frozen down to their blackened stalks for, oh, let's count: three months. Sometimes at this time of year the grocery has little packages containing approximately six leaves of the stuff (young and mild flavored?) for three bucks. If I hauled a big bag of money out to my car and spent the next two days on icy roads foraging the produce aisles of this and the neighboring counties, I might score enough California-grown basil leaves to whip up a hundred-dollar-a-plate pesto meal by the weekend. Gee, thanks for the swell idea.

Okay, I know, it's a free country, and I'm a grouch. (Just two weeks later this chef took off for other work in a distant city where he remains safe from my beetle-browed scrutiny.) But if Arizona children have to cut out snowflakes in winter, maybe cooking-school students could be held to a similar standard, cutting out construction-paper asparagus in springtime, pumpkins in the fall, basil in *summer*. Mightn't they even take field trips to farms, four times a year? In our summer garden they'd get a gander at basil bushes growing not as a garnish but a crop. When the leaves begin releasing their fragrance into the dry heat of August, we harvest whole plants by the bushel and make pesto in large batches, freezing it in pint-sized bags. At farmers' markets it starts showing up by the snippet in June and in bulk over the next two months: fresh, fragrant, and inexpensive enough for nongardeners to put up a winter's supply.

Pesto freezes beautifully. When made in season it costs just a fraction of what the grocery or specialty stores charge for pestos in little jars. It takes very little space when frozen flat in plastic bags, then stacked in the freezer like books on a shelf. A pint bag will thaw in a bowl of warm water in less time than it takes to boil the pasta. Tossed together with some pecans or olives, dried tomatoes, and a grind of Parmesan cheese, it's the best of easy meals. But the time to think of bruising those leaves with our fingers to release the oils would be August. Those of us who don't live in southern California or Florida have to plan ahead, not just for pesto but for local eating in general. That seems obvious. But apparently it isn't, because in public discussions of the subject, the first question that comes up is always the same: "What do you eat in January?"

I wish I could offer high drama, some chilling tales of a family gnawing on the leather uppers of their Birkenstocks. From childhood I vividly recall a saga of a family stranded in their car in the Mojave Desert who survived by eating the children's box of Crayolas. I hope in those days crayons were made of something yummy like rendered lard, rather than petroleum. In any case, my childish mind fretted for years about the untold *bathroom* part of the tale. Our family's story pales by comparison. No Chartreuse or Burnt Sienna for us. We just ate ordinary things like pasta with pesto, made ahead.

In the winter we tended more toward carnivory, probably in answer to the body's metabolic craving for warm stews with more fats and oils. Our local meat is always frozen, except in the rare weeks when we've just harvested poultry, so the season doesn't dictate what's available. A meat farmer has to plan in spring for the entire year, starting the Thanksgiving turkeys in April, so that's when the customer needs to order one. But the crop comes in, and finishes, just as vegetables do. When our farmers' market closed for the winter we made sure our freezer was stocked with grass-finished lamb chops and ground beef, crammed in there with our own poultry. And we would now have fresh eggs in every month, thanks to Lily's foresight in raising good winter layers.

People who inhabit the world's colder, darker places have long relied on lots of cold-water ocean fish in their diets. Research on this subject has cracked open one more case of humans knowing how to be a sensible animal, before Little Debbie got hold of our brains. Several cross-cultural studies (published in *Lancet* and the *American Journal of Psychiatry,* among others) have shown lower rates of depression and bipolar disorder in populations consuming more seafood; neurological studies reveal that it's the omega-3 fatty acids in ocean fish that specifically combat the blues. These compounds (also important to cardiovascular health) accumulate in the bodies of predators whose food chains are founded on plankton or grass—like tuna and salmon. And like humans used to be, before our food animals all went over to indoor dining. Joseph Hibbeln, M.D., of the National Institutes of Health, points out that in most modern Western diets "we eat grossly fewer omega-3 fatty acids now. We also know that rates of depression have radically increased, by perhaps a hundred-fold."

In the long, dark evenings of January I had been hankering to follow those particular doctor's orders. We badly missed one of our imported former mainstays: wild-caught Alaskan salmon. We'd found no local sources for fish. Streams in our region are swimming with trout, but the only trout in our restaurants were the flying kind, we'd discovered, shipped on ice from Idaho. And we weren't going to go ice fishing. But instead of plankton eaters our local food chain had grass-eaters: pasture-finished beef has omega-3 levels up to six times higher than CAFO beef; that and Lily's egg yolks would get us through. Steven threw extra flax seeds (also rich in omega-3s) into his loaves of bread, to keep the troops happy.

Legumes were one of our mainstays. Our favorite meal for snow days starts with a pot of beans simmering all afternoon on the woodstove, warming the kitchen while it cooks. An hour before dinnertime I sauté a skillet of chopped onions and peppers until they sweetly melt; living half my life in the Southwest won me over to starting chili with a *sofrito*. Apart from that, my Kentucky chili recipe stands firm: to the bean pot I add the sautéed onions and peppers, two jars of our canned tomatoes, a handful of dried spicy chilies, bay leaves, and a handful of elbow macaroni. (The macaroni is not negotiable.)

Winter is also the best time for baking: fruit pies and cobblers, savory vegetable pies, spicy zucchini breads, shepherd's pies covered with a lightly browned crust of mashed potatoes. The hot oven is more welcome now than in summertime, and it recaptures the fruits and vegetables we put away in season. We freeze grated zucchini, sliced apples, and other fillings in the amounts required by our pie and bread recipes.

So many options, and still that omnipresent question about what local fare one could possibly eat in January. I do understand the concern. Healthier eating generally begins with taking one or two giant steps back from the processed-foods aisle. Thus, the ubiquitous foodie presumptions about fresh-is-good, frozen-is-bad, and salads every day. I've enjoyed that program myself, marking it as progress from the tinned green beans and fruit cocktail of my childhood era when produce aisles didn't have so much of everything all the time.

While declining to return to the canned-pear-half-with-cottage-cheese cookery I learned in high school Home Ec, I've reconsidered some of my presumptions. Getting over the frozen-foods snobbery is important. The broccoli and greens from our freezer stand in just fine for fresh salads, not just nutritionally but aesthetically. I think creatively in winter about using fruit and vegetable salsas, chutneys, and pickles, all preserved back in the summer when the ingredients were rolling us over. Chard and kale are champion year-round producers (ours grow through the snow), and will likely show up in any farmers' market that's open in winter. We use fresh kale in soups, steamed chard leaves for wrapping dolmades, sautéed chard in omelets.

Another of our cold-weather saviors is winter squash, a vegetable that doesn't get enough respect. They're rich in beta-carotenes, tasty, versatile, and keep their youth as mysteriously as movie stars. We grow yellow-fleshed hubbards, orange butternuts, green-striped Bush Delicata, and an auburn French beauty called a *potimarron* that tastes like roasted chestnuts. I arranged an autumnal pile of these in a big wooden bread bowl in October, as a seasonal decoration, and then forgot to admire them after a while. I was startled to realize they still looked great in January. We would finally use the last one in April. I've become a tad obsessive about collecting winter squash recipes, believing secretly that our family could live on them indefinitely if the world as we know it should end. My favorite so far is white beans with thyme served in a baked hubbard-squash half. It's an easy meal, impressive enough for company.

With stuff like this around, who needs iceberg lettuce? Occasionally we get winter mesclun from farming friends with greenhouses, and I have grown spinach under a cold frame. But normal greens season is spring. I'm not sure how lettuce specifically finagled its way, in so many households, from special-guest status to live-in. I tend to forget about it for the duration. At a January potluck or dinner party I'll be taken by surprise when a friend casually suggests, "Bring a green salad." I'll bring an erstwhile salad of steamed chard with antipasto tomatoes, crumbled goat cheese, and balsamic vinegar. Or else everybody's secret favorite: deviled eggs.

In our first year of conscious locavory (locivory?) we encountered a lot of things we hadn't expected: the truth about turkey sex life; the recidivism rate of raccoon corn burglars; the size attained by a zucchini left unattended for twenty-four hours. But our biggest surprise was January: it wasn't all that hard. Our winter kitchen was more relaxed, by far, than our summer slaughterhouse-and-cannery. November brought the season of our Thanksgiving for more reasons than one. The hard work was over. I'd always done some canning and freezing, but this year we'd laid in a larder like never before, driven by our pledge. Now we could sit back and rest on our basils.

"Driven" is putting it mildly, I confess. Scratch the surface of any mother and you'll find Scarlett O'Hara chomping on that gnarly beet she'd yanked out of the ground. "I'll never go hungry again" seems to be the DNA-encoded rallying cry for many of us who never went hungry in the first place. When my family headed into winter my instincts took over, abetted by the Indian Lore books I'd read in childhood, which all noted that the word for February in Cherokee (and every other known native tongue) was "Hungry Month."

After the farmers' market and our garden both closed for the season, I took an inventory of our pantry. During our industrious summer we'd canned over forty jars of tomatoes, tomato-based sauces, and salsa. We'd also put up that many jars of pickles, jams, and fruit juice, and another fifty or so quarts of dried vegetables, mostly tomatoes but also soup beans, peppers, okra, squash, root vegetables, and herbs. In pint-sized freezer boxes we'd frozen broccoli, beans, squash, corn, pesto, peas, roasted tomatoes, smoked eggplants, fire-roasted peppers, cherries, peaches, strawberries, and blueberries. In large ziplock bags we froze quantities of our favorite snack food, whole edamame, which Lily knows how to thaw in the microwave, salt, and pop from the pod straight down the hatch. I do realize I'm lucky to have kids who prefer steamed soybeans to Twinkies. But about 20 million mothers in Japan have kids like that too, so it's not a bolt out of the blue.

Our formerly feisty chickens and turkeys now lay in quiet meditation (legs-up pose) in the chest freezer. Our onions and garlic hung like Rapunzel's braids from the mantel behind the kitchen woodstove. In the mudroom and root cellar we had three bushels of potatoes, another two of winter squash, plus beets, carrots, melons, and cabbages. A pyramid of blue-green and orange pumpkins was stacked near the back door. One shelf in the pantry held small, alphabetized jars of seeds, saved for starting over—assuming spring found us able-bodied and inclined to do this again.

That's the long and short of it: what I did last summer. Most evenings and a lot of weekends from mid-August to mid-September were occupied with cutting, drying, and canning. We'd worked like wage laborers on double shift while our friends were going to the beach for summer's last hurrah, and retrospectively that looks like a bum deal even to me. But we had taken a vacation in June, wedged between the important dates of Cherries Fall and the First of Tomato. Next summer maybe we'd go to the beach. But right now, looking at all these jars in the pantry gave me a happy, connected feeling, as if I had roots growing right through the soles of my shoes into the dirt of our farm.

I understand that's a pretty subjective value, not necessarily impressive to an outsider. It's a value, nonetheless. Food security is no longer the sole concern of the paranoid schizophrenic. Some of my very sane friends in New York and Washington, D.C., tell me that city households are advised now to have a two-month food supply on hand at all times. This is advice of a different ilk from the duct-tape-and-plastic response to terrorist attacks, or the duck-and-cover drills of my childhood. We now have looming threats larger than any cold-hearted human's imagination. Global climate change has created dramatic new weather patterns, altered the migratory paths of birds, and shifted the habitats of disease-carrying organisms, opening the season on catastrophes we are ill-prepared to predict.

"It's not a matter of 'maybe' anymore," my friend from D.C. told me over the phone. A professional photographer, she had been to New Orleans several months after Hurricane Katrina to document the grim demise of a piece of our nation we'd assumed to be permanent. "I'm starting to feel disaster as a real thing—that it's not *if* but *when*. And I feel helpless. When they say you should be keeping that much food on hand, all I can think to do is go to Costco and buy a bunch of cans! Can't I do better than that?" We made a date for the end of next tomato season: she would drive down for a girlfriend weekend and we could can stuff together. Tomato therapy.

Our family hadn't been bracing for the sky to fall, but we now had the prescribed amount of food on hand. I felt thankful for our uncommon good luck. Or if not *luck,* then the following of our strange bliss through the labors it took to get us here, like the industrious ants of Aesop's fable working hard to prepare because it's their nature. Our luck was our proximity to land where food grows, and having the means to acquire it.

Technically, most U.S. citizens are that lucky: well more than half live within striking distance of a farmers' market (some estimates put it at 70 percent), and most have the cash to buy some things beyond their next meal. One of those things could be a thirty-pound bag of tomatoes, purchased some Saturday in July and taken home to be turned into winter foods. Plenty of people have freezers that are humming away at this moment to chill, among other things, some cardboard. That space could be packed with some local zucchini and beans. Any garage or closet big enough for two months' worth of canned goods from Costco could stash, instead, some bushels of potatoes, onions, and apples, purchased cheaply in season. Even at the more upscale farmers' markets in the D.C. area, organic Yukon Gold potatoes run $2 a pound in late summer. A bushel costs about the same as dinner for four in a good restaurant, and lasts 2,800 times as long. Local onions and Ginger Gold apples at the same market cost less than the potatoes, and the same or less than their transported counterparts at a nearby Whole Foods.

It doesn't cost a fortune, in other words. Nor does it require a pickup truck, or a calico bonnet. Just the unique belief that summer is the right time to go to the fresh market with cash in hand and say to some vendors: I'll take all you have. It's an entirely reasonable impulse, to stock up on what's in season. Most of my farming and gardening friends do it. Elsewhere, Aesop is history. Grasshoppers rule, ants drool.

Three-quarters of the way through our locavore year, the process was becoming its own reward for us. We were jonesing for a few things, certainly, including time off: occasionally I clanged dirty pot lids together in frustration and called kitchen strikes. But more often than not, dinnertime called me into the kitchen for the comfort of predictable routines, as respite from the baked-on intellectual residue of work and life that is inevitably messier than pots and pans. In a culture that assigns nil prestige to domestic work, I usually self-deprecate when anyone comments on my gardening and cooking-from-scratch lifestyle. I explain that I have to do something brainless to unwind from my work, and I don't like TV. But the truth is, I enjoy this so-called brainless work. I like the kind of family I can raise on this kind of food.

Still, what kind of person doesn't ask herself at the end of a hard day's work: Was it worth it? Maybe because of the highly documented status of our experiment, I now felt compelled to quantify the work we had done in terms that would translate across the culture gap—i.e., moolah. I had kept detailed harvest records in my journal. Now I sat at my desk and added up columns.

Between April and November, the full cash value of the vegetables, chickens, and turkeys we'd raised and harvested was $4,410. To get this figure I assigned a price to each vegetable and the poultry per pound on the basis of organic equivalents (mostly California imports) in the nearest retail outlet where they would have been available at the time we'd harvested our own. The value-added products, our several hundred jars of tomato sauce and other preserved foods, plus Lily's full-year egg contribution, would add more than 50 percent to the cash value of our garden's production.

That's retail value, of course, much more than we would have earned from selling our goods wholesale (as most farmers do), but it's the actual monetary value *to us,* saved from our annual food budget by means of our own animal and vegetable production and processing. We also had saved by eating mostly at home, doing our own cooking, but that isn't figured into the tally. Our costs, beyond seeds, chicken feed, and our own labor, had been minimal. Our second job in the backyard, as we had come to think of it, was earning us the equivalent of some $7,500 of annual income.

That's more than a hill of beans. In my younger days I spent a few years as a freelance journalist before I hit that mark. And ironically, it now happens to be the median annual income of laborers who work in this country's fields and orchards. We who get to eat the literal fruits of our labors are the fortunate ones.

I harbored some doubts that our family of four could actually consume (or give away as gifts) this dollar-value of food in a year. But that is only $1.72 per person, per meal; that we'd spent that much and more was confirmed by the grocery receipts I'd saved from the year before we began eating locally. As I sat at my desk leafing through those old receipts, they carried me down an odd paper trail through a time when we'd routinely bought things like BAGGED GALA APP ORG, NTP PANDA PFF, and ORNG VALNC 4#BG (I have no idea, but it set me back $1.99), with little thought for the places where these things had grown, if in fact they had grown at all.

We were still going to the supermarket, but the receipts looked different these days. In the first six months of our local year we'd spent a total of $83.70 on organic flour (about twenty-five pounds a month) for our daily bread and weekly pizza dough, and approximately the same amount on olive oil. We'd spent about $5 a week on fair-trade coffee, and had also purchased a small but steady supply of nonlocal odds and ends like capers, yeast, cashews, raisins, lasagna noodles, and certain things I considered first aid: energy bars to carry in the purse against blood sugar emergencies; boxes of mac-and-cheese. Both my kids have had beloved friends who would eat nothing, literally, except macaroni and cheese out of a certain kind of box. I didn't want anybody to perish on my watch.

Still, our grocery-store bill for the year was a small fraction of what it had been the year before, and most of it went for regionally produced goods we had sleuthed out in our supermarket: cider vinegar, milk, butter, cheese, and wines, all grown and processed in Virginia. About $100 a month went to our friends at the farmers' market for the meats and vegetables we purchased there. The market would now be closed for the rest of our record-keeping year, so that figure was deceptively high, including all the stocking up we'd done in the fall. In cash, our year of local was costing us well under 50¢ per meal. Add the $1.72-per-meal credit for the vegetables we grew, and it's still a bargain. We were saving tons of money by eating, in every sense, at home.

Our goal had not really been to economize, only to exercise some control over which economy we would support. We were succeeding on both counts. If we'd had to purchase all our vegetables as most households do, instead of pulling them out of our back forty, it would still be a huge money-saver to shop in our new fashion, starting always with the farmers' market and organizing meals from there. I know some people will never believe that. It's too easy to see the price of a locally grown tomato or melon and note that it's higher (usually) than the conventionally grown, imported one at the grocery. It's harder to see, or perhaps to admit, that all those NTP PANDA PFFs do add up. The big savings come from a habit of organizing meals that don't include pricey processed additions.

In some objective and measurable ways, we could see that our hard work had been worth the trouble. But the truth is we did it for other reasons, largely because it wasn't our day job. Steven and I certainly could have earned more money by putting our farming hours into teaching more classes or meeting extra deadlines, using skills that our culture rewards and respects much more than food production and processing. Camille could have done the same via more yoga classes and hours at her other jobs. Lily was the only one of us, probably, who was maximizing her earning potential through farm labor.

But spending every waking hour on one job is drudgery, however you slice it. After an eight-hour day at my chosen profession, enough is enough. I'm ready to spend the next two or three somewhere else, preferably outdoors, moving my untethered limbs to a worldly beat. Sign me up on the list of those who won't maximize their earnings through a life of professionally focused ninety-hour weeks. Plenty of people do, I know, either perforce or by choice—overwork actually has major cachet in a society whose holy trinity is efficiency, productivity, and material acquisition. Complaining about it is the modern equivalent of public prayer. "Work," in this context, refers to tasks that are stressful and externally judged, which the worker heartily longs to do less of. "Not working" is widely coveted but harder to define. The opposite of work is play, also an active verb. It could be tennis or bird-watching, so long as it's meditative and makes you feel better afterward.

Growing sunflowers and beans is like that, for some of us. Cooking is like that. So is canning tomatoes, and making mozzarella. Doing all of the above with my kids feels like family life in every happy sense. When people see the size of our garden or the stocks in our pantry and shake their heads, saying "What a lot of work," I know what they're really saying. This is the polite construction in our language for "What a dope." They can think so. But they're wrong.

⁂

This is not to say family life is just la-di-dah around here. Classes and meetings and deadlines collide, mail piles get scary. Children forget to use their inside voices when charging indoors to demand what's for dinner. Mothers may also forget to use them when advising that whoever's junk is all over the table has three seconds to clear it up or IT'S GOING TO THE LANDFILL. Parent-teacher night rises out of nowhere, and that thing I promised to make for Lily's something-or-other is never, ever due any time except *tomorrow morning, Mama!* Tears happen. On the average January weeknight, I was deeply grateful that I could now just toss a handful of spaghetti into a pot and reach for one of the quart jars of tomato sauce that Camille and I had canned the previous August. Our pantry had transformed. No longer the Monster Zucchini roadhouse brawl (do not enter without a knife), it was now a politely organized storehouse of healthy convenience foods. The blanched, frozen vegetables needed only a brief steaming to be table-ready, and the dried vegetables were easy to throw into the Crock-Pot with the chicken stock we made and froze after every roasted bird. For several full-steam-ahead weeks last summer, in countless different ways, we'd made dinner ahead. What do we eat in January? Everything.

But when the question comes up, especially when winter is dead upon us, I feel funny about answering honestly. Maybe I'm a little embarrassed to be a dweeby ant in a grasshopper nation. Or I'm afraid it will come as a letdown to confess we're not suffering as we should. Or that I'll sound as wacky as Chef January Pesto, to folks trying to eat locally who are presently stuck with farmers' markets closed for the season.

If you're reading this in midwinter and that is your situation, put the thought away. Just never mind, come back in six months. Eating locally in winter is easy. But the time to think about that would be in August.

Getting Over the Bananas

BY CAMILLE

Many summers ago my best friend Kate, from Tucson, came out to visit our farm in Virginia for the first time. She was enamored of our beautiful hills, liked working in our garden, and happily helped pick blackberries from the sea of brambles that skirt the surrounding fields. She did her part, carrying with us the armloads of beans, cucumbers, squash, and tomatoes that came out of our garden on a daily basis. But one day, on a trip to the grocery store, we hit a little problem. When my parents asked if there was anything in particular she would like to eat, she replied, "Let's get some bananas!"

My parents exchanged a glance and asked her for another suggestion.

"Why not bananas?" she asked, feeling really baffled by their refusal. My mother is not the type to say no to a guest. She waited until we were in the car to explain to Kate that it seemed wasteful to buy produce grown hundreds of miles away when we had so much fresh fruit right now, literally in our backyard. We'd picked two gallons of blackberries that very morning. She didn't want to see them get moldy while we were eating bananas.

"Plus, think of all the gasoline it takes to bring those bananas here," Mom pointed out. My friend was quiet while the wheels turned inside her head. "I never thought about that before," she admitted.

Kate has grown up to become passionate about farming and eating organic food. Since the banana incident, she has volunteered on small farms and developed a sincere interest in agricultural methods that preserve biodiversity. She looks back at that conversation in the grocery store as a life-changing moment. Some things you learn by having to work around the word "no."

Of course local eating gets trickier in wintertime. Fresh fruits and vegetables are rare or just gone then, for most of us. In the colder months we have to think roots, not fruits. Potatoes, carrots, beets, turnips, parsnips, and celariac cover the full spectrum of color. Winter squash are delicious too; most people I know have never eaten one, probably because they never needed to. They couldn't see them for all the bananas in the way.

Each season requires thinking about food in a different way. In her book *Local Flavors,* Deborah Madison writes that when she teaches a fresh summer eggplant dish in her cooking classes, students always say two things: "Wow, I never liked eggplant before, but I love this!" and "What kind of eggplants should I buy in December, to make this for Christmas?" It takes a while for them to realize that these particular fresh eggplants were so yummy *because* they weren't buying them in the winter. Most of us agree to put away our sandals and bikinis when the leaves start to turn, even if they're our favorite clothes. We can learn to apply similar practicality to our foods. Here are some delicious winter-vegetable recipes, along with a week of dinners for the cold months at our house.

❋ BRAISED WINTER SQUASH

2 pounds winter squash, peeled, halved, and sliced into ½-inch rounds
2 tablespoons butter
2 cups apple cider
1 teaspoon salt
Rosemary and pepper to taste

Melt butter in skillet with rosemary; after a few minutes add the squash, salt, and cider. You may need to add some additional cider (or water), enough to cover the squash. Bring to a boil, then lower heat and braise for 20 minutes or until tender. At this point the juice should be reduced to a glaze. If not, raise heat for a few minutes until excess liquid evaporates. Add pepper and a splash of balsamic vinegar if you like.

❋ BUTTERNUT BEAN SOUP

(serves 4)

1½ cups dried white beans, soaked overnight and drained
3 medium portobello mushroom caps, sliced (optional)
6 garlic cloves, finely chopped
1 tablespoon thyme
1 tablespoon sage
4 teaspoons rosemary

Combine beans and spices in a large saucepan, add water to cover amply, and simmer for 30 to 40 minutes, until beans are tender and most water has cooked off. Add mushrooms toward the end.

2 butternut or hubbard squash, halved lengthwise and seeded
Olive oil

While beans are cooking, drizzle a large roasting pan with olive oil and arrange squash skin side down. Cook at 400° for about 40 minutes, until fully tender when pierced with a fork. Remove from oven and serve each half squash filled with a generous scoop of bean soup.

❋ VEGETARIAN CHILI

1 pound dry kidney beans, soaked overnight and drained
1 cup chopped carrots
2 large onions, chopped
1 cup frozen peppers (or ½ cup dried)
3 cloves garlic, minced
Olive oil
28 ounces canned tomatoes, undrained
4 cups vegetable stock or tomato juice
3–5 tablespoons chili powder

4–5 bay leaves
1 tablespoon cumin

Sauté garlic, peppers, and onions in olive oil until golden, add chopped carrots, and cook until tender. Combine with beans and remaining ingredients; stir well. Thin with extra water, stock, or tomato juice as needed. Cover and simmer for one hour. If you are related to my mother, you have to add 8 ounces of elbow macaroni, 15 minutes before serving.

❋ SWEET POTATO QUESADILLAS

2 medium sweet potatoes
½ onion
1 clove garlic
1 tablespoon oregano
1 tablespoon basil
1 teaspoon cumin
Chile powder to taste
Olive oil for sauté

Cut sweet potatoes into chunks, cook in steamer basket or microwave until soft, then mash. Chop and sauté garlic and onion in a large skillet. Add spices and sweet potato and mix well, adding a little water if it's too sticky. Turn burner very low to keep warm without burning.

4 flour tortillas
4 ounces Brie or other medium soft cheese
2–3 leaves Swiss chard (or other greens)

Preheat oven to 400°. Oil a large baking sheet, spread tortillas on it to lightly oil one side, then spread filling on half of each. Top with slices of Brie and shredded chard, then fold tortillas to close (oiled side out). Bake until browned and crisp (about 15 minutes); cut into wedges for serving.

Download these and all other *Animal, Vegetable, Miracle* recipes at www.AnimalVegetableMiracle.com

WINTER MEAL PLAN

Sunday ~ Roast leg of lamb with mint jelly, baked Yukon Golds, and multi-colored beet salad

Monday ~ Vegetarian chili

Tuesday ~ Pasta with pesto, olives, and grated cheese

Wednesday ~ Steak with winter potato salad or baked sweet potatoes

Thursday ~ Bean burritos with sautéed onion and dried tomatoes

Friday ~ Crock-Pot chicken soup with vegetables, served with warm bread

Saturday ~ Cheese and mushroom tortellini with tomato sauce

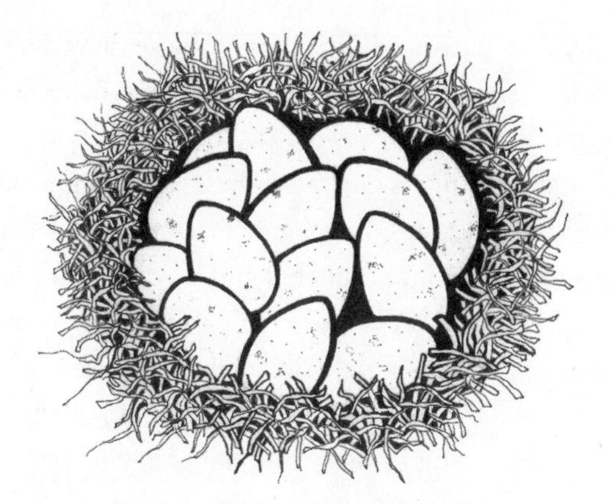

19 · HUNGRY MONTH

February–March

As I grow older, more of my close friends are elderly people. I suppose I am auditioning, in some sense, to join their club. My generation will no doubt persist in wearing our blue jeans right into the nursing homes, kicking out Lawrence Welk when we get there and cranking up "Bad Moon Rising" to maximum volume. But I do find myself softening to certain features of the elder landscape. Especially, I'm coming to understand that culture's special regard for winter. It's the season to come through. My eighty-four-year-old neighbor is an incredibly cheerful person by all other standards, but she will remark of a relative or friend, "Well, she's still with us after the winter."

It's not just about icy sidewalks and inconvenience: she lost two sisters and a lifelong friend during recent winters. She carries in living memory a time when bitter cold and limited diets compromised everyone's immunities, and the weather forced people to hunker down and share contagions. Winter epidemics took their heartbreaking due, not discriminating especially between the old and the young. For those of us who have grown up under the modern glow of things like vaccinations, penicillin, and central heat, it's hard to retain any real sense of this. We flock indoors all the time, to work and even to exercise, sharing our germs in all seasons. But vitamins are ready at hand any time, for those who care, and antibiotics mop up the fallout.

Tying my family's nutritional fortunes to the seasons did not really involve any risk for us, of course. But it did acquaint us in new ways with what seasons mean, and how they matter. The subtle downward pulse of temperatures and day lengths created a physical rhythm in our lives, with beats and rests: long muscles, long light; shorter days, shorter work, and cold that drew us deeper into thoughts and plans, under plaster ceilings instead of an open sky. I watched the rank-and-file jars in our pantry decline from army to platoon, and finally to lonely sentries staggered along the shelves. We weren't rationing yet, but I couldn't help counting the weeks until our first spring harvests and the happy reopening day of the farmers' market. I had a vision of our neighbors saying of us, "Well, they're still with us after the winter."

In late February, official end of "Hungry Month," I was ready to believe we and all our animals had come through the lean times unscathed. And then one of our turkey hens took to standing around looking droopy. She let her wings drop to the ground instead of folding them on her back in normal turkey fashion. Her shoulders hunched and her head jutted forward, giving her a Nixonesque air—minus the eyebrows and crafty agenda. This girl just looked dazed.

Oh, no, I thought. Here we go. Farm animals lucky enough to live on pasture must deal with winter eventually, and health challenges similar to those faced by people in previous generations: less fresh air, more indoor congregation and risk of contagion, and the trial of surviving on stored hay or grain instead of fresh greens and hunted protein. In the realm of contagious maladies, poultry husbandry is notoriously challenging. And turkeys are even more disease-prone than chickens. "You never see it coming," a turkey-experienced friend of mine had warned when we first got our poults. "One day they're walking around looking fine. Next thing you know, drop-dead Fred!" The list of afflictions that can strike down a turkey would excite any hypochondriac: blackhead roundworm, crop bind, coccidiosis, paratyphoid, pullorum disease, and many more. In one of my poultry handbooks, the turkey chapter is subtitled, "A Dickens to Raise."

So far, though, my turkeys had stayed hale and hearty and I'd taken all the doom-saying with a grain of salt. Virtually all turkey-info sources in existence (including my friend with her drop-dead Freds) refer to the Broad-Breasted White, the standard factory-farm turkey that's also the choice of most hobby farmers and 4-H projects, simply because the alternatives aren't well known. My heirloom Bourbon Reds were a different bird, not bred for sluggardly indoor fattening but for scrappy survival in the great outdoors. They retain a genetic constitution for foraging, flying, mating, and—I hoped—resisting germs.

Even so, my goal of keeping these birds alive through the winter and into their second year for breeding was statistically audacious. The longer a bird is kept, the greater its chances of being overwhelmed by pathogens. The great majority of modern turkeys can expect an earthly duration of only four months before meeting their processor. Free-range turkeys may take as long as six months to reach slaughter size. But any bird that lives past its first Thanksgiving inhabits a domain occupied by fewer than one-half of one percent of domestic turkeys. At nine months, my flock had now entered that elite age bracket, among the oldest living turkeys in America. When I undertook to keep a naturally breeding flock, I hadn't thought much about what I was up against.

Nor did I have any clue, now, which possible turkey ailment my poor droopy hen might have. The drear blackhead roundworm topped my worry list, since its inventory of symptoms began with "droopy aspect," proceeding from there into "aspects" too unpleasant to mention. I immediately removed Miss Droop from the rest of the flock, assuming she'd be contagious. I've had kids in preschool—I know *that* much. I ushered her out of the big outdoor pen attached to the poultry house, and escorted her several hundred yards into an isolation room in the cellar of our big barn. Yes, it's the same one where we sequester the death-row roosters; I didn't discuss that with her.

In fact, when I shut her in there by herself she immediately perked up, raising her head and folding her wings onto her back, shaking her fluffed feathers neatly back into place, looking around brightly for what she might find to do next. I hardly trusted this miracle recovery, but an hour later when I came back to check on her she was still perky and now calling desperately for her friends. (Turkey hens hate being alone.) I decided I must have worried the whole thing out of thin air.

I led her back up to the turkey pen. She walked into the midst of her brethren, heaved a great turkey sigh, and drooped back down again: head hunched, wings dragging the ground. Okay, I wasn't imagining it, she really had something. I shepherded her back out and down to the barn again. And once again, watched in astonishment as she lifted her head brightly and began to walk around, looking for something to eat.

I stared at her. "Are you goldbricking?" I demanded.

It was a sunny Saturday, our first little sneak preview of spring. Steven was down in the new orchard working on the fence. I decided my poultry patient could use a mental health day. I let her out of the barn and we walked together down the road toward the orchard. She could get some sunshine and fresh greens, and I'd see if Steven needed help with the fence. He saw us coming down the lane together, and laughed. Turkey herders are not a respected class of people.

"She needs some fresh pasture," I said defensively. "She has some kind of droopy sickness."

"She looks fine," he said (which was maddeningly true), and went back to the fence that's meant to discourage deer from eating our young pear orchard. For a few minutes I watched Ms. Turkey happily foraging among the trees, pecking at seed heads, alerting to any small movement of insects among the clumps of grass. She seemed as healthy as the day she was born. Some of our trees, on the other hand, showed signs of deer damage. I inspected them closely and considered going back up to the shed to get the lop shears. This winter day would be a good time for pruning the fruit trees.

Steven yelled, "Hey, knock it off!"

I looked up to see he wasn't talking to me. My charge had wandered over to him, approaching from behind and reaching up with her beak to give his jacket a good tug, issuing a turkey mandative I would translate as: *Hey, look at me!*

He nudged her away, but she persisted. After several more tugs, he turned to face her directly, planted his feet, and made a very manly sort of huff. On that cue she coyly turned her tail toward him, jutted her neck, and dropped her wings to the ground.

Oh, my goodness. It wasn't *Hey, look at me.* It was *Hey, sailor, new in town?* That's what she had: love sickness. Steven shot me a look I will not translate here.

"Stop that," I yelled. "He is *so* not your type!" I ran to interrupt her, in case she meant to move their relationship to the next level.

Poor thing, how would she know? She was raised by humans, with no opportunity to imprint on adult turkeys of either gender or observe proper turkey relations. As far as she knew, I was her mother. It's only logical that the person I married would strike her as a good catch.

As quickly as possible I ushered her back to the turkey pen, putting the kibosh on her plot to win away my husband. But now what? We'd kept two males and six females for breeding purposes, with no real logic behind this number beyond a hope that we'd still have enough, in case we lost any birds over the winter. Were they now all about to come into season? Would our two toms suddenly wake up and start killing each other over this droopy Lolita? And what about the other hens? Who needed to be separated from whom, for how long? Would every hen need her own nest, and if so, what would it look like?

I had assumed I'd cross all these bridges when I came to them. I remember harboring exactly this kind of unauthorized confidence before I had my first baby, also, only to look back eventually upon my ignorance and bang my head with the flat of my hand. Now, suddenly, long before I'd ever expected any shenanigans, like parents of turkey teens everywhere, I was caught by surprise. *They're too young for this, it's only February!* I went indoors to check our farm library for anything I could find about turkey mating behavior.

I spent way too much of a beautiful day inside, on the floor, with books stacked all around me. Our poultry husbandry manuals contained a total of *nothing* about turkey sex. I kept looking, checking indices for various barnyard euphemisms: nothing. Honestly, our kids' bookshelves had over the years been furnished with more literature in the "Now That You're Growing Up" department. You'd think some turkey fundamentalists had been in here burning books.

The real problem, of course, was that I was looking for a category of information nobody has needed for decades. The whole birds-and-bees business has been bred out of turkeys completely, so this complex piece of former animal behavior is now of no concern to anyone. Large-scale turkey hatcheries artificially inseminate their breeding stock. They extract the eggs in a similarly sterile manner and roll them into incubators, where electric warmth and automatic egg-turning devices stand in for motherhood. For the farmers who acquire and raise these hatchlings, the story is even simpler: fatten them as quickly as possible to slaughter size, then off with their heads. That's it. Poultry handbooks don't go into mating behavior because turkey mating has gone the way of rubberized foundation garments and the drive-in movie.

To restore some old-fashioned sex to our farm, I was going to have to scour my sources for some decent sex ed. The Internet was no more help. A search for "turkey mating" scored 670,000 hits, mostly along the lines of this lively dispatch from the Missouri Department of Conservation: "More excitement this week—hunters statewide will find gobblers more responsive to calls! The key to success is sounding like a lovesick turkey hen."

I already *had* a lovesick turkey hen, no need to fake that one. I tried limiting my search to domestic turkeys rather than wild ones. I still got thousands of hits, but not one shred of fact about turkey hokey-pokey. I did learn that the bright blue-and-pink growths on a male turkey's neck are called his "caruncle." I learned that the name "turkey" for this solely North American bird comes from a 400-year-old geographic mistake made by the English. I learned that the French know this bird as a *dindon sauvage*. That is when I fled from the electronic library, returning to my limited but reassuring paper pages where I could feel safe from the random onslaught of savage ding-dongs.

Finally there I hit pay dirt. My spouse has a weakness for antique natural history books. His collection of old volumes covers the gamut from Piaget and Audubon to William J. Long, an early-twentieth-century ethologist who attributed animal communications to a telepathic force he called "chumfo." You may gather that I was desperate, to be plumbing these depths for help around the farm. But I found a thick tome by

E. S. E. Hafez called *The Behavior of Domestic Animals.* Published mid-twentieth century, it's probably the most modern entry in Steven's collection, but for my purposes that was exactly the right era: animal science had advanced beyond chumfo, but had not yet taken the tomfoolery out of the toms.

What caught my eye as I flipped through the book was a photograph with this caption: "Female turkey giving the sexual crouch to man . . ." Bingo! The text confirmed my worst suspicions: turkeys who had imprinted on humans, as hatchlings, would be prone to batting for the hominid team. But given the chance, the book said, they would likely be open-minded about turkey partners as well.

Oh, good! Reading on, I learned that the characteristic droopy "crouch" is the first sign of sexual receptiveness in girl turkeys. Soon we could expect to see a more extended courtship interaction that would include stomping (boy), deeper crouch (girl), then mounting and much treading around as the male manipulated the female's "erogenous area along the sides of the body," followed by the complicated "copulatory sequence." Domestic turkeys are promiscuous, I learned, with no inclination toward pair bonding. Egg laying would begin in about two weeks. A turkey hen's instinct for sitting on the nest to brood the eggs, if that happened, would be triggered when enough of them accumulated in the nest. The magic number was somewhere between twelve and seventeen eggs.

Eggs and nest were all theoretical at this point, but what concerned me most was the broody instinct getting switched on. These mothering instincts have been bred out of turkeys. For confinement birds the discouragement has been purposeful, and even heirloom breeds are mostly sold by hatcheries that incubate mechanically, so nobody is selecting for good maternal behaviors. Genes get passed on without regard for broody or nonbroody behavior. If anything, it's probably a bother to hatchery operations when a mother gets possessive about her young.

If I wanted to raise turkeys the natural way, I understood now that I was signing up for a strong possibility of failure, not to mention a deep involvement in the sexual antics of a domestic bird. My interests weren't prurient (though you may come to doubt this later in the chapter). As a biologist *and* a PTA member, I have a healthy respect for the complex parameters of motherhood. The longer I think about a food industry organized around an animal that cannot reproduce itself without technical assistance, the more I mistrust it. Poultry, a significant part of the modern diet, is emblematic of the whole dirty deal. Having no self-sustaining bloodlines to back up the industry is like having no gold standard to underpin paper currency. Maintaining a naturally breeding poultry flock is a rebellion, at the most basic level, against the wholly artificial nature of how foods are produced.

I was the rebel, that was my cause. I had more than just sentimental reasons for wanting to see my turkey hens brood and hatch their own babies, however unlikely that might be. I plowed on through my antique reference for more details on nesting and brooding, and what I might do to be a helpful midwife, other than boiling water or putting a knife under the bed. My new turkey-sex manual got better and better. "Male turkeys," I read, "can be forced to broodiness by first being made drowsy, e.g. by an ample dose of brandy, and then being put on a nest with eggs. After recovery from the hangover, broodiness is established. This method was used extensively by farmers in Europe before incubators were available."

I don't think of myself as the type to ply turkey menfolk with brandy and hoodwink them into fatherhood. But a girl needs to know her options.

⚘

Six quarts of spaghetti sauce, four jars of dried tomatoes, four onions, one head of garlic at the end of a long, skinny, empty braid—and weeks to go. January is widely held to be the bugbear of local food, but the hungriest month is March, if you plan to see this thing through. Your stores are dwindling, your potatoes are sending pale feelers out into the void, but for most of us there is nothing new under the sun of muddy March, however it might intend to go out like a lamb. A few spring wildflowers, maybe, but no real eats. Our family was getting down to the bottom of our barrel.

Which was a good thing for the chest freezer. I know people who layer stuff in there year after year, leaving it to future archaeologists, I gather, to read the good and bad green-bean years like tree rings. I've taken microbiology, and honestly, ick. I'm pretty fanatical about emptying the freezer completely before starting over. A quick inventory found our frozen beans long gone, but we still had sliced apples, corn, one whole turkey, and some smoked eggplant from last fall. Also plenty of zucchini, *quelle surprise*. We would not be the Crayola Family, then, but the one that survived on zucchini pie. Pretty cushy, as harrowing adventures go.

Maybe March doesn't get such a bad rap because it doesn't *feel* hungry. If it's not the end of winter, you can see it from here. Lily and I were now starting our vegetable and flower seeds indoors, puttering in earnest under the fluorescent lights of our homemade seedling shelves. She had given up all hope of further snow days. And one fine afternoon she bounced off the bus with the news that the fourth-graders were going to study *gardening* at school. For a kid like Lily, this was an unbelievable turn of events: Now, children, we are going to begin a unit on recess!

It wasn't just the fourth-graders, as it turned out. The whole school was that lucky, along with three other elementary schools in our county. School garden programs have lately begun showing up in schools from the trend-setting Bay Area to working-class Durham, North Carolina. Alice Waters founded the Berkeley programs, developing a curriculum that teaches kids, alongside their math and reading, how to plant gardens, prepare their own school lunches, and sit down to eat them together in a civil manner. She has provided inspiration nationwide for getting fresh-grown food into cafeterias.

But most of the garden-learning programs scattered through our country's schools have been created independently, as ours was. A local nonprofit helps support it, the school system has been cooperative, but our Learning Landscapes curriculum is the dream and full-time project of a green-thumbed angel named Deni. She helps the kindergartners grow popcorn and plant a rainbow of flowers to learn their colors. Second-graders make a special garden for hummingbirds, bees, and butterflies, while learning about pollination. Third-graders grow a pizza garden that covers the plant kingdom. Lily's class was starting seeds they planned to set out in a colonial herb garden, giving some life to their Virginia history lessons. Each grade's program is tied to concrete objectives the kids must know in order to pass their state-mandated testing.

Virtually everyone I know in the school system feels oppressed by these testing regimes hanging over everything. Teachers sense them as huge black clouds on the horizon of April. For the kids it's more like a permanent threat of air attack. In our state—no kidding—they are called Standards of Learning, or "SOLs." (I don't think anyone intended the joke.) But Learning Landscapes works because it gets kids outdoors studying for tests while believing they are just playing in dirt.

Deni knows how to get the approval of a school board, but she has a larger game plan for these kids than just passing the next exam. "One of the key things gardens can teach students is respect: for themselves, for others, and the environment," she says. "It helps future generations gain an understanding of our food system, our forests, our water and air, and how these things are all connected."

From a biological perspective, the ultimate act of failure is to raise helpless kids. Not a parent I know who's worth the title wants to do that. But our operating system values Advanced Placement Comparative Politics, for example, way, way ahead of Knowing How to Make Your Own Lunch. Kids who can explain how supernovas are formed may not be allowed to get dirty in play group, and many teenagers who could construct and manage a Web site would starve if left alone on a working food farm.

Legislating Local

The epidemic of childhood obesity in the United States has incited parents, communities, and even legislators to improve kids' nutrition in one place they invariably eat: schools. Junk foods have been legally banned from many lunchrooms and school vending machines. But what will our nation's youth eat instead—fresh local produce? *As if!*

Dude, it's going down. In 2004, in a National School Lunch Act amendment, Congress authorized a seed grant for the Farm to Cafeteria Program, promoting school garden projects and acquisition of local foods from small farms. The Local Produce Business Unit of the Department of Defense actually procures produce. Benefits of these programs, above and beyond the food, include agricultural education through gardening, farm visits, presentations by local farmers, and modest economic gains for the community. More than one-third of our states now have active farm-to-school programs; farm-to-college alliances are also growing.

That's hardly their fault. We all may have some hungry months ahead of us, even hungry years, when a warmed-up globe changes the rules of a game we smugly thought we'd already aced. We might live to regret some of our SOL priorities. But the alumni of at least one Appalachian county's elementary schools will know how to grow their own pizzas, and I'm proud of them. If I could fit that on a bumper sticker, I would.

My pupils in the turkey coop were not such quick studies. The first hen who'd come into season was getting no action from either of the two males, whom we had lately been calling Big Tom and Bad Tom. These guys had been fanning their tails in urgent mating display since last summer, more or less constantly, but they directed the brunt of their show-off efforts toward me, each other, or any sexy thing I might leave sitting around, such as a watering can. They really tried hard with the watering can. Lolita kept plopping herself down where they'd have to trip over her, but they only had eyes for some shiny little item. She sulked, and I didn't blame her. Who hasn't been there?

I determined to set a more romantic scene, which meant escorting

---

The USDA Special Supplemental Nutrition Program for Women, Infants, and Children (WIC) has a Farmers' Market Nutrition Program for purchasing local food. It provides coupons good for fresh produce purchased from farms, farmers' markets, and roadside stands. In 2006, some $20 million in government funds provided these benefits to more than 2.5 million people.

In a strong legislative move, Woodbury County, Iowa, mandated in 2006 that the county (subject to availability) "shall purchase . . . locally produced organic food when a department of Woodbury County serves food in the usual course of business." Even the prisons are serving local food, in a county that truly recognizes the value of community support.

For more information visit www.foodsecurity.org and www.farmtoschool.org.

STEVEN L. HOPP

Lolita and one of the toms into their own honeymoon suite, a small private room inside the main barn, and removing any watering cans from his line of sight. She practically had to connect the dots for him—no bras to unhook, heaven be praised—but finally he started to get the picture. She crouched, he approached, and finally stopped quivering his tailfeathers to impress her. After all these many months, it took him a couple of beats to shift gears from "Get the babe! Get the babe!" to "O-oh *yess!*" Inch by inch he walked up onto her back. Then he turned around in circles several times, s-l-o-w-l-y, like the minute hand of a clock, before appearing to decide on the correct orientation. I was ready to hear the case for artificial insemination. But it looked now like he was giving it a go.

The final important event after all this awkward foreplay is what bird scientists call the "cloacal kiss." A male bird doesn't have anything you would call "a member," or whatever you call it at your house. He just has an orifice, or cloaca, more or less the same equipment as the female except that semen is ejected from his, and eggs come out of hers later on. Those eggs will be fertile only if the two orifices have previously made the prescribed kind of well-timed contact.

I watched, I don't mind saying. Come on, wouldn't you? Possibly you would not have stooped quite as low as I did for the better view, but geez, we don't get cable out here. And this truly was an extraordinary event, something that's nearly gone from our living world. For 99.9 percent of domestic turkeys, life begins in the syringe and remains sexless to the end. Few people alive have witnessed what I was about to see.

Cloacal kiss is exactly the right name for it. The male really has to extend that orifice, like puckering up for a big smooch. Try to picture this, though: he's standing on her back, tromping steadily and clutching his lady so as not to fall off. The full complement of her long tail-feather fan lies between his equipment and hers. The pucker has to be heroic to get around all that. Robert Browning said it perfectly: *Ah, but a man's reach should exceed his grasp, or what's a Heaven for?*

Paradise arrives when a fellow has kneaded his lady's erogenous wing zones for a long, long time with his feet, until she finally decides her suitor has worked himself up to the necessary fervor. Without warning, quick as an eyeblink, she flips up her tail feathers and reaches upward to meet him. Oh, my gosh! I gasped to see it.

It was an *air kiss.*

They really did miss. *Mwah!* —like a pair of divas onstage who don't want to muss their lipstick. (Not Britney and Madonna.) But rare is the perfect first attempt, I know as well as the next person who has ever been young.

She wandered off, slightly dazed, to a corner of the dark little room. He stared after her, his feathers all slack for once in his life, divining that this was not the time to put on a tail-shaking show. He knit his caruncled brow and surely would have quoted Shakespeare if he'd had it in him: *Trip no further, pretty sweeting; journeys end in lovers meeting . . .*

She pecked listlessly at some grain on the dirt floor. Probably she'd been hoping for better room service.

What's to come is still unsure: in delay there lies no plenty. Then come kiss me, sweet and twenty, Youth's a stuff will not endure.

I left them there, to love again on the morrow. Or maybe in fifteen minutes. After all, they were kids.

⚜

Animal behaviorists refer to a mating phenomenon called the "Coolidge effect," a term deriving from an apocryphal story about the president and first lady. On an official visit to a government farm in Kentucky, they are said to have been impressed by a very industrious rooster. Mrs. Coolidge asked her guide how often the cockerel could be expected to perform his duty, and was informed: "Dozens of times a day."

"Please tell that to the president," she said.

The president, upon a moment's reflection, asked, "Was this with the same hen each time?"

"Oh, no, Mr. President," the guide replied. "A different one each time."

The president smiled. "Tell that to Mrs. Coolidge."

Two weeks after our Lolita came down with lovesickness, the rest of the hens followed. Now we recognized the symptoms. Scientific as always in our barnyard, we applied the Coolidge effect, separating either Big Tom or Bad Tom with a new hen each day in the romantic barn room while the other tom chased the rest of the girls around the pasture. We had to keep the boys apart from one another, not so much because they fought (though they did), but because any time one of them managed to mount a hen, the other would charge like a bowling ball down the lane and topple the lovers most ungracefully, ka-*pow*. Nothing good was going to come of that.

But after the February of Love dawned over our barnyard, it was followed by the March of the Turkey Eggs. We hoped this was good, although the first attempts looked like just one more wreck along the love-train track. It's normal for a young bird to need a few tries, to get her oviduct work in order. But to be honest I didn't even recognize the first one as an egg. I went into the turkey coop to refill the grain bin and almost stepped on a weird thing on the floor. I stooped down to poke at it: a pale bag of fluid, soft to the touch, teardrop-shaped with a rubbery white corkscrew at the pointy end. Hmmm. A small visitor from another planet? I tentatively decided it was an egg, but did not uncork the champagne.

Soon, real eggs followed: larger and more pointed than chickens' eggs, light brown with a cast of reddish freckles. I was thrilled with the first few. Then suddenly they were everywhere, dropped coyly on the floor like hankies: hither and yon about the coop, outside in the caged run, and even splat on the grass of the pasture. When the urge struck these girls, they delivered, like the unfortunate mothers one hears about having their babies in restaurant foyers and taxicabs.

I had fashioned what I thought to be a respectable turkey nest on the floor in one corner of the coop, but no one was using it. Clearly it didn't look right to them, maybe not cozy enough. We built a big wooden box with open sides to set over the nest for protection. The turkeys roosted at night on high rafters inside their coop, and always flew around rambunctiously before going to bed. Bourbon Reds have wings and are not afraid to use them. Maybe the nest on the floor would have more appeal, I reasoned, if I made it safe from aerial assault.

This struck some chord in the turkey psyche, but not the right one: the hens immediately began laying eggs on top of the plywood platform, about three feet off the ground. My reference book insisted turkeys will only use floor nests. My turkeys hadn't heard about that. Within days I had no more eggs on the floor, but nearly two dozen in a precarious clutch on a plywood platform where they could easily roll off and smash. I cut down the sides of a big cardboard box to make a shallow tray, filled it with straw and leaves, and put the eggs in there.

Finally I'd guessed right. The sight of this cozy pile of eggs in a computer-monitor shipping carton was just what it took to throw the hormonal switch. One by one, the turkey hens began sitting on the nest. After a fashion. They would lay an egg, sit just a few minutes longer, and leave. Soon I had more than thirty eggs in this platform nest and no mother worth a corsage, to say the least.

Most of them did try, a little. As time passed and the pile grew to ridiculous proportions, they seemed to feel some dim sense of obligation. A hen would sit on the eggs for an hour. Then she'd hop up, wander away, and go get a snack. Or she would land on the nest, lay one egg, tromp around on the pile until she'd broken two, then eat them and go bye-bye. Often two hens would sit on the eggs together, amiable for awhile until they'd begin to tussle with one another. Sometimes this would escalate until they were fanning their tail feathers and displaying at one another, exactly the way the males do. Then, suddenly, they'd quit and go do lunch together. I became hopeful when one hen (not always the same one) would stay on the eggs until late morning, when I usually let all the turkeys out into the pasture. She would stay behind as the sisters nattered out the door, but always after a while she would decide she'd had all she could take of *that,* and scream to be let out with her friends for the rest of the day.

With all due respect for very young mothers who are devoted to their children, I began to think of my hens as teen moms of the more stereotypical kind. "I'm not ready to be tied down" was the general mindset. "Free bird" was the anthem. Nobody was worrying over this growing pile of eggs, except me. I fretted as they strolled away, scolding each slacker mother: You turkey! *Dindon sauvage,* pardon my French. You've made your nest, now sit on it.

My nagging had the predictable effect, i.e. none. I felt bereft. Most nights were still below freezing. What could be more pitiful than a huge nest of beautiful eggs sitting out in the cold? Potentially viable, valuable eggs left to die. That many heirloom turkey eggs, purchased mail-order for incubation, would cost about three hundred dollars, and that is nothing compared with the real products of awkward, earnest turkey love. But what was I supposed to do, sit on them myself?

That, essentially, is what the professionals do. Our feed store carried several models of incubators, which I'd scrutinized more than once. This would be the simple answer: put the eggs in an electric incubator, watch them hatch, and raise baby turkeys myself, one more time. Turkeys that would, once again, grow up wanting to mate with something like me.

Is it possible to rear eggs in an incubator and slip them under a female adult after they've hatched? Easy answer: Yes, and she will kill them. Possibly eat them, as horrifying as that sounds. Motherhood is the largest work of most lives, and natural selection cannot favor a huge investment of energy in genes that are not one's own. It's straightforward math: the next generation will contain zero young from individuals whose genes let them make that choice. In animals other than humans, adoption exists only in rare and mostly accidental circumstances.

In the case of turkeys, the mother's brain is programmed to memorize the sound of her chicks' peeping the moment they hatch. This communication cements her bond with her young, causing her to protect them intensely during their vulnerable early weeks, holding her wings out and crouching to keep the kids hidden under something like a feathery hoopskirt, day and night, while they make brief forays out into the world, learning to find their own food.

Early-twentieth-century experiments (awful ones to contemplate) showed that deafened mother turkeys were unable to get the all-important signal from their young. These mothers destroyed their own chicks, even after sitting on the eggs faithfully for weeks.

My hens seemed to have good ears, but the faithful sitting was not their long suit. Still, I didn't buy an incubator. I wanted turkey chicks raised by turkey mothers, creatures that would literally know how to be true to their own kind. The project allows no shortcuts. If we could just get a first generation out of one of these mothers, the next ones would have both better genes and better rearing.

The alternative possibility, a lot of botched hatchlings, made me sad. The temptation is to save the individual that pulls on your heartstrings, even at the cost of the breed. When I'd signed on to the small club of heritage animal breeders, part of the deal was refraining from this kind of sentimentality. Poor mothering instincts, runts, and genetic weaklings all have to be culled. In a human-centric world that increasingly (and wisely, in my opinion) defines all humans as intrinsically equal, it's hard not to color this thinking outside of the lines. But the rules for healthy domestic animal populations are entirely unlike those we apply to ourselves. I came up against this when trying to explain to my nephew why we can't let the white rooster mate with the brown hens. I decided to drop the subject for a few years. But I'll bring it up again if he asks, because it's important information: respect takes different forms for different species. The apple tree gains strength from strict breeding and regular pruning. So does the herd.

Our purpose for keeping heritage animals is food-system security, but also something else that is less self-serving: the dignity of each breed's true and specific nature. A Gloucester Old Spots hog in the pasture, descended from her own ancient line, making choices, minute by minute, about rooting for grubs and nursing her young, contains in her life a sensate and intelligent "pigness." It's a state of animal grace that never even touches the sausages-on-hooves in an industrial pig lot. One can only hope they've lost any sense of the porcine dignities stolen from them.

If it seems a stretch to use the word *dignity* in the same sentence with *pig,* or especially *turkey,* that really proves my point. It was never their plan to let stupid white eunuchs take over, it was ours, and now the genuine, self-propagating turkeys with astute mothering instincts are all but lost from the world. My Bourbon Reds and I had come through hard times together, and I was still rooting for them. They had grown up handsome and strong, disease free, good meat producers, efficient pasture foragers.

I found myself deeply invested in the next step: I wanted them to make it to the next generation on their own. Natural Childbirth or Bust. All my eggs were in one basket now. If they dropped it, we'd have pumpkin soup next Thanksgiving.

Taking Local On the Road

BY CAMILLE

I have a confession to make. Five months into my family's year of devoted local eating, I moved out. Not because the hours of canning tomatoes in early August drove me insane or because I was overcome by insatiable cravings for tropical fruit. I just went to college. It was a challenging life, getting through chemistry and calculus while adjusting to a whole new place, and the limited dining options I had as a student living on campus didn't help. I suppose I could have hoed up a personal vegetable patch on the quad or filled my dorm room with potted tomato and zucchini plants, but then people would *really* have made fun of me for being from Appalachia. Instead, I ate lettuce and cucumbers in January just like all the other kids.

Living away from home, talking with my family over the phone, gave me some perspective. Not having fresh produce at my disposal made me realize how good it is. I also noticed that how I think about food is pretty unusual among my peers. When I perused the salad bar at my dining hall most evenings, grimly surveying the mealy, pinkish tomatoes and paperlike iceberg lettuce, I could pick out what probably came from South America or New Zealand. I always kept this information to myself (because who really cares when there are basketball games and frat parties to talk about?), but I couldn't help noticing it.

I suppose my generation is farther removed from food production than any other, just one more step down the path of the American food industry. More than our parents, we rely on foods that come out of shiny wrappers instead of peels or skins. It still surprises a girl like me, who actually lives on a real farm with *real animals* and stuff growing out of the *ground*, that so many young adults couldn't guess where their food comes from, or when it's in season where they live. It's not that my rising generation is unintelligent or unworldly—my classmates are some of the smartest, most cultured people I know. But information about food and farming is not very available. Most of the people I know have never seen a working farm, or had any reason to do so. Living among people my age from various cities across the United States made me realize I actually know a lot about food production, and I don't take that for granted.

I also won't forget to appreciate how much better local food tastes. Next to getting a good night's sleep on a comfortable mattress, cooking good food became my main motivation for coming home from school to visit. Of course seeing my family was nice, but priorities are priorities, right? It was great after weeks of dorm life to eat eggs with deep golden yolks, and greens that still had their flavor and crunch. I loved being able to look at a table full of food and know where every vegetable was grown, where the meat lived when it was still a breathing animal.

During my first year of college I found two campus eateries that use organic, locally grown produce in their meals, and one that consistently uses free-range meat. For the most part, these vendors did not widely advertise the fact that they were participating in the local food economy. I only found out because I cared, and then tried to buy most of my food from those places.

My generation, I know, has the reputation of sticking iPods in our ears and declining to care about what might happen in ten years, or even next week. We can't yet afford hybrid vehicles or solar homes. But we do care about a lot of things, including what we eat. Food is something real. Living on the land that has grown my food gives me a sense of security I'm lucky to have. Feeling safe isn't so easy for people my age, who face odious threats like global warming, overpopulation, and chemical warfare in our future. But even as the world runs out of fuel and the ice caps melt, I will know the real sources of my sustenance. My college education may or may not land me a good job down the road, but my farm education will serve me. The choices I make now about my food will influence the rest of my life. If a lot of us felt this way, and started thinking carefully about our consumption habits just one meal at a time, we could affect the future of our planet. No matter how grave the predictions I hear about the future, for my peers and me, that's a fact that gives me hope.

20 · TIME BEGINS

Years ago, when Lily was not quite four, we were spending one of those perfect mother-daughter mornings in the flower garden: I planted pansies while she helped by picking up the bugs for closer looks, and not eating them at all. Three is a great age. She was asking a lot of questions about creature life, I remember, because that was the day she first worked up to the Big Question. I don't mean sex, that's easy. She wanted to know where *everything* comes from: beetles, plants, us. "How did dinosaurs get on the earth, and why did they go away?" was her reasonable starting point.

How lovely it might be to invoke for my child in just one or two quotes the inexplicable Mystery. But I went to graduate school in evolutionary biology, which kind of obligates me to go into the details. Lily and I talked about the millions and millions of years, the seaweeds and jellyfish and rabbits. I explained how most creatures have many children (some have thousands!) with lots of small differences between them. These specialties—things like quick hiding or slow, picky eating or just shoveling everything in—can make a difference in whether the baby lives to be a grown-up. The ones that survive will have children more like themselves. And so on. The group slowly changes.

I've always thought of this as a fine creation story, a sort of quantifiable miracle, and was pleased to think I'd rendered such a complex subject comprehensible to a toddler. She sat among the flowers, pondering it. At length she asked, "Mama, did *you* get born, or are you one of the ones that evolved from the tree primates?"

I'm not eight million years old. But I am old enough to know I should never, ever, trust I've explained anything perfectly. Some part of the audience will always remain at large, confused or plain unconvinced. As I wind up this account, I'm weighing that. Is it possible to explain the year we had? I can tell you we came to think of ourselves, in the best way, as a family of animals living in our habitat. Does that reveal the meaning of our passage? Does it explain how we're different now, even though we look the same? We are made of different stuff, with new connections to our place. We have a new relationship with the weather. So what, and who cares?

All stories, they say, begin in one of two ways: "A stranger came to town," or else, "I set out upon a journey." The rest is all just metaphor and simile. Your high school English teacher was right. In *Moby-Dick,* you'll know if you were half awake, the whale was not just an aquatic mammal. In our case, the heirloom turkeys are not just large birds but symbols of a precarious hold on a vanishing honesty. The chickens are secondary protagonists, the tomatoes are allegorical. The zucchini may be just zucchini.

We set out upon a journey. It seemed so ordinary on the face of things, to try to do what nearly all people used to do without a second thought. But the trip surprised us many times, because of all the ways a landscape can enter one's physical being. Like most of the other top-heavy hominids walking around in shoes, failing to notice the forest for the mashed trees reincarnated as our newspapers and such, I'd nearly forgotten the truest of all truths: we are what we eat.

As our edible calendar approached its arbitrary conclusion, we were more than normally conscious of how everything starts over in the springtime. All the milestones that had nudged us toward the start of our locavore year began to wink at us again. Our seedlings came up indoors. The mud-ice melted, and the spicebushes in the lane covered themselves with tiny yellow pompoms of flower. The tranquils bloomed. On April 3, the secretive asparagus began to nose up from its bed.

What were we doing when *the day* finally came? Standing by our empty chest freezer at midnight, gnawing our last frozen brick of sliced squash, watching the clock tick down the seconds till we could run out and buy Moon Pies? No. I'm sorry, but the truth is so undramatic, I can't even find "the day" in my journal.

The best I can do is recall a moment when I understood I had kept some promise to myself, having to do with learning to see the world differently. It was a day in early April when three little trees in our yard were covered with bloom—dark pink peach blossoms, pale pink plum, and white pear, filling the space like a Japanese watercolor. The air smelled spicy; the brown pasture had turned brilliant green. From where I stood on the front porch I could see my white-winged turkeys moving slowly through that emerald sea, nibbling as they went. I pictured how it would be in another month when the grass shot up knee-deep. I was struck, then, with a vivid fantasy of my family being in the turkeys' place, imagining what a thrill it would be to wander chest-deep in one's dinner as an ordinary routine. I mean to say I pictured us wading through piles of salad greens, breast-stroking into things like tomatoes, basil, and mozzarella.

I snapped out of it, recognizing this was not a very normal daydream. This was along the lines my astute children would diagnose as *wackadoo.* I took myself to be a woman changed by experience.

But I'd noticed the kids had changed too. One day at the farmer's market a vendor had warned us there might be some earworms in the corn because it was unsprayed. He pointed out a big one wriggling in the silks of one of the ears in our bag, and reached in to pluck it off. Lily politely held out her hand: that was our worm, we'd paid for it. She would take that protein to her chickens, and in time it would be eggs. Camille used similar logic to console me after my turkeys raided the garden and took some of the nicest tomatoes. "Mom," she said, "you'll eat them eventually." And I did.

It wasn't just our family, either, that had changed in a year. Food was now very much a subject of public conversation—not recipes, but issues. When we'd first dreamed up our project, we'd expected our hardest task would be to explain in the most basic terms what we were doing, and why on earth we'd bother. Now our local newspaper and national ones frequently had local-food feature stories on the same day. Every state had it going on, including Arizona, the food scene we feared we had left for dead. *Alaska* was experiencing a farmers' market boom, with the "Alaska Grown" logo showing up on cloth shopping bags all over Anchorage. Tod Murphy's Farmers Diner, in order to accommodate more diners, had relocated south to Quechee Village, Vermont (near Hanover, NH). Other like-minded eateries now lay in the path of many a road trip. Hundreds of people were signing up online and reporting on their "Locavore Month" experiences. We had undertaken a life change partly as a reaction against living in a snappily-named-diet culture; now this lifestyle had its own snappy diet name: "The 100-Mile Diet Challenge!" What a shock. We were trendy.

As further proof that the movement had gained significance, local eating now had some official opposition. The standard criticisms of local food as Quixotic and elitist seemed to get louder, as more and more of us found it affordable and utterly doable. The *Christian Science Monitor* even ran a story on how so much local focus could breed "unhealthy provincialism." John Clark, a development specialist for (where else) the World Bank, argued that "what are sweatshop jobs for us may be a dream job" for someone else—presumably meaning those folks who earn a few dreamy bucks a day from Dole, Kraft, Unilever, or Archer Daniels Midland—"but all that goes out the window if we only buy local." He expressed concern that local-food bias would lead to energy waste, as rabidly provincial consumers drove farmers in icy climes to grow bananas in hothouses.

That's some creative disapproval, all right—a sure sign the local-food movement was getting worrisome to food industrialists who had heretofore controlled consumer choices so handily, even when they damaged our kids' health and our neighborhoods. Shoppers were starting to show some backbone, clearly shifting certain preferences about what foods they purchased, and from where. An estimated 3 percent of the national supply of fresh produce had moved directly from farmers to customers that year.

The "why bother" part of the equation was also becoming obvious to more people. Global climate change had gone, in one year, from unmentionable to cover story. "The end of the oil economy" was now being discussed by some politicians and many economists, not just tree huggers and Idaho survivalists. We were starting to get it.

But it's also true what the strategists say about hearts and minds—you have to win them both. We will change our ways significantly as a nation not when some laws tell us we *have* to (remember Prohibition?), but when we *want* to. During my family's year of conscious food choices, the most important things we'd learned were all about that: the wanting to. Our fretful minds had started us on a project of abstinence from industrial food, but we finished it with our hearts. We were not counting down the days until the end, because we didn't want to go back.

A few days after my momentary chest-deep-in-food fantasy, we had dinner with our friends Sylvain and Cynthia. Sylvain grew up in the Loire Valley, where local food is edible patriotism, and I sensed a kindred spirit from the way he celebrated every bite of our salad, inhaling the spice of the cut radishes and arugula. He told us that in India it's sometimes considered a purification ritual to go home and spend a year eating everything from one place—ideally, even to grow it yourself. I liked this name for

The Blind Leading the Blind

Critics of local food suggest that it's naive or elitist, whereas industrial agriculture is for everybody: it's what's for dinner, all about feeding the world. "Genetically modified, industrially produced monocultural corn," wrote Steven Shapin in the *New Yorker,* "is what feeds the victims of an African famine, not the gorgeous organic technicolor Swiss chard from your local farmers' market."

The big guys have so completely taken over the rules of the game, it's hard to see how food systems really work, but this criticism hits the nail right on the pointy end: it's perfectly backward. One of industrial agriculture's latest feed-the-hungry schemes offers a good example of why that's so. Exhibit A: "golden rice." It's a genetically modified variety of rice that contains beta-carotene in the kernel. (All other parts of the rice plant already contain it, but not the grain after it is milled.) The developers of this biotechnology say they will donate the seeds—with some strings attached—to Third World farmers. It's an important public relations point because the human body converts beta-carotene to vitamin A; a deficiency of that vitamin affects millions of children, especially in Asia, causing half a million of them every year to go blind. GM rice is the food industry's proposed solution.

what we had done: a purification ritual, to cultivate health and gratitude. It sounds so much better than wackadoo.

Over the years since I first acquired children and a job, I've often made reference to the concern of "keeping my family fed." I meant this in the same symbolic way I'd previously used (pre-kids, pre-respectable job) to speak of something "costing a lot of bread." I was really talking about money. Now when I say bread, I mean bread. I find that food is not symbolic of anything so much as it is real stuff: beetroot as neighbor to my shoe, chicken as sometime companion. I once read a pioneer diary in which the Kansas wife postponed, week after week, harvesting the last hen in her barren, windy yard. "We need the food badly," she wrote, "but I will miss the company."

I have never been anywhere near that lonely, but now I can relate to the relationship. When I pick apples, I miss the way they looked on the tree. Eggplants look like lightbulbs on the plant, especially the white and

But most of the world's malnourished children live in countries that already produce surplus food. We have no reason to believe they would have better access to this special new grain. Golden rice is one more attempt at a monoculture solution to nutritional problems that have been *caused* by monocultures and disappearing diversity. In India alone, farmers have traditionally grown over 200 types of greens, and gathered many more wild ones from the countryside. Every single one is a good source of beta-carotene. So are fruits and vegetables. Further, vitamin A delivered in a rice kernel may not even help a malnourished child, because it can't be absorbed well in isolation from other nutrients. Throwing more rice at the problem of disappearing dietary diversity is a blind approach to the problem of blindness. "Naïve" might describe a person who believes agribusinesses develop their heavily patented commodity crops in order to feed the poor. (Golden rice, alone, has seventy patents on it.) Technicolor chard and its relatives growing in village gardens—that's a solution for realists.

STEVEN L. HOPP

neon purple ones, and I observe the unplugging of their light when I toss them in the basket. My turkey hens have names now. I do know better, but couldn't help myself.

⚜

At the end of March, one of my turkey mothers found her calling. She sat down on the platform nest and didn't get up again for a week. Then two, then three. This was Lolita, the would-be husband-stealer—the hen who had been first to show mating behaviors, and then to lay eggs. Now she was the first to begin sitting with dedication. We expunged "Lolita" from her record and dubbed her "Number One Mother."

Underneath the platform where she now sat earning that title, we fixed up two more nests to contain the overflow. Together the hens had now produced more than fifty eggs. While Number One Mother incubated about two dozen of them, Numbers Two, Three, and Four were showing vague interest in the other piles. Number Two had started to spend the nights sitting on eggs, but still had better things to do in the daytime. Three and Four were using the remaining nest the way families use a time-share condo in Florida.

But something inside the downy breast of Number One had switched on. Once she settled in, I never saw her get up again, not even for a quick drink of water. With her head flattened against her body and a faraway look in her eyes, she gave herself over to maternity. I began bringing her handfuls of grain and cups of water that she slurped with desperation. I apologized for everything I'd said to her earlier.

I was the free bird now, out in the sunshine as much as possible, walking into the open-armed embrace of springtime. A balmy precipitation of cherry petals swirled around us as we did our garden chores. The ruddy fiddleheads of peony leaves rolled up out of the ground. The birthday garden made up of gift plants I'd received last year now surprised me like a series of unexpected phone calls: the irises bloomed; the blue fountain grass poured over the rocks; I found the yellow lady's slipper blossoms when I was weeding under the maple. One friend had given me fifty tulip bulbs, one for each of my years, which we planted in a long trail down the driveway. Now they were popping up with flaming red heads on slender stalks like candles on a birthday cake. The groundhog that dug up some bulbs over the winter had taken a few years off. I would try to remain grateful to the groundhog later on, when he was eating my beans.

Spring is made of solid, fourteen-karat gratitude, the reward for the long wait. Every religious tradition from the northern hemisphere honors some form of April hallelujah, for this is the season of exquisite redemption, a slam-bang return to joy after a season of cold second thoughts. Our personal hallelujah was the return of good, fresh food. Nobody in our household was dying for a Moon Pie, but we'd missed crisp things, more than we'd realized. Starting the cycle again was a heady prospect: cutting asparagus, hunting morels, harvesting tender spinach and chard. We'd made it.

Did our year go the way we'd expected? It's hard to say. We weren't thinking every minute about food, as our family life was occupied front and center by so many other things. Devastating illnesses had darkened several doors in our close family. We'd sent a daughter off to college and missed her company, and her cooking. If our special way of eating had seemed imposing at first, gradually it was just dinner, the spontaneous background of family time as we met our fortunes one day, one phone call, one hospital visit, wedding, funeral, spelling bee, and birthday party at a time. It caused us to take more notice of food traditions of all kinds—the candy-driven school discipline program, the overwhelming brace of covered dishes that attend a death in the family. But in the main, our banana-free life was now just our life. So much so, in fact, I sometimes found myself a bit startled to run across things like bananas in other people's kitchens—like discovering a pair of Manolo Blahnik sandals in the lettuce bed. Very nice I'm sure, just a little bit extravagant for our kind.

We pressed ourselves to pronounce some verdicts on our year. Our planning and putting-by for the winter had passed muster, as we still had pesto and vegetables in our freezer to last comfortably till the abundances of June. We'd overplanted squash, could have used more garlic, but had enough of everything to stay happy. The Web site of the local-eating Vancouver couple said they'd ended their year fifteen pounds lighter (despite what they described as "a lot of potatoes"), whereas we all weighed out of the year right about where we'd weighed in, and hoped to remain—except for Lily, who had gained twelve pounds and grown nearly five inches. Obviously we never went hungry, and you can't raise that much good kid on potatoes alone. The Canadians had been purists, though, and really we weren't; we'd maintained those emergency rations of mac-and-cheese. (And anyone giving up coffee gets a medal we weren't even in the running for.) But frankly, any year in which no high-fructose corn syrup crosses my threshold is pure enough for me.

Our plan to make everything from scratch had pushed us into a lot of great learning experiences. In some cases, what we learned was that it was too much trouble for everyday: homemade pasta really is better, but we will always buy it most of the time, and save the big pasta-cranking events for dinner parties. Hard cheeses are *hard.* I never did try the French-class mayonnaise recipe. I'd also imagined at some irrational moment that I would learn to make apple cider and vinegar, but happily submitted to realism when I located professionals nearby doing these things really well. On the other hand, making our daily bread, soft cheeses, and yogurt had become so routine we now prepared them in minutes, without a recipe.

Altered routines were really the heart of what we'd gained. We'd learned that many aisles of our supermarket offered us nothing local, so we didn't even push our carts down those: frozen foods, canned goods, soft drinks (yes, that's a whole aisle). Just grab the Virginia dairy products and organic flour and get out, was our motto, before you start coveting thy neighbor's goods. A person can completely forget about lemons and kiwis once the near occasion is removed.

As successful as our sleuthing into local markets had been, we never did find good local wheat products, or seafood. I was definitely looking forward to some nonlocal splurges in the coming months: wild-caught Alaskan salmon and bay scallops and portobellos, hooray. In moderation, of course. I had a much better sense of my options now and could try for balance, buying one bottle of Virginia wine, for example, for every import.

The biggest shock of our year came when we added up the tab. We'd fed ourselves, organically and pretty splendidly we thought, on about fifty cents per family member, per meal—probably less than I spent in the years when I qualified for food stamps. Of course, I now had the luxury of land for growing food to supplement our purchases. But it wasn't a *lot* of land: 3,524 square feet of tilled beds gave us all our produce—that's a forty-by-twenty-two-foot spread, per person. (It felt a lot bigger when we were weeding it.) We appreciate our farm's wooded mountainsides for hiking and the rare morel foray—and for our household water supply—but in the main, one doesn't eat a nature preserve. Adding up the land occupied by our fruit trees, berry bushes, and the pasture grazed by our poultry brings our land-use total for nutritional support to about a quarter acre—still a modest allotment. Our main off-farm purchases for the year were organic grain for animal feed, and the 300 pounds of flour required for our daily bread. To put this in perspective, a good wheat field yields 1,600 pounds of flour per acre. In total, for our grain and flour, pastured meats and goods from the farmers' market, and our own produce, our family's food footprint for the year was probably around one acre.

By contrast, current nutritional consumption in the U.S. requires an average of 1.2 cultivated acres for every citizen—4.8 acres for a family of four. (Among other things, it takes space to grow corn syrup for that hypothetical family's 219 gallons of soda.) These estimates become more meaningful when placed next to another prediction: in 2050, the amount of U.S. farmland available per citizen will be only 0.6 acres. By the numbers, the hypothetical family has change in the cards. By any measure, ours had discovered a way of eating that was more resourceful than I ever could have predicted.

In the coming year, I decided, I would plant fewer tomatoes, and more flowers. If we didn't have quite such a big garden, if we took a vacation to the beach this summer, we'd do that thanks to our friends at the farmers' market. The point of being dedicated locavores for some prescribed length of time, I now understand, is to internalize a trust in one's own foodshed. It's natural to get panicky right off the bat, freaking out about January and salad, thinking we could never ever do it. But we did. Without rationing, skipping a meal, buying a corn-fed Midwesternburger or breaking our vows of exclusivity with local produce, we lived inside our own territory for one good year of food life.

❦

"I can't exactly explain what we're looking for," I told our guests, feeling like a perfectly idiotic guide. "Your eye kind of has to learn for itself."

We were back on Old Charley's Lot, scanning the dry-leaf-colored ground for dry-leaf-colored mushrooms. Steven found the first patch, a trio tilted at coy angles like garden gnomes. We all stood staring, trying to fix our vision. The color, the shape, the size, everything about a morel resembles a curled leaf lying on the ground among a million of its kind. Even so, the brain perceives, dimly at first and then, after practice, with a weirdly trenchant efficiency. You spot them before you know you've seen them.

This was the original human vocation: finding food on the ground. We're wired for it. It's hard to stop, too. Our friends Joan and Jesse had traveled a long way that day, and their idea of the perfect host might not be a Scoutmaster type who makes you climb all over a slick, pathless mountainside with cat briars ripping your legs. But they didn't complain, even as rain began to spit on our jackets and we climbed through another maze of wild grapevines and mossy logs. "We could go back now," I kept saying. They insisted we keep looking.

After the first half hour we grew quiet, concentrating on the ground, giving each other space for our own finds. It was a rare sort of afternoon. The wood thrushes and warblers, normally quiet once the sun gets a good foothold, kept blurting out occasional pieces of song, tricked into a morning mood by the cool, sunless sky. Pileated woodpeckers pitched ideas to one another in their secret talking-drum language. These giant, flamboyant woodpeckers are plentiful in our woods. We all took note of their presence, and were drawn out of our silence to comment on the remarkable news about their even more gigantic first cousins, the ivory-billed woodpeckers. These magnificent creatures, the "Lord God Birds" as they used to be called in the South, had been presumed extinct for half a century. Now a reputable research team had made an unbelievable but well-documented announcement. Ivorybills were still alive, deep in a swamp in Arkansas. Lord God.

Was it true? A mistake or a hoax? Was it just one bird, or a few, maybe even enough for the species to survive? These were still open questions, but they were headliner questions, inspiring chat rooms and T-shirts and a whole new tourist industry in swampy Arkansas. People who never gave a hoot about birds before cared about this one. It was a miracle, capturing our hopes. We so want to believe it's possible to come back from our saddest mistakes, and have another chance.

"How do you encourage people to keep their hope," Joan asked, "but not their complacency?" She was deeply involved that spring in producing a film about global climate change, and preoccupied with striking this balance. The truth is so horrific: we are marching ourselves to the maw of our own extinction. An audience that doesn't really get that will amble out of the theater unmoved, go home and change nothing. But an audience that *does* get it may be so terrified they'll feel doomed already. They might walk out looking paler, but still do nothing. How is it possible to inspire an appropriately repentant stance toward a planet that is really, really upset?

I was as stumped on the answer to that as I'd been earlier on the mushroom guidance. However much we despise the monstrous serial killer called global warming, it's hard to bring charges. We cherish our fossil-fuel-driven conveniences, such as the computer I am using to write these words. We can't exactly name-call this problem, or vote it away. The cure involves reaching down into ourselves and pulling out a new kind of person. The practical problem, of course, is how to do that. It's impossible to become a fuel purist, and it seems like failure to change our ways only halfway, or a pathetic 10 percent. So why even try? When the scope of the problem seems insuperable, isn't it reasonable just to call this one, give it up, and get on with life as we know it?

I do know the answer to that one: that's called child abuse. When my teenager worries that her generation won't be able to fix this problem, I have to admit to her that it won't be up to her generation. It's up to mine. This is a now-or-never kind of project.

But a project, nevertheless. Global-scale alteration from pollution didn't happen when human societies started using a little bit of fossil fuel. It happened after unrestrained growth, irresponsible management, and a cultural refusal to assign any moral value to excessive consumption. Those habits can be reformed. They *have* been reformed: several times in the last century we've learned that some of our favorite things like DDT and the propellants in aerosol cans were rapidly unraveling the structure and substance of our biosphere. We gave them up, and reversed the threats. Now the reforms required of us are more systematic, and nobody seems to want to go first. (To be more precise, the U.S.A. wants to go last.) Personally, I can't figure out how to give up my computer, but I'm trying to get myself onto a grid fueled by wind and hydro power instead of strip-mined coal. I could even see sticking some of the new thin-film photovoltaic panels onto our roof, and I'm looking for a few good congressmen or -women who'd give us a tax credit for that. In our community and our household we now have options we didn't know about five years ago: hybrid vehicles, geothermal heating. And I refused to believe a fuel-driven food industry was the only hand that could feed my family. It felt good to be right about that.

I share with almost every adult I know this crazy quilt of optimism and worries, feeling locked into certain habits but keen to change them in the right direction. And the tendency to feel like a jerk for falling short of absolute conversion. I'm not sure why. If a friend had a coronary scare and finally started exercising three days a week, who would hound him about the other four days? It's the worst of bad manners—and self-protection, I think, in a nervously cynical society—to ridicule the small gesture. These earnest efforts might just get us past the train-wreck of the daily news, or the anguish of standing behind a child, looking with her at the road ahead, searching out redemption where we can find it: recycling or carpooling or growing a garden or saving a species or *something*. Small, stepwise changes in personal habits aren't trivial. Ultimately they will, or won't, add up to having been the thing that mattered.

We all went crazy over finding the ivorybill because he is the Lord God's own redheaded whopper of a second chance. Something can happen for us, it seems, or *through* us, that will stop this earthly unraveling and start the clock over. Like every creature on earth, we want to make it too. We want more time.

Natural cycles persist in being predictable, despite all human caprice. It probably happened by the grace of biology, rather than magic, that the very date Lily had circled on her calendar one year earlier got circled now on mine, for the same reason. When Number One Mother sat down on her clutch of eggs, I'd made note of it in my journal. Now I counted forward like expectant mothers everywhere: *My babies due!*

I was on pins and needles, watching the date approach. Having done it myself twice, I knew the expectant-mother gig: focus, summon strength for the task at hand. But now I found myself in a role more along the lines of expectant dad: dither uselessly. I could do absolutely nothing to help, which increased my need to hover.

On the actual due date I walked down to the poultry barn to check on Mom. Maybe, oh, about sixteen times. She raised her hackles and hissed at me to go away. This was a whole different demeanor from her glassy-eyed hunker of yesterday and the twenty-six days previous. I took her fussy defensiveness to be a good sign. Chicks begin peeping from inside the egg a day or so before they hatch. This mother must be hearing that, I thought, getting ready for the blessed event.

The outcome of Sunday, April 23, however, was a big nothing. Monday brought more of the same. Has any anxious person ever *really* respected the warning about watched pots that never boil? Well, good for you, is all I can say, because I checked that nest morning, noon, and night, hoping for little fluffy chicks that did not appear. After all we'd been through together, Number One and I, what if nothing hatched at all?

On Tuesday I went back through my journal and recalculated the due date, thinking I might be off by a day. I wasn't. They would hatch by the end of Tuesday then, I concluded reasonably, and they did not. That night I double-checked my reference books, which all agreed the incubation period for domestic turkey eggs is "about 27 days." What does *about* mean? Twenty-nine? Forty? On Wednesday I checked on the poor mother until she was visibly fed up. I even poked my hand under her to feel the eggs. She stuck tightly to the nest, but became so accustomed to my prodding that she began to ignore me rather than hissing. Possibly she was slumping into post-due-date despair.

Looking for Mr. Goodvegetable

How small is a small farm? How nearby does local have to be? Is organic more important than local? Which of these should we favor, and when?

Eco-gastronomy isn't just a minimum-distance food-buying contest. The three basic components of responsible eating are to favor food grown in an environmentally responsible way, delivered with minimal petroleum use, in a manner that doesn't exploit the farmers. Most of us won't have a diversified farm located within walking distance, or a Local Foods-R-Us opening nearby anytime soon. Here are some guidelines that can help define responsible food choices.

Begin by visiting a nearby farmers' market to see what's available. Don't go in with the goal of buying anything in particular, but simply to learn. Is it local? Most farmers' markets have rules about how recently vegetables can have been picked. Are they from a small farm? Probably; if not, they likely wouldn't be sold there. Are they organic? Likely; while certification is not always required at these markets, most small market growers have assumed sustainability as part of their identity. It's probably the most common question they hear, so ask. Pay attention to what's available, what is in season. Every region has its strengths and weaknesses. Some are obvious (seafood near the coasts, or citrus in Florida). Others you'll need to learn. Buy a good supply of what you can use.

Now, armed with what you learned at the farmers' market, you can visit your conventional grocery store. Applying ethics in a modern grocery store can be daunting, but here are a few general rules that may help sort out the whole equation.

- If items are available regionally, and are in season, get them from a farmer or ask a grocer to obtain them from a local source.
- Do as much as possible of your own cooking or preparation. Make meal plans for the seasons, rather than starting with a recipe and having a treasure hunt for its ingredients.
- Food processing uses energy in two main ways: (1) extracting, dicing, mixing, and cooking the ingredients; (2) transporting each individual ingredient. Products with fewer ingredients have probably burned less gas. For example, the oatmeal box on our pantry shelf lists one ingredient: rolled oats. With some local walnuts and honey, it makes a great breakfast. By contrast, our Free-range Happy 75% Organic Cereal Chunks box lists

seventeen ingredients, all of which had to be transported to the processing plant. Who even knows how much fossil fuel it took to make it 75% Happy?

- For fresh fruits and vegetables, consider travel distance. On an autumn trip to our grocery I found apples grown in a neighboring state (North Carolina), Washington State, and New Zealand. That choice is easy. If we lived in Oregon, that would be a different easy choice.
- Consider how you feel about using energy to move water. All fresh produce contains a lot of it. Apparent differences between more and less juicy items can be deceiving: watermelon is 92 percent water; cucumbers 96 percent; tomatoes 95 percent, while on the firmer side, carrots are 92 percent; peppers 94 percent; and broccoli 91 percent. All watery. If you care about this, when considering world travelers, favor dried fruits or vegetables, dried spices, nuts, coffee beans, dry beans, and grains.
- If produce or a processed item needs to be refrigerated (or frozen), energy was used to keep it cool from its point of origin to you. How can you tell? It's refrigerated (or frozen) in the store!
- Should you buy industrial organics? By shifting to organic methods, corporate farms are reducing the pesticide loads in our soil and water, in a big way. This should be one of many considerations, along with everything listed above.

How local is local? Our friend Gary Nabhan, in his book *Coming Home to Eat,* defined it as a 250-mile-radius circle for the less-productive desert Southwest. By contrast, the Bay Area group Locavores (www.locavores.com) recommends a 100-mile-radius circle for the more fertile California valleys. It depends on the region, and the product. For us, in Appalachia, seasonal vegetables are literally next door, but our dairy products come from about 120 miles away. That's better, we think, than 1,200, which is also an option in our store. We bear in mind our different concerns: fuel use, pesticide use, quality, and support for farms. By pushing the market with our buying habits, we continually shape our buying choices, and the nature of farming.

STEVEN L. HOPP

Or maybe she was starting to sense what I hated to admit: that these eggs were dead. A hundred things can go wrong with the first breeding attempt of an animal that was not even selected, to begin with, for its reproductive wits. Infertility is common in first-year males, compounded by the incompetent mating attempts of creatures reared by humans. Bacteria in the nest can stealthily destroy the embryos. Improper incubation is also fatal. Had the mother left the nest, on one of the freezing nights we were still having in late April? She seemed dedicated, but lacked experience. She might, just once, have flown up to the roost to get warm among her child-free peers. One hour of that kind of exposure could kill the developing chicks.

Increasingly glum, I had no good news to report on Thursday morning when I came back to the house from my crack-of-dawn nest check. And now we had to leave the farm. I was due that evening in North Carolina for an event that had been scheduled for a year. Steven and Lily were going along too, since we planned a quick visit with Camille at college. At noon, in my earrings and dress shoes, I was still dithering in the poultry house, postponing our drive till the last minute on the grounds that a hatch would probably happen in the warmest part of the day. In truth, I couldn't bear to leave my expectant mother, though I knew the feeling was not at all mutual. Steven assured me that she could manage without me. Kindly, he did not say, "Honey, *it's a turkey*." I sighed, threw my overnight bag into the car, and off we went.

The event went without a hitch. I delivered my lecture from the pulpit of a magnificent gothic chapel and did not even once mention poultry. The book signing afterward went on until midnight, but still I was up before dawn the next day, pacing in our hotel room. As soon as the hour seemed forgivable I roused Steven and insisted on an early return to the farm.

The drive back was endless. I felt like a dope for my impatience, aware that if current trends held, I was rushing us all back to a surly turkey hen sitting on a mound of dead eggs. Even so, as we approached our interstate exit and Steven suggested going on into town to run some errands, I snapped a panicky *"No!"* Looking straight ahead, I adjusted my tone. "Can we go straight back to the farm, please?"

Before we even pulled up to the house I was out of the truck, making a beeline for the poultry barn. When I stepped inside I thought I heard a new sound—a *peeping* sound. It was probably the sparrows that always hung around the barn looking for spilled grain. "Don't be disappointed," I counseled myself, and then repeated the warning aloud because Lily was right behind me. I opened the door to the turkey coop and we slipped inside, approaching the nest-corner slowly, letting our eyes adjust to the dimness inside the slatted turkey room. Number One Mom still sat on her nest. She looked different, though, with her wings held out oddly from her body. We stood still and watched.

There, under her wing, was it something? Lily squeezed my hand and uttered a high-pitched squeak like a baby mouse. It *was* something. A tiny dark eye, as small as a hatpin head, peeked out at us. A fluffy head emerged. Two heads!

One of them wiggled out from under Mama, and it was the real thing: a ball of fluff just like a marshmallow peep, honey blond with a dark spot on top of its head. We could see the white egg-tooth still on the end of its beak. This chick was still damp from the egg, its fluff a bit spiky and its walk adorably uncoordinated. Lily looked at me with huge eyes and whispered: *"We have babies."*

"*She* has babies," I said. This time they would be raised right, by a turkey mother, ending once and for all in our barnyard the indignity of unnatural intervention. But my heart was on Lily's side: we had babies. This was about the youngest creature we had ever seen, tottering on wobbly legs, falling over its feet.

It was hard to resist the temptation to scoop it up in our hands, but we didn't. We were dying to know how many more she had, how old, whether the hatch was finished. But when we approached she lowered her head and hissed at us, snakelike, rumpling her auburn feathers to make herself twice her normal size. Then she looked away. Number One Mother had bigger things on her mind now, and the instincts to do them perfectly.

She had been so faithful to her nest, she had to be hungry and thirsty. Bribery might be just the ticket. Lily ran outside to gather a handful of grass while I approached with a cup of water, holding it close enough for her to get a long drink. She accepted détente and settled down. When Lily came back with the grass, she gobbled it.

While she was distracted by the food, I reached underneath her breast feathers. I could feel a considerable number of eggs under there, smooth to my fingertips. Their heat was almost shocking. One of them felt less smooth. I touched its surface carefully and decided it was slightly cracked. As slowly as I could manage, I drew it out from under her and took a close look. Near the pointed end, a spiderweb of cracks had begun.

The egg began to tremble and thump in my palm, a sensation so animate and peculiar. I put it to my ear and heard a sturdy, high-pitched peeping. I held it to Lily's ear, and watched her eyes grow wide. This egg was alive, though it looked for all the world like an ordinary breakfast food. The effect was wildly unsettling. My heart raced as I tucked the warm egg back under Mama.

We'd gone the whole circle, raising our mail-order hatchlings into the most senior demographic of American turkeys. Now, just after her first birthday, one of the nation's eldest had begat its newest. Only a few times in my life have I actually seen lives begin, and *never* had I held in my palm that miracle caught in the act.

The chick that had come out now dived back into the feather security blanket, disappearing completely under Mom. But we kept staring. We couldn't help it. She glared back—I suppose she couldn't help that either. After another minute, a whole crowd of little black eyes appeared under their mother's wing—two, four, six, eight, ten.

It's hard to explain how irrationally proud I felt of this success. *Their* success, a mother's and, in his clumsy way, a father's too, but most of all these creatures who had pecked themselves heroically into the bright wide world to give this life a go.

Lily and I backed away and slipped out of the turkey coop into the grain room. I thought of that day when I'd tried to explain to Lily the beginning of everything. However I might have bungled it, I hadn't undone for her the Beautiful Mystery. That part tells itself. Crazed and giddy, there in the dusty barn, we held hands and danced: *Babies!* That was all, and that was enough. A nest full of little ding-dongs, and time begins once more.

EPILOGUE: 2017

❧

AVM Plus Ten

THE GROUND UNDERNEATH US

by Barbara Kingsolver

This is what a decade looks like, I think as I take up pruning shears against my overgrown flower borders and clear a path to the vegetable garden. Ten years have passed since the storied birthday party when my friends brought little plants to fill the barren yard of our farmhouse, and we eked out a local feast from our frontier ground. Those gifts of peony divisions and rose slips have grown into a wild ramble beyond all common cottage-garden decency. The little purple butterfly bush is now a butterfly filling-station megastore, nectaring up the neighborhood and shading the whole front porch. The sapling pear tree my parents gave me is taller than the barn.

This farm has hosted and fed a procession of birthdays since that one, along with memorable back-to-school and graduation blowouts and, best of all, Camille's wedding in the front yard among the dahlias and hollyhocks. I get to remember forever the way a young man's face was struck with light when my daughter walked through the garden gate in a white dress and sunflower smile. And now, every time they walk up the porch steps to our front door, Reid and Camille get to recover the ground where they spoke their vows. It seems to me our family is not just people but a place. Alongside the pear trees and peonies we've grown to include a beloved son-in-law and, soon, a new baby. When we sit down together in our dining room it's still this place that feeds us. The foodways we explored in this book have become routine to us now, we don't dwell on the story of this leg of lamb, the broccoli, the garlic, the apples, the tomatoes, and the green beans. Our farm gives these foods up with abandon, fills our summers and our freezers with them. We spent a year consciously learning a habit of eating what we produced ourselves or found at the farmers' market, and it was long enough for a habit to grow into a preference.

We like eating this way, that's all. Now that this book is history for our family—and for some of us, relatively ancient history—none of us thinks of food choice as our defining characteristic. That would strike us as weird and a little boring (except maybe for a spokesperson for a diet program or restaurant chain). When we travel or find ourselves pressed for time we'll happily eat whatever, without apology or remorse. But when we're at leisure, in this particular house, we enjoy cooking with the ingredients that grow from this particular farm.

Now in midlife, I take some of my greatest pleasures from our family's food culture. It's not just the meals that make me happy, but the life. No matter how crummy a day I've had, I know it will get better if I haul myself into the kitchen; even better if Steven is there to listen to my grievances while we make dinner. Better yet if we're there with our daughters and son-in-law, all scooting easily around each other as we manage our pieces of a complex operation we all know by heart. There are never too many cooks in my kitchen. Nor too many hands in the garden; this farm still owns us. Camille and Reid live just down the road. Lily is in college a few hours away, but her face still lights up and says "home" when she walks through the front gate between the hollyhocks. We're what you'd call embedded.

Steven and I constantly talk about scaling back our operation, while steadily doing the opposite. Since we wrote this book our vineyard has matured, we've learned how to make wine and cider, and planted a new apple orchard. We've added a dozen Dexter cattle, a small-framed breed that's beautifully suited to pasture finishing. Also, a flock of Icelandic sheep for meat and wool, which we chose for their admirable hardiness and spectacular natural colors from white and pinkish cream to silver, brown, and black. Our flock ranges from twenty to forty sheep, depending on the time of year. Our vegetable garden is still immense. And, because we really needed something else to do, Steven launched an unusual restaurant and community-development project (more about that later).

Obviously ours is a working farm that feeds more than just our family, and employs a few extra hands, especially in summer. Among the products that go off the farm are specialty vegetables, lamb, beef, and wool that gets spun into yarn. This place is not just our domicile, but a piece of ground that is well suited to producing food. It would feel wrong to occupy that kind of land and let it lie fallow. The ethical choice is to manage it for food production, ideally in a way that maintains productivity, improves the health of its soils and watershed, and sequesters more carbon than it burns. If we weren't willing to do this, I think we would need to move out and let somebody else do it.

But I'm more than willing. I *like* putting on my muck boots and traipsing up to the garden in springtime to see what's come up overnight. When I discover little curve-necked bean sprouts emerging in perfectly even rows, I am flooded with a warm glow of predictable order imposed on a disorderly planet. It will evaporate as soon as I come back inside and read the newspaper. But that's part of the deal; hope is a renewable option. Farming is renewal by definition. I love watching the curly-haired lambs the first minute after they're born as they find their wobbly legs, stand up, and stagger after mama, doggedly bunting a nose against her front legs, back legs, belly, the wall of the paddock, and me—if I'm in there with them—until they finally latch onto the bliss of colostrum and milk. When I pick up these fresh-born creatures they're damp and surprisingly hot, with little hearts pounding like the engines of life they are. In lambing season we stay close to the barn because sometimes they'll need help, not just finding the teat but getting through the mortal doorway. I've had to deliver stuck lambs, breech lambs, tangled triplets, and revive two or three that were born not breathing. If I had any chance of pausing first to consider whether I knew what I was doing, I would have said, in every case, "heck no!" And any livestock farmer will tell you that's a regular April morning: save a couple of lives and then go in the house, wash up, and make your oatmeal.

If those are ordinary days, they refuel me nevertheless with a sense of extraordinary possibilities. I'm moved by these gentle lives under my care, ridiculously proud of myself when I save one, frustrated when I can't, attentive as we vaccinate and nurse the sick, authentically grateful when we harvest them for meat, and generally unsentimental when we make our practical choices. (As I mentioned, our flock ranges in size from twenty to forty and back again. Not forty, eighty, one hundred sixty.) But I got a little teary-eyed this spring when we buried Old Meg, one of the three founding mothers of our flock, who died of old age after more than a decade of turning our good pastures into thirteen lambs and many dozen armloads of ink-black wool. These sheep and cattle are a model of frugal enterprise: solar power and rain make the grass, the animals do the manufacturing, and we just do our best to direct the proceedings. Still, I feel so clever when I cook a roast or knit a sweater that my farm made entirely out of sunshine.

Clever and sometimes really tired. It's not just the animals that tax us. We also pull off dramatic rescues of the vegetable kind, involving frantic sprints with row cover fabric on nights before an unseasonal freeze. We love this life, mostly, except on the days when it threatens to break our hearts or our bones. We've put in grueling days and still lost hay crops to badly-timed rain, lost grape crops to fungus, lost whole seasons of pears and apples to one ruthless late-spring frost. We've gotten up at all hours to bottle-feed an orphaned lamb that still didn't make it. I thank my stars that I have another source of income, but then again, in modern times most farmers do. They have to. In addition to art and work, farming is luck. Between the whims of the market and the fickle cruelties of a changing climate we forge our partnership with the land, always hoping for the best, and sometimes—this goes for all of us—we ask ourselves why we keep doing it?

It's a question anyone might ask, farmer or not: why pay attention? Why not just let an unknown person on the far side of the world do this hard work, enabling our illusion of food as the magically inexhaustible, invisibly created resource we require three times a day? What good reason can we find for keeping up a relationship with our own local food system? It would be so easy to run off with the first plump Peruvian asparagus or Chinese apple that catches our eye. And, of course, less costly to opt for the calorie-dense, processed packaged foods that are all (somewhat covertly) subsidized with tax dollars in a way that fresh vegetables are not. For plenty of people it's not a question because it's not even a choice.

Ten years ago, we wrote this book about our decision to get serious with eating local, more or less forsaking all others. It was such a novel idea that we had to make up our own language for it, although the word "locavore" germinated so naturally from the lexicon that it was invented simultaneously by many people at around the same time. Now it's in the dictionary and on menus all over the place. What we'd thought of as our eccentric family notion broke out into something of a scene. I don't mean that we created a scene—not at all—but that we happened to publish our family's story at a moment when lots of families were ready to examine and take more control of their own food stories. Our book tour took us from coast to coast, San Francisco to Vermont, Chicago to Dallas. And then from Montreal to Vancouver, then to the United Kingdom, and even to France where we were very surprised that anyone might want to hear anything at all about food from people of the American persuasion.

For us, the pleasure of these tours was not talking but listening. We encountered an amazing number of people who wanted to talk about food, think about it in new ways, and renew their own dedication to a long-forgotten relationship between personal hungers and the land around them. In Philadelphia, in Alberta, in Devonshire, good citizens of their food webs wanted to give us samples of the local fare, take us to their farmers' markets, show us their urban garden projects in cohousing communities and low-income neighborhoods, and treat us to dinner in great new farm-to-table restaurants. (This doesn't get old. Let me just say two words: *Chez Panisse*.) Food-wise, it was the book tour to die for. Our only regret is all the fabulous homemade stuff people gave us—chutneys, cheeses, moonshine!—that we had to leave behind because we were getting on airplanes. We made some taxi drivers really happy.

In much greater numbers and far beyond the scope of our own experience, people wanted to explore this animal-vegetable miracle on their own terms. They wanted to dig up their suburban front yards, for example, and plant them chockablock with corn and tomatoes, and send us photos of their adorable locavore toddler out there with his big tomato-juicy grin. We loved these photos and put them up on our website. In the limited way we could manage, we used our website as a local-food forum and posted a rolling photo-farm-tour to give interested readers an ongoing acquaintance with our project in all its seasons. We did our best to keep up our end of a conversation we never expected.

Over time, of course, we've gotten languid about updating that website. We've had crops to take in and other work to do, all of us. I've had novels to write. I had to exercise discipline with myself and others in remembering my place as an author, not a spokesperson for a food and farming movement. Sometimes it felt painful or even unkind to decline these respectful invitations, but the movement has leaders infinitely more qualified than I am to speak for it. I'm not an expert on anything by trade, except being a writer. That word means applying myself not to books I've already published, but to the one that's coming next. It will only get written if I stop parading around and stay at my desk, focused on something new. Everyone in our family has, likewise, needed to get on with life and projects, defining himself or herself in ways quite apart from the question of what we eat.

But, of course, we still eat, and of course we still care. This thing we now call "AVM" for short wasn't just a book but a profoundly influential year of our family life. When we took our locavore vows we did it as a family. When we told our story aloud, readers asked us all sorts of questions ranging from the obvious "How's the turkey-sex going these days?" to the more obscure "How do you keep flea beetles from shredding your eggplants?" But these two questions began nearly every interview and were called out from every audience: "What did you find hardest to give up?" And "Are you still eating this way?" At the time, those answers were straightforward: we set up this project in a way that would encourage us to succeed, so nothing about it was terribly hard and we didn't feel deprived. And barely a year later when the book came out, we were still in the same place, very much in locavore mode. (And yes, we still have Bourbon Red turkeys; Old Tom eventually learned to be quite good with the ladies, and graduated to Lothario Emeritus.)

When we were invited to write an anniversary update ten years after the first publication of *Animal, Vegetable, Miracle*, it struck us as a useful enterprise not just for readers but for ourselves. Has our love affair with local foodways blossomed or faded? Have we been faithful? What does "faithful" even mean in this context? Are we going to keep doing these things? Why or why not? If you could join us at our dinner table you'd see we are a family of passions and strong opinions. These questions have a lot of possible answers.

The world has turned in ten years, inside and outside of our household. We've had weddings and funerals, we've grown up, gotten married, gone to school, gone gray. We've remained ourselves, as people do, by remembering what we love best. The world as we now know it is the world into which we first launched this book, and then some. A global reckoning on climate change is visible on the horizon as islands drown, super-hurricanes pound the coasts, and governments race turtle-wise to reach agreements. Global energy use has changed drastically since Steven wrote the essay "Oily Food," with economic and social factors shifting the balance away from petroleum and coal, toward natural gas and, little by little, renewable energy sources. Sustainable agriculture has come of age, offering resilience and food security in unpredictable climates, and valuable potential for carbon sequestration. Slow food has gone viral in its own gentle way. Locavore was the New Oxford American Dictionary Word of the Year in 2007. It's no longer considered impolite to use the word "farm" at the dinner table. Some of the most unlikely purveyors, including big box stores and fast-food chains, are jumping on the bandwagon.

One evening this past summer, as I sat on an airplane midway over the Atlantic, I was amazed to see the words "ingredients from local farmers and artisans" on the in-flight menu. I understood that the word "local" was being stretched to new heights. But I also understood how drastically the ground has shifted under our feet since those cold March days in Virginia when we first waited for our asparagus.

TURNING TABLES

by Steven Hopp

Some people might expect a best-selling book to bring them fame and fortune. But I never saw this coming: a restaurant.

I wasn't looking for a new career, a source of income, or more to do. I'm still a college professor and a farmer. I appear to have no good business sense: I serve as the director of this restaurant but have made no money from it, despite the hard work I've put into it year after year. It's a business founded on unconventional principles. When I first described my idea to a friend in the industry he offered these gentle words of encouragement: "Did a poleax fall on your head?"

Let me back up and start at the beginning.

Picture a town square, maybe half the size of a typical city block, a parking lot in the center surrounded by wooden-framed, early-1900s commercial buildings with pillared front porches. An empty shell of an old railroad depot stands alongside the quiet railroad track. Stand in the middle of the square and turn around slowly and you'll see mostly empty buildings in advanced states of disrepair, their front display windows covered with plywood. This is my little town, really just a village, without enough of a population to support any type of town government. We have no downtown, no stoplights, and fewer than a thousand people.

During the year we were writing this book something big came to our little village of Meadowview, Virginia: a national company proposed building a service station and truck stop at our town's interstate exit. This proposal opened a community discussion around job opportunities. Before this, nobody had put much effort into reviving our little town, probably for decades, so the idea of some jobs was appealing. But would a national truck stop chain likely be a community-minded employer? How else might we turn things around and create meaningful jobs for a town so small and so far down on its luck?

Meadowview isn't unique. All over our country small towns described as "once-thriving" are now home to too many empty buildings and declining communities. People are quick to write these towns off, calling them the doomed victims of the global economy. The cause of this rural decline involves a multitude of forces: tax laws that favor corporations, consolidation of services to centralized facilities, outsourcing at many levels, and a long history of advertising to foster big-name brand loyalty. Small businesses weren't nimble enough to compete with the big-box stores that popped up conspicuously on the outer edges of town. There are arguably pluses and minuses to this. Shopping at big-box stores offers bargains in the short run. But corporate businesses managed from far away have no interest in a store's local community beyond financial gain, and do not share in its suffering. The money we spend in these stores mostly gets drained away from us and our neighborhoods. Researchers who study these trends estimate that locally owned businesses selling imported goods, such as a small bookstore or a florist, redirect around 50 percent of their money back into their own community. A locally owned business that sells locally made things, such as a craft shop, can redirect from 75 to 85 percent of its money back into the community. Corporate and big-box stores are nearly the reverse: such businesses drain up to 85 percent of the money spent there out of the community. In terms of supporting a local economy, the worst is the huge trend of ordering online; the only benefit is the small amount designated to the local delivery. The revenue drain from the community also drains its viability, and that flow tends to keep going in one direction. In recent history in our country, the smaller the community, the greater the financial drain, with the worst-hit towns being those with fewer than a thousand residents.

This brings us back to Meadowview. In 2007 when we were preparing to hand in our manuscript, I had just spent two years reading about building local economies and regional food systems, and I began hatching the idea of creating a local business in our village. My goal was to maximize the number of people who could benefit. The question was, what kind of enterprise? Specifically, what did Meadowview have to offer? How could we effectively get money flowing in to the community rather than out? Like many rural towns, the economic base of our region is historically rooted in small farms, but for decades that livelihood has slowly declined, following the same pattern of corporate consolidation we describe in this book. Tobacco farming has now lost its support. The main agricultural product of our region these days is cattle, but not for local consumption. Calves are raised on our grassy rolling hills, then shipped off to be finished and processed in huge midwestern feedlots, with the bulk of the profits also rolling out on those cattle trucks.

We still had small, local vegetable growers who needed consumers. I wondered if we could use the ideas from our book to help connect more local diners with the food of our region. After lots of talk and a roundup of community and family support, we decided to pull the plywood off the face of one of those derelict buildings on the square, modernize the electricity and plumbing, and hang up a sign.

That's the simple version of how we opened the Harvest Table Restaurant. The restaurant is led by Chef Philip Newton and a loyal staff committed to promoting local and regional goods as creatively as possible. Farm-to-table restaurants are becoming popular across the country, but the Harvest Table is working at the front edge of this trend, building its whole menu around local sourcing. We depend on local growers in every way possible, and change our menu daily, not just seasonally. The restaurant continually tries to find, add, and feature new local items whether from seasoned organic growers or from those we can encourage to broaden their range of products. At our specific request, various farmers have grown sweet corn, winter squash, parsnips, turnips, heirloom tomatoes, mushrooms, and Appalachian heirloom dent corns for grits, to name a few examples.

Serving local foods in restaurants moved from fringe to mainstream in the past decade. It's no longer unusual for chefs to partner with farmers or create their own gardens. That's part of a broader American movement toward increasing the contribution of local foods into our diet. Other efforts include rebuilding human-scale production facilities like grain mills, butcher shops, cideries, and small-scale canneries. Local breweries and brewpubs have flourished with a record high number in the United States. Beer, while not strictly food (although some might call it breakfast), made with locally sourced ingredients has introduced thousands of people to that preference, and many of these pubs also serve local food, creating a new generation of lubricated locavores.

At the time we released this book, most local foods changed hands via direct sales, with farmers serving as both growers and salespeople. Since then, many creative entrepreneurs invented novel ways to collect and distribute local foods. Existing conventional food distributors have been happy to add local products to their inventories. Food hubs pool products from local or regional producers and make them available to restaurants, schools, hospitals, country clubs, or private dining events. More than 40 percent of school districts in our country now have some version of a farm-to-school program. Food hubs have been most successful in urban areas, and the growth of urban markets has led to an explosive growth of small-scale sustainable farm operations surrounding cities and even embedded in city centers. This has reduced the food miles for lots of our population. We have a regional food hub in rural, southwestern Virginia, called Rooted in Appalachia, helping to connect growers in our region with interested consumers.

Diners visiting our restaurant can see on our menu that we're committed to our region's small farms, and they're often curious about how we define local. It seems like it should be easy to specify that all or most of our food comes from within a 100-mile radius, but we have goals beyond just reducing our food miles. First, we want to promote farming systems that respect the environment, soil and animals. This means partnering with our neighbors who follow sustainable practices, including organic production, and other methods that sequester carbon in the soil. Many people are aware that agriculture in general, and animal production in particular, contributes to the climate-warming excess of atmospheric carbon dioxide. What they may not realize is that those blanket statements refer to industrial forms of agriculture: chemical intensive commodity cropping and large-scale feedlot animal production. Sustainable farming practices can actually do the opposite, removing excess carbon dioxide from the atmosphere. Climate scientists call this regenerative agriculture, and we support it as a top priority. Ironically, this is why we're not a vegetarian restaurant, though we gladly accommodate vegans and vegetarians. But we occupy a steep, hilly region where cultivating tilled crops, like grains and soybeans, is likely to send the disturbed topsoil streaming downhill into watersheds. In contrast, sheep, pigs, and cattle raised and finished on well-managed pasture-based farms allow plant roots to grow deeply and extensively under their hooves, like an underground forest. This root network removes carbon from the atmosphere, and sequesters it more permanently than an equal-sized plot of forest. To promote these carbon-negative systems, we never serve feedlot meats, only pasture-based. These methods also improve the soil quality rather than robbing it of nutrients, so grass-fed meat on our table is also like money in the bank for the planet's future. In fact, these agroecological practices are so effective in sequestering carbon that they were proposed as one of the climate solutions—known as the 4 per 1,000 Initiative—by the French government at the recent United Nations Climate Change conference in Paris, COP21. More than thirty countries have agreed to adopt their proposal.

Our second criterion is that we prefer to source from families rather than companies. Our commitment isn't only to quality food, but to the community as a whole, and the people we know personally. We often visit their farms and see their growing practices. We work together to deepen our commitment to sustainability. In committing ourselves to food sources that haven't traveled a lot of miles, we're not just participating in the local challenge but truly working with our neighbors, helping to build a better community and economy.

What we've created over the years is a modern restaurant with a neighborhood feel. We serve lunch, dinner, and Sunday brunch. The menu is rewritten virtually every day based on what's in season, and especially featuring what we have in abundance. We try to find a happy, creative balance between casual and refined. So you can find a pasture-finished burger of the day, or step it up to an oven-roasted pork tenderloin with herbed fingerling potatoes, squash, butternut squash butter, and fennel flowers. Or a vegan dish of roasted yellow squash with oyster mushrooms, curtido (quick pickled cabbage), Carolina Gold rice, sour gherkins, and purslane. Those were actually on our menu the day I wrote this. Every day our farmers bring us something new, challenging us to reinvent ourselves.

When we opened, we relied on growers we knew, using what they already grew. But soon our restaurant's needs outpaced the local suppliers. In 2010, we opened Harvest Table Farm to boost supplies, extend the normal growing season, and provide items that other local growers weren't producing, mainly because of low market value. A restaurant uses, for example, a lot of onions and potatoes, but local growers can't get much for these at farmers' markets so they haven't been excited about them. Our own farm could help fill in these gaps, and provide a reliable supply of greens all winter using innovative, low-tech methods like unheated hoop houses. We didn't want to replace our local suppliers, but rather to supplement what was available. Our farm also serves as an education center. Restaurant customers can visit if they're interested, and we frequently give tours to school and faith groups. The farm also brings in WWOOFers (World Wide Opportunities on Organic Farms), interns, and agricultural volunteers from all over the country, and helps to train a new generation of sustainably-minded farmers.

The number of suppliers in our region has grown since we opened, giving us a remarkable diversity of local goods from gourmet mushrooms to cured cheeses, local wines and beers, reflecting the same creative growth across the whole country. This trend included a tripling of the national number of farmers' markets since we started our year of local eating, with many communities building dedicated structures for these markets, and a proliferation of produce stands and farm stores. Other cooperative farm endeavors pool land, talent, and equipment, organize urban and community gardens, and otherwise creatively reinvent sustainable food systems appropriate to their own neighborhoods.

Food represents about 10 percent of the American economy, and that's a starting point for a small community without much else. We decided to expand our contribution by creating a general store featuring not just food but other local goods: We have an open invitation to anyone in our community who makes anything they think is worth selling. The general store's shelves are stocked with arts and crafts, hand-knitted caps and scarves, ceramics, soaps, jewelry, and the occasional children's tin-can art (which we allow to stay for at least a few months). More than two hundred of our community members have earned money from the general store. We offer the restaurant walls as gallery space for local fine artists to exhibit their work, and showcase our best local musicians in performances during our weekly live music nights.

Between all these enterprises—the restaurant, general store and farm—we've managed to redirect more than two million dollars into our formerly boarded-up village and local community. If we'd opened a conventional restaurant or other business based on outsourced goods, that amount of money would have been taken out of the community. As an academic, farmer, and sometime writer, I never thought of myself as a business owner. But as a citizen of Meadowview, I'm proud to be part of a dedicated group working to bring life and cash flow into my community. We attract nearly 20,000 visitors a year to a little town square where people come for a farm-to-table experience and leave money behind to be spread among chefs, servers, farmers, and artisans. We keep prices affordable while paying a fair wage to workers and suppliers, which means the profit margin is slim, and the director (that's me) volunteers his time. Our goal is to make profits that will flow into the community, not to me or my family since we have other sources of income. Maybe a poleax did fall on my head, but I believe people matter more than the bottom line.

Now when I visit Meadowview I'm likely to see the town square buzzing, and cars with license plates from anywhere in the country parked alongside those belonging to our regular restaurant customers. We're located close to the interstate, and, thanks to modern mobile devices, it's hard to predict who might find us and walk in the door. We've served local, state, and national legislators; famous musicians and film celebrities; environmental activists; and we're a mecca for book clubs. We've also fed vanloads of tattoo artists heading for a convention, and convoys of motorcycle enthusiasts out for a joyride on a sunny afternoon. There's no telling who might pull in next to the pickup truck of a farmer bringing produce to the restaurant, or a local church group coming in for brunch after Sunday service. To me, that looks like success.

HOME AGAIN

by Camille Kingsolver

I guess I've come full circle. Now that I'm expecting my first child, one of the few taste aversions I've developed is to asparagus. As the platter goes around the table, I take a sniff and feel like wrinkling my nose as I did when I was little. It took me years to learn to love asparagus and now here I am, ironically, back to reluctantly trying a single bite.

I've also come back to southwestern Virginia. My husband, Reid, and I moved here last year after spending the first decade of our adulthood in North Carolina. He grew up just an hour from my family's farm, so the move was a homecoming for both of us. Leaving our home of six years and many dear friends in Asheville was difficult, but felt right. I'd just finished graduate school and our conversations about starting a family were becoming more frequent, less hypothetical. We realized we want our children to grow up in a place where they can play in the woods, see their grandparents regularly, and learn an appreciation for the hard work that goes into growing and making food.

Had you asked 18-year-old me if I imagined myself returning to small-town southwestern Virginia in ten years, I probably would have laughed. At that age, I never dreamed I'd fall in love with a Virginia boy, nor did I realize my Appalachian heritage would become such an important part of my adult identity. Of course, I haven't always lived in these mountains, and a big chunk of my heart will always belong to the endless skies and fresh tamales of Tucson, Arizona. But I've come to appreciate the value of having roots in a place. I feel grounded knowing people with my last name lived here before my parents were even an idea. In all the hours I've spent digging in Washington County dirt, I may have planted a piece of my soul here. It wasn't until I'd been gone awhile that I realized something was missing.

During the year we documented in *Animal, Vegetable, Miracle* I went off to college at Duke University. College life made me a better student, shaped the way I understand the world, and challenged me to examine values I'd taken for granted. Up until then, I'd mostly been around people whose lifestyles and backgrounds were not too different from mine. My close friends and I may not have looked alike or gone to the same church, but we shared the same basic expectations about our day-to-day lives. When I moved my Indian-print bedspread and suitcases of hand-me-down clothes into my freshman dorm, I was suddenly swimming in a sea of people with whom I shared little more than a birth year. Many of my new peers had grown up in big cities and gone to private schools. Few had ever worked the minimum-wage jobs I'd taken for granted as an inescapable part of teenage life, like pimples and first-date jitters. I met plenty of wonderful people and befriended those who shared my world view, as well as those who didn't. The experience of living as an outsider to my college's majority culture taught me about myself, and about privilege.

In fact, a lot of my life experiences over the past decade have deepened my understanding of privilege. I graduated from Duke in the spring of 2009, at the peak of an economic crisis. I determinedly applied to jobs that required a college degree in the sciences, but didn't land a single interview because I was in competition with applicants who had doctorates and decades of field experience. After spending just enough time back in my parents' house to get extremely restless, I decided to move to Asheville, North Carolina with no promise of a job. Somehow I just felt confident I would like it there. I rented a room in a house downtown and was amazed to get two job offers by the end of my first week, both in food service. I'd worked in restaurant kitchens throughout college because my upbringing gave me the knife skills and knowledge of ingredients necessary to hold my own on a line. For all those years, I'd enjoyed the familiar rhythms of food prep, and as a student I appreciated the escape from academia. I didn't love the condescension of customers, including fellow students who'd sat across from me in a seminar just hours earlier but didn't recognize me in my work uniform. But the work kept me humble, something I valued almost as much as my biweekly paychecks. It hadn't necessarily been my plan to continue working in restaurants after earning a bachelor's degree, but in a new, post-crash economy those were the jobs

I could find and I was grateful. I avoided discussing my education with coworkers because they tended to assume, once they found out I had a degree from Duke, that I considered our daily work to be beneath me. But I never felt that way. Cooking in a restaurant kitchen is hard, and it's important, whether or not you have a degree from a fancy college.

After a couple of years making food for a living, I found an entry-level job working with children and adolescents with mental health problems. I was ready to make use of some of my other skills and knowledge. I had started college thinking I'd go into nutrition science, maybe as a natural outgrowth of my family's interest in healthy food systems. But along the way I accidentally fell in love with the psychology courses I took as electives, and I graduated with hopes for a career in the mental health field. When I finally took my first step in that direction, in a nonprofit mental health agency in Asheville, I learned that working with young people in treatment for severe emotional and behavioral issues is not for everyone. The work is exhausting physically and emotionally, the bureaucracy involved can be extremely frustrating, the paperwork is tedious, the pay is pretty lousy, the stories you hear are often deeply disturbing, the progress can be slow to undetectable, and there is no other work I'd rather do.

My experience in that world inspired me to keep going. While working towards my master's degree in clinical mental health counseling, I also got married, adopted the world's sweetest dog, and grew onions, greens, tomatoes, squash, peppers, eggplant, and even some fennel in our tiny front yard. Reid and I quickly learned the urban gardener's nemesis—our lovely raised beds appealed to the neighborhood cats as luxury litter boxes. Cats or no cats, we managed to harvest a decent crop of veggies for several summers.

Now we've moved back to a rural life and are struggling to outwit the healthy population of deer and groundhogs who share our property. Though my transition back home has been a little rocky in the gardening department, my career has fallen into place as I immediately found a job with our county's community mental health agency. I work with 11- to 18-year-olds and their families, most of whom receive Medicaid. The challenges faced by the young people in my care vary widely, but many are common to a low-income, rural community. I'm inspired daily by the resilience and resourcefulness of the adolescents and families I see.

Working with low-income families, I see up close how hard it is to eat in a healthy way. Ten years ago I wasn't able to see that food choice is a type of privilege, but now that couldn't be clearer to me. For a lot of hardworking, conscientious people in this country, eating locally is not really an option. Low-income families have a hard time finding and affording fresh produce, local or not. Here in southwestern Virginia, getting to the grocery store is an ordeal for some people. In more remote parts of this region, the nearest grocery store can be twenty miles away. If that family is sharing a vehicle between two or more adults who require transportation to work, and has a budget that allows for only one or two tanks of gas per month, trips to the grocery store can't be frequent. For this family's best hope of having food on the table every day, they have to rely on low-cost items with long shelf lives like potted meat and boxed macaroni and cheese. Alternatively, if fast food establishments are closer than the grocery store to their home and work, the family might look to the dollar menu to help keep everyone fed.

An idealized view of Appalachia probably puts all of us out in our gardens. And it's true, many households in this region do participate in the tradition of a summer garden, but it's not an option for everyone. Even those living furthest from town may not own the land they're living on, or be able to count on staying in one spot for an entire growing season. And many more have physically demanding jobs, in mines or factories for example, that leave them with little energy at the end of the day for the hard work of growing food. It's one thing to go out in the heat and hoe weeds after sitting at a desk in an air-conditioned room all day, but finding the will for it after spending a long shift at a cash register or on the factory floor is another story. Some folks enjoy their gardens enough to put in that kind of effort, but I wouldn't pass an ounce of judgment on those who don't.

And there are barriers that run deeper than simple time and effort. According to a report released by the National Alliance on Mental Illness in 2014, about 80 percent of unemployed adults in the state of Virginia were currently diagnosed with some type of mental illness. From the outside it might be tempting to look at an unemployed adult living on a decent plot of land in southern Appalachia and think, "That person has plenty of time to do something productive, like growing a garden." But anyone who has experienced depression understands that even a project as simple as a backyard garden can feel insurmountable. Planning, gathering information and resources, and following through with a garden's demands all require a level of psychological wellness that, sadly, not everyone enjoys.

The good news is that in the ten years since we wrote our book about food choices, substantial efforts have been made to bring better options to more people in the United States. The most recent revision of the United States Department of Agriculture's (USDA) Farm Bill allots more funds than ever to healthier crops and public health efforts. Food stamp incentive programs have been launched at farmers' markets all over the country, including the one closest to where I live. Families receiving nutritional assistance can double their EBT (electronic benefits transfer) points, or food-stamp dollars, up to $25 at each farmers' market. Since ours takes place on Saturdays and Tuesdays, that's potentially $50 worth of free, healthy local food per week. Of course, transportation and scheduling are still challenges, but this newly expanded program has huge potential rewards for those most in need. Farm-to-school programs have also been on the rise in urban and rural communities across the nation, teaching kids about sustainable agriculture and bringing fresh, healthy foods into the schools so these choices might lead kids into healthier adulthoods.

One big problem that persists, especially in rural areas, is regional food infrastructure. Corner store initiatives have begun to address the issue of access to healthy food in cities. By stocking fresh, local produce at a reasonable price in high-volume convenience stores, these initiatives bring food choices to urban people (especially kids) who rely on corner stores for most of their meals, for lack of transportation. Unfortunately this model doesn't work for a smaller population spread over a larger geographic area. Rural communities don't have an equivalent to the corner store. Small, independent grocery stores once filled a similar niche, but the little country store is a dying breed. I don't have a solution to the problem of food accessibility in regions like mine, but identifying the problem can bring us a step closer.

As I make my final edits to this essay, I'm less than a week from my due date and quite preoccupied by what's to come. I'm thrilled to meet this little person I've been growing for nine months. I hope my son will get to see the local food movement flourish in his lifetime. I hope as he gets older, more of his friends and their families will have access to healthy, sustainable food, regardless of their socioeconomic status. As a parent I'll do my best to pass on my appreciation for both the fragility and bounty of this place we call home. I can't wait to witness my child's sense of wonder the first time he sees a family of deer skipping through our front yard, even if they're the same deer that wiped out our sweet potato crop. I know his father and I will make time to show him how to mix biscuit dough from scratch, and to push seeds in neat rows into the soil. I imagine the three of us giggling at the squish of ground meat between our fingers as we shape sausage patties. We'll teach him these things so we can feel confident he'll always know how to feed himself and never take his meals for granted. I won't get to choose where our little boy will go when he grows up, or what kind of work he'll be drawn to. But if I can raise a person who sees himself as one piece of a much bigger picture, I think I'll be doing my job.

BUILDING A PANTHEON

by Lily Hopp Kingsolver

I never did quite what I planned to do with my egg money. I saved the $2.50 each time I sold a dozen eggs from my lovely flock, and had big plans for my small savings account. By the time my family published our book, my dreams had grown even bigger than a horse. Our conversations about sustainable food at the dinner table had gotten through to me, and I began to form new ideas about how I wanted to exist in the world. The horse dreams of my nine-year-old mind grew into something that would carry me through my next ten years into the beginning of my adult life.

The autumn after we began our adventure in seasonal eating, I was off to fifth grade with a blue-flowered backpack and a lot of new things to think about. I'd always been drawn to the natural world, but now that I was paying closer attention to my food, I took the biosphere more seriously. Pulling weeds for hours at a time and carefully coaxing seedlings out of the earth felt spiritual to me, and I yearned to belong more thoughtfully to the ecological system I'd uncovered. I started cutting out pictures of animals from magazines and taping them to my walls, the same way some of my friends taped up pictures of celebrities. The more pictures I collected, the more I found myself in awe, especially of tropical creatures. The rainforest was my own personal celebrity, and a pantheon of dart frogs and leopards and birds of paradise spread over my walls and ceiling like ivy. I would lie in bed and stare, promising to devote my life to the preservation of these Hieronymus Bosch-esque creatures with their watercolor brilliance and strangeness. They howled and hummed and haunted me until one day I got off the school bus and announced to my mother that I wanted to donate all my egg money to the rainforest.

As a ten-year-old, of course, I didn't have the foggiest idea how to do that. My proclamation conjures up images of spider monkeys spending my hard-earned cash on fruit, or iguanas counting out the dollars to buy themselves hammocks. I started doing research, religiously poring over the websites of various conservation groups. I made a few small donations to ones I liked, but the more I read about this work, the more I understood that what I really wanted was to get my hands on the rainforest. I wanted to be out somewhere in the heat counting trees or tagging birds. I memorized the encyclopedia entries for biodiversity and conservation, convinced in my fiercely devoted little mind that being able to recite this information word-for-word was an essential part of the biological discipline. To put myself completely into my dreams of conservation ecology, I decided to become an honest-to-goodness biologist.

Now I stand with one foot still in childhood and the other foot in grown-up life, and find my love of nature has been refined but has never lessened. My family's practice of mindful eating has fed my belief that my own small contribution to this world has meaning, as long as I put my whole self into it. So I threw what was left of my egg earnings into the pocket of the American educational system and headed off to college with another brand-new backpack, this time with a schedule full of science classes, a shiny new dissection kit, and a plan to graduate with a degree in environmental science. I can still recite some encyclopedia passages I'll never need, in fact I can recite quite a few new ones now, and I'm a step closer every day to realizing my dream.

Now I spend my days trying to fit in meals between geology labs and classes on climate change. During school breaks I've started working for the American Chestnut Foundation where I get to spend my days with an amazing endemic tree species that's disappearing from North America due to chestnut blight. In my free time I care for my pets, a lovely corn snake and a hamster I rescued from euthanasia after it was used for a biology lab study. I try hard to resist buying new pets, even though I've been known to sneak the occasional clutch of frog eggs into my dorm room to hatch. I plan to study abroad next semester in a Central American ecology program, where I'm thrilled to be meeting my lifelong true love for the first time. I spent uncountable hours sitting on the leaf-printed rug in my bedroom and cutting pictures out of magazines to add to my growing pantheon of jungle creatures. I said their names to myself over and over as I fell asleep, mouthing hoatzin and capuchin like they were my secrets. As I got older I came to know these creatures more intimately through my studies, but I still admired them from afar. Now I feel as if I'm prepping for a first date, but instead of fixing my hair I'm buying plane tickets and filling out forms for a visa.

In the meantime, I've found it can be tough to save the world when you're a twenty-year-old with a minimum wage job. Once while I was babysitting a sweet little girl, she asked me earnestly if I was a baby or a mommy. I was surprised at how difficult it was for me to answer. The truth is, I'm a little of both. I have my days, as most college students do, when I eat pizza in bed and tell myself tomato sauce is a vegetable. I also have days when I shop and cook thoughtfully and feel like a grown-up. The trick for me is finding a healthy balance between a mindless college-cafeteria diet and my own expression of the emotional maturity my family encouraged me to develop when we all pushed ourselves to eat in a way that made our bodies and hearts feel full. I've figured out that sharing tasks with a friend helps me find the motivation to cook, shop for groceries more carefully, and be more creative with meal planning. There's also some bonding in the forced intimacy of cooking together in my apartment kitchen, which is approximately the size of a cereal box. Discussing and creating food with another person is a beautiful way of connecting, and one that I came to love in the kitchen where I grew up.

I've also learned to give myself the space to flounder a little, especially when my life is in such a dynamic period of growth. I'm living in a city on a college student's budget. Sometimes my only realistic shot at a locally grown meal is the jar of Mom's applesauce I brought back from my last visit home. Self-respect means accepting that even when I'm struggling to align my life with my values, I am truly doing okay.

This self-respect, like so many things, has been a journey for me. It's a little strange to grow up feeling like thousands of other people are sitting around the dinner table with your family. I've accidentally stumbled across online posts by people who've named their chickens Lily after me, and I've sat down next to total strangers who introduced themselves by saying "I feel like I already know you." These people aren't unkind, in fact they are usually very lovely, but it's still odd to grow up feeling like there's a window into your fridge. In the years after our book became widely read, I started to feel nervous about whether I was eating the right things and representing my family the right way. The way people sometimes feel about a pimple, I felt about food—I knew nobody was looking *that* closely, but I was self-conscious anyway.

Throughout high school I avoided making food decisions, electing to let my family lead the way when cooking, or just blurting out the first thing I saw when I ordered off a menu. When I entered college I discovered that people really do need to make food choices in order to survive, and I was going to have to figure out how to feel good about my own. It's been very good for me to be in charge of my own eating, even as I've had to come to terms with some cultural choices I was lucky enough not to encounter as a child. Most obviously, of course, there's the matter of cost.

I work full-time every summer and part-time during the semester, but still there's a limit to how much I can save. I know the whole story of the industrial food complex and misplaced government subsidies, but it doesn't change the fact that those organic carrots are a bit steep compared to their industrial counterparts. I deal with this by trying to stay informed, sending my dad texts from the grocery store asking things like, "More fuel to ship fruit or fish? Standing in line, need answer NOW please." I'm not always able to buy the most natural food available, but if I can feel aware and in control of my food choices, I feel like I'm honoring my family's culture as I meet my own life's challenges.

My friends tease me for checking the ingredients of my food, as if buying one less package of crackers will save Indonesia from being deforested for oil palm plantations. They know me as that girl who might bring up GMOs at parties, or burst out of my room to cry, "Did I hear someone say *factory farm*?" (I swear I'm more fun than this makes me sound!) But I do it anyway because it helps me remember I haven't given up on this beautiful living planet. I feel connected to my environment in a way I often have trouble explaining to my peers, especially those who haven't had the opportunities I've had to grow my own food and live on a farm. People tend to laugh in disbelief when I tell them where I grew up. Behind their surprise at my agricultural background I can almost hear *but you're so clean*, or *how did you get into a school like this?* I'm proud, though, and not apologetic for my knowledge of food production and comfort with manual labor. These things gave me a window into the biological world, and help define the culture I choose for myself.

Biologically (for me, at least) there's no better way to understand an organism than by looking at it with your own eyeballs. For all the incredible things I've read about in textbooks, I'm more excited to find a *real live turtle in torpor!* I do know comatose turtles aren't on everybody's list of life highlights, but I'm a person who grew up planting seeds and battling groundhogs—not symbolically in a videogame but with fences and persistence—for my dinner. I appreciate the interconnectedness and the balance between what we consume, what we grow, what we harvest, and what we provide for the world (even the groundhogs). So yes, I might get a horrified look from my boyfriend when I get on my hands and knees on the downtown sidewalk of my hip college town to pick up a particularly radiant cicada, but I am never happier than when I encounter another life, especially one so distinct from my own.

I was always expected to pull my own weight on the farm, and I'm willing to bet I was the only person in my first-year dorm who knew how to castrate a sheep or how to treat bumblefoot in an elderly hen. I'm glad I was taught the value of hard physical work, even if I wouldn't always have said so at the time. I grew up hauling hay bales and chasing lambs through the mud, and that's lucky for me because it turns out that environmental field work mostly involves things like climbing trees and walking upstream in rushing rivers, all of which strike me as a good time. Growing food also taught me about how people connect. We live in a time when one of the most common ways to form a relationship is by sitting near another person while you both eat. Food represents cultural background, social status, even personality. My boyfriend isn't a fan of cicadas, but he and I delight in going to a market called C'ville Oriental, one of those places that sells everything from gourd juice to an unidentifiable brown "pickled vegetable." We cook together often, trying to transform our scant time and budget into something worth eating. This feels like a natural next step, since I grew up watching food turn from a patch of soil and some seeds into a meal over which people might fall in love or have a business meeting. Food is so important. I'm glad my family taught me how it gets that way.

The values I drew from our local food experience have stayed with me, shaping my life more than anything else in my twenty years. I carry it with me into my classes, my personal relationships, my view of myself, and my attachments to the world. It's a moral position that became the fulcrum of my studies and my drive to contribute what I can to the many species struggling to stay alive on this earth. What I learned from watching the seasons and letting them dictate my food choices, as they do for other creatures, became the basis of my love for the rainforest. I learned at an early age about the vast effect of the biosphere on human life: how much of our society is based on the carbon sinks of the oceans and rainforests, the pharmaceutical resources of the South American tropics, the water filtration systems of mollusks native to American lotic environments, and a thousand other impossibly complicated webs of coexistence. I learned enough facts from my dad about the American food economy to fill a feedlot. I also learned about camaraderie and all the ways a food community can unite people. I learned how to disembowel a turkey and how to explain why I don't eat feedlot beef to my school friends without either lecturing or shrugging it off.

My body is made of food of course, but much of my soul is made of food too, built on an emotional platform of self-discipline, mindfulness, and love. I'm created of food, and through food I may begin to create change. Eating locally has helped me understand how I'm bound to this world through the choices I make about what I consume. It helped me realize my passions, taught me how my choices affect my health, and drove me to care about all the lives around me from weeds to birds to people, and cormorants to Congressmen. It turned my big, crazy childhood dreams into a journey of devotion to conservation.

See, isn't that better than a horse?

LEARNING TO BE HUMAN

by Barbara Kingsolver

I write this in August, the month when Lily once declared me the Tomato Queen and used the whole red end of her Crayola spectrum to make my portrait. Now she understands this is not so much a coronation as a collision, something that happens when a frugal adult work-ethic meets the extravagant eruption of an August garden. She comes home sun-cooked and tired from her summer job as a field biologist and still helps me cut up tomatoes. We crank up a high-energy playlist while we work and talk about everything under the sun. This week our tomato-sauce-canning soundtrack is Watsky and Steely Dan.

These late-summer kitchen marathons have grown poignant, August after August, as Camille first and then Lily spent those same weeks packing up to leave for college. Ten years ago when it was happening for the first time I felt stricken, believing some sweet mother-daughter era was coming to an end. Now I've lived through a decade of these partings and am reassured that sturdy family cultures don't end. They surprise us with their persistence.

I have no way to count the hours I've spent cutting up fruits and vegetables with Lily and Camille, Steven, my girlfriends, and my mother-in-law, Joann, who still enjoys extended stays at our farm in summertime. She says it reminds her of her childhood in the Colorado enclave of immigrant farmers (she was Giovanna then) where she grew up helping her Sicilian father in his vegetable fields. On her most recent visit with us, at the age of ninety-one, she was still game to help me put up tomato sauce. She assumed the tedious role of cutting and coring tomatoes while I bounced around measuring other ingredients into the gigantic bubbling pot, sterilizing jars, and timing the canning kettle. It takes concentration. The afternoon's passing escaped me until I looked at the clock and then across the counter island at my darling mother-in-law: barely tall enough to get her elbows on the cutting board, her arthritic knuckles curled around the knife, she seemed lost in her own thoughts while she cored and tossed red fruits into a bowl piled higher than her shoulders. She'd been standing there more than three hours.

"I'm sorry!" I sort of yelped. "Let's take a break. I can't believe your stamina."

She shrugged her shoulders and smiled. "Well, I'm Italian."

I could spend years diagramming that sentence. That shrug, that smile, that unapologetic identity with labor. I am not, technically, Italian. But my fondest hope might be to reach the age of ninety-one and still be so reliably useful that my progeny will forget I am ancient and keep me standing at the counter for hours cutting up vegetables.

On my good days, I feel like I have a decent shot.

It's not that I'm wildly optimistic about the state of the world, or of me. I've known enough sickness and loss to claim mortality as my familiar, and I know how one unlucky half-second can cost a year, or even forever. I've broken each leg twice; to save my dignity I should just declare broken bones a hobby. And as for the world, oh man. In my lifetime I've witnessed the end of the notion of endless: there will *not* always be more fish in the sea, more bread in the basket, more coal in the ground, more ice on the poles. When my kids were small and I fretted about their future, my thoughts gelled around things like college tuition and equal respect for women. Now I'm considering *their* children's prospects and have to fend off images of the ocean swallowing Florida. The world already has too many mouths to feed and not enough arable acres or good water tables. And that same fever-stressed planet is poised on the brink of showing us, I mean *really*, who's boss.

These are bigger worries than I've ever known. Size matters: we're into the domain of oceans, glaciers, rainforests, species, the Ogalalla Aquifer, the desertified continents of Africa and Australia, and all the people fed by all the rivers fed by the vanishing Himalayan icecaps. That kind of thing.

The wisest people in my life have always taught me that worry is either a disease or an engine: you can get paralyzed or you can get moving. Where to go, is the question, when one feels like an ant attempting to carry the proverbial six hundred times its own weight. Anything I might do today feels effete in the face of global catastrophe. But then, nobody's actually asking me at this moment to scrub the face of global catastrophe. It's a point to consider. Humans aren't designed to have any practicable thoughts about the entire world at once, much less behave or affect consequences on that scale, any more than an ant or elephant is naturally gifted at global supremacy. We're animals too, we have a niche. We evolved in small groups. Children are raised by families and, as they say, villages, but not international coalitions. Towns are run by town councils and schools are run by school boards, both of which you have a shot at influencing if you're paying attention. The spectacular global news cycle notwithstanding, the unsung local is where we live, and eat, and go to work. It's where we are born and buried, and do virtually everything in between, barring the odd beach vacation. We're duty-bound as citizens to vote, pester our legislators and try to weigh our influence on the national and global scale, but local problems are relatively comprehensible and also deserve some regard. Little choices are the only kind most of us get to make, and over time they add up to big ones.

This sounds like bumper-sticker anodyne, I know, but it's also true, most obviously in family life where spongelike little people observe their parents' every choice and snap it into the blueprint. When my children were small this tended to freak me out, especially when I was too tired to be exemplary, or some bonehead cut me off in traffic. I enriched my daughters' vocabularies in some unintended ways. But moments, noble or otherwise, get swamped by quotidian routines. I can see now that it was my everyday intentions that my daughters have subtly translated into aspects of their own adult personalities and passions.

As a sphere of potential influence, a community is only slightly bigger than a family—especially where I live. Ten years ago, anybody passing through my town would have called it an empty place, if they noticed it at all. Our public face was a weedy one of vacant lots, broken windows, chained-up dogs, and tattered Confederate flags hanging from the odd front porch. Hit the gas and get out of there! I wouldn't blame you. But my family wanted to stay, for a lot of good reasons that weren't visible to outsiders. Now, some of them might actually catch your eye: two well-kept buildings on the town square, for starters, painted barn red and aquamarine. These buildings are the restaurant and the general store; the white building next to them is a community center and sliding-scale health clinic. Where once there was nothing much, my town is now a place where people can meet for lunch, drinks, or dinner; go to the doctor; get help with filing taxes or citizenship papers; attend a community meeting; and go shopping for garden seeds or a birthday present. Students from the nearby college can comfortably hang out here, and every Wednesday night people can listen to live music while they have dinner. This week we went to hear a really good jazz quartet. The house was packed, the kitchen and waitstaff hopping, and the experience, for me, transporting, maybe because jazz is a little outside our bailiwick here. Most Wednesdays it's likely to be a singer-songwriter or a bluegrass band. But as I sat looking around at all the contented, mostly familiar faces bobbing subtly to the beat of *Take Five*, I thought: Holy cow! This is where I live. The bass player is a geography professor, the saxophonist is a farmer. Every person I saw in that room, save for the few out-of-towners, knows something useful and applies it for the good of our little town. I swear there is no such thing as an empty place.

In the darkness outside that room, of course, I knew that dogs still barked on the ends of their chains and rebel flags hung limp in their faded colors. In a culturally polarized America, country music fans rarely give a jazz quartet half a chance, or vice versa, and that's everybody's loss. But beyond any question of preference, some of our neighbors just couldn't afford to walk in the door that evening, not even for a five dollar bowl of soup. We have things to talk about and a lot more to do in our town, and some of it makes me uneasy. I'm an extreme introvert, I hate conflict, and I'm no good at persuading people to do things. I became a novelist, I think, because I'm good at seeing why every single person in a hornet's nest of disagreement believes he or she is right, and I tend to believe all of them. It's so easy to concede to a stalemate. Ten years ago when my husband and a handful of other optimists laid out proposals to shake up our economically comatose town, I was skeptical. Maybe even afraid. I wouldn't have put it this way, but I felt something immutable about the way things already were, cemented inside thick walls of class and culture. I had no idea. Food brings people together; where bread is broken, trust follows. Nothing stood between that earlier Meadowview and this one except years of workdays, one at a time, and our everyday good intentions.

Work is a language I understand. If worry is an engine, I got the Corvette 454. Sometimes I can only calm down by doing two things at once: knitting a sweater through a tedious meeting, making a good chicken soup while I commiserate with a friend's heartbreak. The odds are, only one of the several things I'm doing will have a productive outcome, so I hedge my bets. When I take stock of my irrefutable animal assets they are these: I live on land where enough rain falls and no rising tides can drown us. I like all of my neighbors and they are kind to me. Alongside the elder farmers we have a tribe of young agrarians who bring their cheerful energy to a shearing or harvest whenever we ask for help. I have hands that still work, elbows I can rest on a cutting board, and an ardor for feeling useful. I've figured out how to do a few things well, and one of them is making food happen.

So that's it, once upon a time we planted seeds, one thing led to another and this weekend before Lily leaves us again for college, she and I will be canning tomatoes. Every evening this week Steven and I poured an after-work glass of cider and debriefed, making each other laugh, while facing off across the counter island with knives in hand and some vegetable mountain between us. I imagine us getting old in this companionable way. On Sunday Camille and Reid will come over and help make a celebration dinner for Steven's birthday. Lucky Steven, to have been born in the season when larders are stocked: right now we're looking at a surplus of filet beans, zucchini, beets, blueberries, tomatoes, cucumbers, and melons. I'll send Camille and Reid home with a giant bag of cukes to advance their experiments in fermented pickles, a territory I've never explored. I love it when the kids do things in the kitchen that I didn't teach them, that I don't know how to do myself. It feels like a progress of civilization.

For my part, I'm experimenting with ways to freeze baby-sized quantities of pureed green beans and squash because at some point this winter we'll be starting the newest member of our family on solid foods. This grandchild constantly inhabits my mind, shape-shifting in the paranormal way of the not-yet born: I see him as a gasping neonate, a wide-eyed toddler, a question mark, a beloved sure thing. By next summer he'll be pulling up, holding onto crib rails, then plunking out beginner steps while keeping a grip on one adult finger. I imagine a proud little promenade down the mulched path between green-scented rows of tomato vines. My mental cinema keeps taking us into the garden. Why? I don't know. Steven says, "Because to get good weeding help, you start training them early."

He's absolutely right, and we've got the lineage to prove it, but I don't think my motives are administrative. They're primordial. When I put my imagination on free range it wants to go where I'm happiest: outside. That's always been true, since I was old enough to tell the difference between carpet and grass, ceiling and sky. My earliest happy memories are of pretending to get lost in the florid, cornstalk jungle of my grandparents' garden. Making squadrons of little dolls from hollyhock flowers that stained my hands red during their exciting, very short lives. Picking and eating blackberries beside some tall person: maybe a parent, a grandparent, I don't even picture the face, just the well-worn jeans I tagged after, and the working hands I watched like a little hawk. Or like *some* kind of little beast, learning how to be a successful human. Babies of every new generation are born to this task. No matter what has happened in the interim. We will keep doing these things.

ACKNOWLEDGMENTS

Every list of gratitudes should begin with the hands that feed us: Anthony and Laurel Flaccavento, Tom and Deni Peterson, Charlie Foster and family, Mike Hubbard, Paul Rizzo, Kirsty Zahnke, Kate Richardson, the Kling family, Will and Charlie Clark, David King, and everyone else at the Abingdon market. People always say, "I couldn't have survived without you," but in our case that's literally true.

Many mentors helped shape this project: Wendell and Tanya Berry were there all along; everything we've said here, Wendell said first, in a quiet voice that makes the mountains tremble. Joan Gussow also did it all ahead of us, and is the kind of friend who'll help with anything, whether it's scholarship or pulling weeds. Gary Nabhan, fellow chile-roaster from the early days, still keeps us smiling from a distance. Wendy Peskin and the Peruvian staff of Heifer International opened extraordinary doors to help us understand sustainable development. Marikler Giron truly saved us. Our debts to other colleagues and writers are as numerous as the books in our library: especially Vandana Shiva, Michael Pollan, Wes Jackson, and Brian Halweil. And the kitchen bookshelf: Alice Waters, Deborah Madison, Mary Beth Lind, and Cathleen Hockman-Wert.

Friendship with a writer—or in this case, a whole family of them—means you may sometimes fall into the pages when you're weren't looking. We're grateful to all those who opened their lives this way: most courageously, David and Elsie Kline, and the Worth-Jones family. Also Ricki Carroll, Tod Murphy, Pam Van Deursen, Anne Waddell and our postal pals, Amy Klippenstein, Paul Lacinski, Wendell and Ginny Kingsolver, Joann Hopp, and the Hopp-Ostiguys. Neta and Joe Findley are not just neighbors but family, and tell the best stories. Bill, Sanford, and Elizabeth are forever with us. Kate Forbes has earned a lifetime pass as our official extra farm kid, along with Abby Worth-Jones, who provided the title for chapter 14. Abby, Eli, Becky, and Roscoe Worth-Jones, Laura and Jerry Grantham, and the Malusa-Norman and Malusa-Froelich families get medals of valor for not running away on harvest day. Kay Hughes didn't run from a hungry crowd. Nancy and Paul Blaney, Sandy Skidmore, Jim Warden, Tandy and Lee Rasnake, Dayle Zanzinger, Fred Hebard, Rob Kingsolver, Ann Kingsolver, and so many others have sustained us with bread and kindness, rain or shine. Will White rose to any challenge; Mary Hanrahan pulled the Devil's Own weeds. The Bobs were fearless and undaunted. Jim Watson uncovered Eden from the brambles, and Cade helped. Our hardworking friends at Appalachian Sustainable Development keep reminding us why farmers matter: Anthony Flaccavento, Tom and Deni Peterson, Robin Robbins, Rebecca Brooks, Kathlyn Chupik, and all the staff.

Richard Houser, Virginia's most talented illustrator-painter-musician-chef-historian, saw how to make our book smile, and did it. Judy Carmichael is so much more than an office manager, we're working on a better title: research ace, rooster wrangler, esteemed colleague, best pal, and guardian angel all come to mind. Amy Redfern organized the chaos with panache. Jim Malusa and Sonya Norman left their fingerprints on the manuscript, for the better. Terry Karten is a champion editor and our very good fortune. No words are big enough to carry our devotion to Frances Goldin, so we'll just use little ones: we love you. Ditto for the whole office: Sam Stoloff, Ellen Geiger, Matt McGowan, Phyllis Jenkins, and Josie Schoel.

We all three thank our parents for putting tools in our hands at an early age and turning us loose on the project of making food happen. And we thank Lily for absolutely everything—plus eggs. If you think she's a charming character in this book, you should see her walk out the front door.

2007—BK, SLH, CHK

REFERENCES

For updates and a complete list of references, see our website: www.animalvegetablemiracle.com.

Ableman, Michael. *Fields of Plenty.* San Francisco: Chronicle Books, 2005.

Berry, Wendell. *What Are People For?* New York: North Point Press, 1990.

Brooks Vinton, Sherri, and Ann Clark Espuelas. *The Real Food Revival.* New York: Jeremy P. Tarcher, 2005.

Chadwick, Janet. *The Busy Person's Guide to Preserving Food.* North Adams, MA: Storey Publishing, 1995.

Charles, Daniel. *Lords of the Harvest: Biotech, Big Money, and the Future of Food.* New York: Perseus Books, 2002.

Cook, Christopher. *Diet for a Dead Planet: How the Food Industry Is Killing Us.* New York: New Press, 2004.

Fallon, Sally. *Nourishing Traditions: The Cookbook That Challenges Politically Correct Nutrition and the Diet Dictocrats.* Washington, DC: NewTrends, 2000.

Fowler, Cary, and Pat Mooney. *Shattering: Food, Politics and the Loss of Genetic Diversity.* Tucson: University of Arizona Press, 1990.

Fox, Michael W. *Eating with Conscience: The Bioethics of Food.* Troutdale, Oregon: NewSage Press, 1997.

Fromartz, Samuel. *Organic, Inc.: Natural Foods and How They Grew.* New York: Harcourt, 2006.

Goodall, Jane, Gary McAvoy, and Gail Hudson. *Harvest for Hope: A Guide to Mindful Eating.* New York: Warner Books, 2005.

Gussow, Joan Dye. *This Organic Life: Confessions of a Suburban Homesteader.* White River Junction, VT: Chelsea Green, 2002.

Guthman, Julie. *Agrarian Dreams: The Paradox of Organic Farming in California.* Berkeley and Los Angeles: University of California Press, 2004.

Hafez, E. S. E. *The Behaviour of Domestic Animals.* Hagerstown, MD: Williams and Wilkinson, 1969.

Halweil, Brian. *Eat Here: Homegrown Pleasures in a Global Supermarket.* New York: W. W. Norton, 2004.

Katz, Sandor Ellix. *The Revolution Will Not Be Microwaved.* White River Junction, VT: Chelsea Green, 2006.

Kimbrell, Andrew. *Fatal Harvest: The Tragedy of Industrial Agriculture.* Sausalito, CA: Foundation for Deep Ecology, 2002.

Kimbrell, Andrew, ed. *The Fatal Harvest Reader.* Washington, DC: Island Press, 2002.

Lambrecht, Bill. *Dinner at the New Gene Café: How Genetic Engineering Is Changing What We Eat, How We Live, and the Global Politics of Food.* New York: St. Martin's Griffin, 2002.

Lind, Mary Beth, and Cathleen Hockman-Wert. *Simply in Season.* Scottdale, PA: Herald Press, 2005.

Lappe, Frances Moore, and Anna Lappe. *Hope's Edge: The Next Diet for a Small Planet.* New York: Jeremy P. Tarcher, 2003.

Lyson, Thomas A. *Civic Agriculture: Reconnecting Farm, Food, and Community.* Medford, MA: Tufts University Press, 2004.

Madison, Deborah. *Local Flavors.* New York: Broadway Books, 2002.

Magdoff, Fred, John Bellamy Foster, and Frederick H. Buttel, eds. *Hungry for Profit: The Agribusiness Threat to Farmers, Food, and the Environment.* New York: New York University Press, 2000.

Manning, Richard. *Against the Grain: How Agriculture Has Hijacked Civilization.* New York: North Point Press, 2005.

Merzer, Glen, and Howard Lyman. *Mad Cowboy: Plain Truth from the Cattle Rancher Who Won't Eat Meat.* New York: Scribner, 1998.

Midkiff, Ken. *The Meat You Eat: How Corporate Farming Has Endangered America's Food Supply.* New York: St. Martin's Griffin, 2005.

Nabhan, Gary Paul. *Coming Home to Eat: The Pleasures and Politics of Local Foods.* New York: W. W. Norton, 2002.

Nestle, Marion. *Food Politics: How the Food Industry Influences Nutrition.* Berkeley and Los Angeles: University of California Press, 2003.

Norberg-Hodge, Helena, Peter Goering, and John Page. *From the Ground Up: Rethinking Industrial Agriculture.* London: Zed Books, 2001.

Norberg-Hodge, Helena, Todd Merrifield, and Steven Gorelick. *Bringing the Food Economy Home: Local Alternatives to Global Agribusiness.* London: Zed Books, 2002.

Robbins, John, and Dean Ornish. *The Food Revolution: How Your Diet Can Help Save Your Life and Our World.* Boston: Conari Press, 2001.

Petrini, Carlo. *Slow Food (The Case for Taste).* New York: Columbia University Press, 2003.

Pfeiffer, Dale Allen. *Eating Fossil Fuels: Oil, Food and the Coming Crisis in Agriculture.* Gabriola Island, Canada: New Society, 2006.

Pollan, Michael. *The Omnivore's Dilemma: A Natural History of Four Meals.* New York: Penguin, 2006.

Pringle, Peter. *Food, Inc.: Mendel to Monsanto—The Promises and Perils of the Biotech Harvest.* New York: Simon & Schuster, 2003.

Pyle, George. *Raising Less Corn, More Hell.* Cambridge, MA: Public Affairs, 2005.

Rifkin, Jeremy. *Beyond Beef.* New York: Plume, 1992.

Robinson, Jo. *Pasture Perfect.* Vashon, WA: Vashon Island Press, 2004.

Rogers, Marc. *Saving Seeds.* Pownal, VT: Storey Press, 1990.

Shepherd, Renee, and Fran Raboff. *Recipes from a Kitchen Garden.* Berkeley, CA: Ten Speed Press, 1993.

Shiva, Vandana. *Stolen Harvest: The Hijacking of the Global Food Supply.* Cambridge, MA: South End Press, 2000.

Smith, Jeffrey M. *Seeds of Deception: Exposing Industry and Government Lies About the Safety of the Genetically Engineered Foods You're Eating.* Fairfield, Iowa: Yes! Books, 2003.

Ticciati, Laura, and Robin Ticciati. *Genetically Engineered Foods.* New York: McGraw-Hill, 1999.

Waters, Alice. *Chez Panisse Vegetables.* New York: HarperCollins, 1996.

Willett, Walter. *Eat, Drink, and Be Healthy: The Harvard Medical School Guide to Healthy Eating.* New York: Free Press, 2005.

Wirzba, Norman, ed. *The Essential Agrarian Reader: The Future of Culture, Community, and the Land.* Emeryville, CA: Shoemaker & Hoard, 2004.

ORGANIZATIONS

Also see our website: www.animalvegetablemiracle.com.

LOCAL FOOD, EATING, AND FOOD SECURITY

Local Harvest
220 21st Avenue, Santa Cruz, California 95062
www.localharvest.org

FoodRoutes
37 East Durham Street, Philadelphia, Pennsylvania 19119
National nonprofit dedicated to reintroducing Americans to their food.
www.foodroutes.org

Slow Food International
Via Mendicità Istruita 8, 12042 Bra (CN), Italy
www.slowfood.com
20 Jay Street, Suite 313, Brooklyn, New York 11201
www.slowfoodusa.org

USDA Food and Nutrition Service
3101 Park Center Drive, Alexandria, Virginia 22302
www.fns.usda.gov/fns

The Community Food Security Coalition
PO Box 209, Venice, California 90294
Dedicated to building strong, sustainable local and regional food systems.
www.foodsecurity.org

Sustainable Table
215 Lexington Avenue, Suite 1001, New York, New York 10016
www.sustainabletable.org

Edible Communities
PMB 441, 25 NW 23rd Place, Suite 6, Portland, Oregon 97210-5599
www.ediblecommunities.com

National Farm to School Network
1600 Campus Road, Mail Stop M1, Los Angeles, California 90041
www.farmtoschool.org

Local Food Works (UK)
40–56 Victoria Street, Bristol BS1 6BY, United Kingdom
www.localfoodworks.org

SUSTAINABLE AGRICULTURE AND FARMING

The Sustainable Agriculture Research and Education (SARE) Program
Waterfront Center, Room 4462, 800 9th Street SW, Washington, DC 20024
www.sare.org

National Sustainable Agriculture Information Service
PO Box 3657, Fayetteville, Arkansas 72702
www.attra.org

The Land Institute
2440 E. Water Well Road, Salina, Kansas 67401
www.landinstitute.org

The Ecological Farming Association (EFA)
406 Main Street, Suite 313, Watsonville, California 95076
Promotes ecologically sound and economically viable agriculture.
www.eco-farm.org

The National Campaign for Sustainable Agriculture
PO Box 396, Pine Bush, New York 12566
www.sustainableagriculture.net

The Rodale Institute Farming Resources
611 Siegfriedale Road, Kutztown, Pennsylvania 19530-9749
www.rodaleinstitute.org

Sustain: The Alliance for Better Food and Farming (UK)
94 White Lion Street, London N1 9PF, United Kingdom
www.sustainweb.org

Organic Farming Research Foundation
PO Box 440, Santa Cruz, California 95061
www.ofrf.org

City Farmer—Canada's Office of Urban Agriculture
Box 74567, Kitsilano RPO, Vancouver, BC V6K 4P4, Canada
www.cityfarmer.org

American Community Gardening Association
1777 East Broad Street, Columbus, Ohio 43203
www.communitygarden.org

GOVERNMENT AGENCIES

The U.S. Department of Agriculture
1400 Independence Avenue SW, Washington, DC 20250
www.usda.gov

The National Agricultural Statistics Service
1400 Independence Avenue SW, Washington, DC 20250
www.nass.usda.gov

USDA Economic Research Service
1800 M Street NW, Washington, DC 20036-5831
www.ers.usda.gov

The U.S. Food and Drug Administration
5600 Fishers Lane, Rockville, Maryland 20857-0001
www.fda.gov

The Food and Agriculture Organization of the United Nations
Viale delle Terme di Caracalla, 00100 Rome, Italy
www.fao.org

The National Agricultural Library (NAL)
Abraham Lincoln Building, 10301 Baltimore Avenue, Beltsville, Maryland 20705-2351
www.nal.usda.gov

FOOD POLICY, CONSUMER AND ADVOCACY ORGANIZATIONS

The Organic Consumers Association (OCA)
Public interest organization campaigning for health, justice, and sustainability.
6771 South Silver Hill Drive, Finland, Minnesota 55603
www.organicconsumers.org

Food and Water Watch
1616 P Street NW, Washington, DC 20036
www.foodandwaterwatch.org

The Center for Food Safety
660 Pennsylvania Avenue SE, #302, Washington, DC 20003
Challenging harmful food production and promoting sustainable alternatives.
www.centerforfoodsafety.org

CropChoice
PO Box 33811, Washington, DC 20033
Alternative news and information source for American farmers and consumers.
www.cropchoice.com

The Food Trust
1201 Chestnut Street, 4th Floor, Philadelphia, Pennsylvania 19107
Works to increase access to affordable and nutritious food.
www.thefoodtrust.org

American Society for Nutrition
9650 Rockville Pike, Suite L-4500, Bethesda, Maryland 20814
www.nutrition.org

Pesticide Action Network North America
49 Powell Street, Suite 500, San Francisco, California 94102
www.panna.org

Union of Concerned Scientists
2 Brattle Square, Cambridge, Massachusetts 02238–9105
www.ucsusa.org

International Food Policy Research Institute
2033 K Street NW, Washington, DC 20006-1002
www.ifpri.org

The Institute for Food and Development Policy/Food First
398 60th Street, Oakland, California 94618
www.foodfirst.org

SIDEBAR RESOURCES

Other web sources are included in the sidebars. More comprehensive resources can be found on our website: www.animalvegetablemiracle.com.

OILY FOOD

David Pimentel, Marcia Pimentel, and Marianne Karpenstein-Machan, "Energy Use in Agriculture: An Overview," dspace.library.cornell.edu/bitstream/1813/118/3/Energy.PDF.

Richard Manning, "The Oil We Eat," *Harper's Magazine,* February 2004, http://harpers.org/archive/2004/02/the-oil-we-eat/

U.S. Energy Information Administration: www.eia.doe.gov/.

HUNGRY WORLD

Thalif Deen, "Tied Aid Strangling Nations, Says U.N.," Inter Press Service News Agency, 2004, http://www.ipsnews.net/2004/07/development-tied-aid-strangling-nations-says-un/

Celia W. Dugger, "Supermarket Giants Crush Central American Farmers," *New York Times,* December 28, 2004, www.organicconsumers.org/corp/walmartca122804.cfm.

Sophia Murphy and Kathy McAfee, *U.S. Food Aid: Time to Get It Right,* Institute for Agriculture and Trade Policy: Trade and Global Governance Program, July 2005, http://www.iatp.org/files/451_2_73512.pdf

Vandana Shiva, "Force-Feeding GMOs to the Poor," www.organicconsumers.org/ge/poor.cfm.

The Food and Agriculture Organization of the United Nations (FAO): www.fao.org.

Physicians and Scientists for Responsible Application of Science and Technology: www.psrast.org/nowohu.htm.

"Tied Aid—Promoting Donors' Self-Interest," *South Bulletin* 57, www.southcentre.org/info/southbulletin/bulletin57/bulletin57-08.htm.

HOW TO FIND A FARMER

New Generation Cooperatives on the Northern Plains, "Declining Farm Value Share of the Food Dollar," www.umanitoba.ca/faculties/afs/dept/agribusiness/media/pdf/ARDI_PDF.pdf.

Missouri Farmers Union: missourifarmersunion.org/coop/ffcenter/about.htm.

THE STRANGE CASE OF PERCY SCHMEISER

Gregory M. Lamb, "When Genetically Modified Plants Go Wild," *Christian Science Monitor,* August 31, 2006, http://www.csmonitor.com/2006/0831/p15s01-sten.html.

E. Ann Clark, "On the Implications of the Schmeiser Decision," University of Guelph, Guelph, Ontario, May 2001, http://www.plant.uoguelph.ca/research/homepages/eclark/percy.htm.

Ron Friesen, "Studies Show Gene Flow in GM Canola Likely Widespread," *Manitoba Co-operator,* July 4, 2002, www.percyschmeiser.com/Gene%20Flow.htm.

"Schmeiser Decision Causes Uproar Around the World," *CNW* (Canada), May 21, 2004, www.mindfully.org/GE/2004/Schmeiser-Uproar-World21may04.htm.

THE GLOBAL EQUATION

Brian Halweil, "Why No One Wins in the Global Food Fight," *Washington Post Sunday,* September 21, 2003, https://www.washingtonpost.com/archive/opinions/2003/09/21/why-no-one-wins-in-the-global-food-fight/502c5c20-6d3e-4629-8733-50894830425d/?utm_term=.3335390ed4cc.

John Otis, "Ruled by Fear, Banana Workers Resist Unions," *Houston Chronicle,* January 19, 2004, www.chron.com/disp/story.mpl/special/04/leftbehind/2095828.html.

USDA Department of Agriculture, Cooperative State Research Service, Office for Small-Scale Agriculture, http://sfp.ucdavis.edu/pubs/brochures/Specialtypotatoes/.

IS BIGGER REALLY BETTER?

Gerard D'Souza and John Ikerd, "Small Farms and Sustainable Development: Is Small More Sustainable?" *Journal of Agricultural and Applied Economics* 28 (1996): 73–83.

Peter M. Rosset, *The Multiple Functions and Benefits of Small Farm Agriculture in the Context of Global Trade Negotiations,* Institute for Food and Development Policy Brief no. 4, September 1999, http://www.map-abcdf.com.ph/documents/submitted%20papers/FOOD%20FIRST%20POLICY%20NO%204.pdf.

Ronald C. Wimberley et al., *Food from Our Changing World: The Globalization of Food and How Americans Feel About It,* 2004, sasw.chass.ncsu.edu/global-food/foodglobal.html.

The National Family Farm Coalition: www.nffc.net/.

THE PRICE OF LIFE

"Press Release: Consumer Reports Finds 71 Percent of Store-Bought Chicken Contains Harmful Bacteria," *Consumer's Union,* February 23, 1998, http://consumersunion.org/news/consumer-reports-finds-71-percent-of-store-bought-chicken-contains-harmful-bacteria/.

Economic and Structural Relationships in U.S. Hog Production, AER-818, Economic Research Service/USDA, https://www.ers.usda.gov/webdocs/publications/aer818/17860_aer818app3_1_.pdf.

PAYING THE PRICE OF LOW PRICES

Christopher D. Cook, "Thanksgiving's Hidden Costs," *AlterNet,* November 23, 2004, www.alternet.org/envirohealth/20556/.

K. Delate, M. Duffy, C. Chase, A. Holste, H. Friedrich, and N. Wantata, "An Economic Comparison of Organic and Conventional Grain Crops in a Long-Term Agroecological Research (LTAR) Site in Iowa," *American Journal of Alternative Agriculture* 18 (2002): 59–69.

Y. O. Ogini, D. P. Stonehouse, and E. A. Clark, "Comparison of Organic and Conventional Dairy Farms in Ontario," *American Journal of Alternative Agriculture* 14 (1999): 122–28.

D. Pimentel, "Environmental and Economic Costs of Pesticide Use," *Bioscience* 42 (1992): 750–60.

D. Pimentel, "Environmental and Economic Costs of Soil Erosion and Conservation Benefits," *Science* 267 (1995): 1117–23.

D. Pimentel, P. Hepperly, J. Hanson, D. Douds, and R. Seidel, "Environmental, Energetic, and Economic Comparisons of Organic and Conventional Farming Systems," *Bioscience* 55 (2005): 573–82.

Brian Riedl, "Still at the Federal Trough: Farm Subsidies for the Rich and Famous Shattered Records in 2001," *Heritage Foundation Backgrounder* #1542, http://www.heritage.org/research/reports/2002/04/farm-subsidies-for-the-rich-amp-famous-shattered-records-in-2001.

J. D. Smolik, T. L. Dobbs, and D. H. Rickert, "The Relative Sustainability of Alternative, Conventional, and Reduced-till Farming Systems," *American Journal of Alternative Agriculture* 16 (1995): 25–35.

The Rural Coalition, "Brief Background and History of the US Farm Bill: 1949 to Present," www.ruralco.org/library/admin/uploadedfiles/Farmbill_History.doc.

USDA Economic Research Service: https://www.ers.usda.gov/topics/farm-economy/farm-commodity-policy/.

SPEAKING UP

The Food Project: www.thefoodproject.org/.

National Sustainable Agriculture Information Service: attra.ncat.org/.

The Community Food Security Coalition: www.foodsecurity.org/.

LOSING THE BUG ARMS RACE

Robert G. Bellinger, *Pest Resistance to Pesticides,* Southern Region Pesticide Impact Assessment Program Report 1996, http://ipm.ncsu.edu/safety/factsheets/resistan.pdf.

Pesticide Action Network, (PAN) International, List of Highly Hazardous Pesticides, https://www.panna.org/sites/default/files/PAN_HHP-List_1101(1).pdf.

Weed Science Society of America, list of herbicide-resistant weeds: http://wssa.net/wssa/weed/resistance/.

HOME GROWN

Travis Beck and Martin F. Quigley, *Edible Landscaping,* Ohio State University Extension Factsheet HYG-1255-02, ohioline.osu.edu/hyg-fact/1000/1255.html.

Ben Sharvy, *Edible Landscaping & Gardening,* http://members.efn.org/~bsharvy/edible.html.

Ron Scherer, "Farmers Markets Boom Across the USA," *Christian Science Monitor,* August 29, 2001, www.organicconsumers.org/Organic/FarmersMarket901.cfm.

Center for Integrated Agricultural Systems, *Your Consumer Food Dollar: How Does It Carve Up?* www.cias.wisc.edu/foodshed/pubsntools/meal2.htm.

SUSTAINING THE UNSUSTAINABLE

Douglass Cassel Jr., "The Great Trade Robbery," *Chicago Daily Law Bulletin,* May 16, 2002.

Jim Goodman, "Bush Team Squeezes Farmers Stifles Dissent," *Capital Times* (Madison, WI), February 26, 2006, www.familyfarmdefenders.org/pmwiki.php/Main/BushTeamSqueezesFarmersStiflesDissent.

Anuradha Mittal, *Giving Away the Farm: The 2002 Farm Bill,* www.foodfirst.org/pubs/backgrdrs/2002/s02v8n3.html.

Environmental Working Group, *Bumper Crop: Concentration of Commodity Loan Subsidies,* http://static.ewg.org/reports/2000/BumperCrop.pdf?_ga=1.147972784.646996897.1485032005.

National Family Farm Coalition, *Food from Family Farms Act: A Proposal for the 2007 U.S. Farm Bill,* http://nffc.net/Learn/Fact%20Sheets/FFFA2007.pdf.

Union of Concerned Scientists, *Industrial Agriculture: Features and Policy,* http://www.ucsusa.org/sites/default/files/legacy/assets/documents/food_and_agriculture/cafos-uncovered.pdf.

Why Family Farmers Need Help, www.farmaid.org.

REALLY, WE'RE NOT MAD

Charles Abbott, "Meatpacker Sues US for Right to Do Mad Cow Tests," *Reuters,* March 24, 2006, http://usatoday30.usatoday.com/money/industries/food/2006-03-23-mad-cow-suit_x.htm.

Libby Quaid, "Government to Scale Back Mad Cow Testing," *Associated Press,* March 15, 2006, http://www.cbsnews.com/news/government-to-reduce-mad-cow-testing/.

Sabin Russell, "USDA Lacks Power to Inform Public, Mandate Returns," *San Francisco Chronicle,* January 6, 2004, http://www.organicconsumers.org/madcow/recall1604.cfm.

"Mad Cow Watch Goes Blind," *USA Today,* August 4, 2006, http://usatoday30.usatoday.com/news/opinion/editorials/2006-08-03-our-view_x.htm.

United States Government Accountability Office, "USDA and FDA Need to Better Ensure Prompt and Complete Recalls of Potentially Unsafe Food," GAO-05-51, October 2004, www.gao.gov/new.items/d0551.pdf.

"DIG! DIG! DIG! AND YOUR MUSCLES WILL GROW BIG"

Abiola Adeyemi, *Urban Agriculture: An Abbreviated List of References & Resource Guide 2000,* National Agricultural Library, https://pubs.nal.usda.gov/sites/pubs.nal.usda.gov/files/urban_0.htm.

Rachel Moscovich, "Grow Your Own, Big City," 4/19/2006, www.zerofootprint.net/green_stories/green_stories_item.asp?type_=50&ID=5019.

William Thomas, "Victory Gardens Can Save Us Again," *Convergence Weekly,* April 28, 2005, www.willthomas.net/Convergence/Weekly/Gardens.htm.

Online Magazine of Metropolitan Agriculture: www.metrofarm.com/.

TRADING FAIR AND SQUARE

Russell Greenberg, Peter Bichier, Andrea Cruz Angon, and Robert Reitsma, "Bird Populations in Shade and Sun Coffee Plantations in Central Guatemala," *Conservation Biology* 11, no. 2 (1997): 448–59.

Russell Greenberg, Peter Bichier, and John Sterling, "Bird Populations in

Rustic and Planted Shade Coffee Plantations of Eastern Chiapas, Mexico," *Biotropica* 29, no. 4 (1997): 501–14.

Adriana Valencia, *Birds and Beans: The Changing Face of Coffee Production,* World Resources Institute, May 2001, earthtrends.wri.org/features/view_feature.php?theme=7&fid=35.

World Fair Trade Organization: http://wfto.com/.

For more on migratory birds and coffee, visit: https://nationalzoo.si.edu/scbi/migratorybirds/coffee/online.cfm.

LEGISLATING LOCAL

Community Food Security Coalition, *Farm to Cafeteria in 2006: Helping Farmers, Kids, and Communities,* http://foodsecurity.org/farm_to_school/.

The National Farm to School Network: www.farmtoschool.org/ and www.farmtocollege.org/.

How Local Farmers and School Food Service Buyers Are Building Alliances, https://naldc.nal.usda.gov/naldc/download.xhtml?id=38355&content=PDF.

Small Farms/School Meals Initiative: www.fns.usda.gov/cnd/Lunch/Downloadable/small.pdf.

The National Association of Farmer's Market Nutrition Programs: www.nafmnp.org/.

For more resources and downloadable recipes, please visit www.animalvegetablemiracle.com.

INDEX

BIOGRAPHIES FOR THE AUTHORS OF *ANIMAL, VEGETABLE, MIRACLE*:

BARBARA KINGSOLVER is the author of fourteen books of fiction, poetry, and creative nonfiction including the novels *The Bean Trees*, *The Poisonwood Bible*, and *The Lacuna*, winner of the Orange Prize for Fiction. Translated into more than twenty languages, her work has won a devoted worldwide readership and many awards, including the National Humanities Medal. Many of her books have been incorporated into the core English literature curriculum of colleges throughout the United States.

As a grower of food, she traces her roots back to the family garden where she was given her own plot on which to grow one vegetable of her choice at the age of six. (She chose squash.) She cultivated the soil in many cities, and on several continents, before settling once and for all on the Virginia farm where she now lives with her husband.

STEVEN HOPP received his PhD from Indiana University and has a background in life sciences. He published academic papers in bioacoustics, ornithology, animal behavior, and more recently in sustainable agriculture. He is the founder and director of the Meadowview Farmers Guild and teaches environmental sciences at Emory and Henry College in Virginia.

CAMILLE KINGSOLVER studied biology at Duke University before pursuing a graduate degree and career in mental health counseling. She now lives with her husband, Reid Snow, and their son, Owen, in Washington County, Virginia, a few minutes away from her parents' farm.

LILY HOPP KINGSOLVER is studying biology at the University of Virginia in Charlottesville. She was too young to sign a publishing contract when *Animal, Vegetable, Miracle* debuted in 2007. Always an integral part of her family's local food efforts and story, she joins *AVM*'s original authors in writing new material for the tenth anniversary edition.

BOOKS BY BARBARA KINGSOLVER

FLIGHT BEHAVIOR
A Novel

Available in Paperback, E-Book, Audio CD, Digital Audio, and Large Print

A suspenseful and brilliant novel about catastrophe and denial which explores the complexities that lead us to believe in our chosen truths.

THE LACUNA
A Novel

Available in Paperback, E-Book, Audio CD, Digital Audio, and Large Print

The poignant story of a man pulled between two nations as well as an unforgettable portrait of the artist—and of art itself.

THE POISONWOOD BIBLE
A Novel

Available in Paperback and E-Book

Kingsolver's bestselling epic novel chronicling three decades in the life of an American family who travel to the Belgian Congo as missionaries in 1959.

SMALL WONDER
Essays

Available in Paperback, E-Book, and Digital Audio

Twenty-three essays that are a passionate invitation to readers to be part of the crowd that cares about the environment, peace, and family.

PRODIGAL SUMMER
A Novel

Available in Paperback, E-Book, Digital Audio, and Large Print

Three stories of human love woven together within a larger tapestry of lives amid the mountains and farms of southern Appalachia.